Advances in Intelligent Systems and Computing

Volume 1218

AF848577

The series “Advances in Intelligent Systems and Computing” contains publications on theory, applications, and design methods of Intelligent Systems and Intelligent Computing. Virtually all disciplines such as engineering, natural sciences, computer and information science, ICT, economics, business, e-commerce, environment, healthcare, life science are covered. The list of topics spans all the areas of modern intelligent systems and computing such as: computational intelligence, soft computing including neural networks, fuzzy systems, evolutionary computing and the fusion of these paradigms, social intelligence, ambient intelligence, computational neuroscience, artificial life, virtual worlds and society, cognitive science and systems, Perception and Vision, DNA and immune based systems, self-organizing and adaptive systems, e-Learning and teaching, human-centered and human-centric computing, recommender systems, intelligent control, robotics and mechatronics including human-machine teaming, knowledge-based paradigms, learning paradigms, machine ethics, intelligent data analysis, knowledge management, intelligent agents, intelligent decision making and support, intelligent network security, trust management, interactive entertainment, Web intelligence and multimedia.

The publications within “Advances in Intelligent Systems and Computing” are primarily proceedings of important conferences, symposia and congresses. They cover significant recent developments in the field, both of a foundational and applicable character. An important characteristic feature of the series is the short publication time and world-wide distribution. This permits a rapid and broad dissemination of research results.

**** Indexing: The books of this series are submitted to ISI Proceedings, EI-Compendex, DBLP, SCOPUS, Google Scholar and Springerlink ****

More information about this series at http://www.springer.com/series/11156

Evangelos Markopoulos ·
Ravindra S. Goonetilleke ·
Amic G. Ho · Yan Luximon
Editors

Advances in Creativity, Innovation, Entrepreneurship and Communication of Design

Proceedings of the AHFE 2020 Virtual Conferences on Creativity, Innovation and Entrepreneurship, and Human Factors in Communication of Design, July 16–20, 2020, USA

Springer

Editors
Evangelos Markopoulos
School of Management
University College London (UCL)
London, UK

Ravindra S. Goonetilleke
Information Services Division (ISD)
HKUST
Kowloon, Hong Kong

Amic G. Ho
School of Arts and Social Sciences,
Department of Creative Arts
The Open University of Hong Kong
Ho Man Tin, Kowloon, Hong Kong

Yan Luximon
The Hong Kong Polytechnic University
Kowloon, Hong Kong

ISSN 2194-5357 ISSN 2194-5365 (electronic)
Advances in Intelligent Systems and Computing
ISBN 978-3-030-51625-3 ISBN 978-3-030-51626-0 (eBook)
https://doi.org/10.1007/978-3-030-51626-0

This Springer imprint is published by the registered company Springer Nature Switzerland AG
The registered company address is: Gewerbestrasse 11, 6330 Cham, Switzerland

Advances in Human Factors and Ergonomics 2020

AHFE 2020 Series Editors

Tareq Z. Ahram, Florida, USA
Waldemar Karwowski, Florida, USA

11th International Conference on Applied Human Factors and Ergonomics and the Affiliated Conferences

Proceedings of the AHFE 2020 Virtual Conference on Creativity, Innovation and Entrepreneurship and the International Conference on Human Factors in Communication of Design, July 16–20, 2020, USA

Advances in Neuroergonomics and Cognitive Engineering	Hasan Ayaz and Umer Asgher
Advances in Industrial Design	Giuseppe Di Bucchianico, Cliff Sungsoo Shin, Scott Shim, Shuichi Fukuda, Gianni Montagna and Cristina Carvalho
Advances in Ergonomics in Design	Francisco Rebelo and Marcelo Soares
Advances in Safety Management and Human Performance	Pedro M. Arezes and Ronald L. Boring
Advances in Human Factors and Ergonomics in Healthcare and Medical Devices	Jay Kalra and Nancy J. Lightner
Advances in Simulation and Digital Human Modeling	Daniel N Cassenti, Sofia Scataglini, Sudhakar L. Rajulu and Julia L. Wright
Advances in Human Factors and Systems Interaction	Isabel L. Nunes
Advances in the Human Side of Service Engineering	Jim Spohrer and Christine Leitner
Advances in Human Factors, Business Management and Leadership	Jussi Ilari Kantola, Salman Nazir and Vesa Salminen
Advances in Human Factors in Robots, Drones and Unmanned Systems	Matteo Zallio
Advances in Human Factors in Cybersecurity	Isabella Corradini, Enrico Nardelli and Tareq Ahram

(continued)

(continued)

Advances in Human Factors in Training, Education, and Learning Sciences	Salman Nazir, Tareq Ahram and Waldemar Karwowski
Advances in Human Aspects of Transportation	Neville Stanton
Advances in Artificial Intelligence, Software and Systems Engineering	Tareq Ahram
Advances in Human Factors in Architecture, Sustainable Urban Planning and Infrastructure	Jerzy Charytonowicz
Advances in Physical, Social & Occupational Ergonomics	Waldemar Karwowski, Ravindra S. Goonetilleke, Shuping Xiong, Richard H.M. Goossens and Atsuo Murata
Advances in Manufacturing, Production Management and Process Control	Beata Mrugalska, Stefan Trzcielinski, Waldemar Karwowski, Massimo Di Nicolantonio and Emilio Rossi
Advances in Usability, User Experience, Wearable and Assistive Technology	Tareq Ahram and Christianne Falcão
Advances in Creativity, Innovation, Entrepreneurship and Communication of Design	Evangelos Markopoulos, Ravindra S. Goonetilleke, Amic G. Ho and Yan Luximon

Preface

Creativity is the key driver of innovation. The purpose of the AHFE 2020 conference track on Creativity, Innovation and Entrepreneurship has been bringing together experts from different areas to show how creativity drives design and innovation, and thus merging a wider spectrum of related topics from engineering design, industrial design, and ergonomics in design. The peer-reviewed contributions to this track, which have been collected in this book, do cover industrial and business topics within and lying in between disciplines such as information technology, computer science, social science, systems engineering, artificial intelligence, cognitive science, and psychology.

Communication of design is concerned with understanding users, creating communication, and engaging experiences. Users expect communication to be an exciting and interactive experience. To provide such enjoyable and stimulating experiences, researchers in the discipline of communication design have been concerned with human factors research and methods enabling them to enrich communication and optimize design experience. By gathering peer-reviewed contributions presented at the AHFE 2020 virtual conference track on Human Factors in Communication of Design, this book merges classical humanistic approaches concerning the role of human needs, emotions, thoughts, and actions, with innovative and technological approaches, such as those improving visual experience, at the purpose of optimizing users' experiences in design consumption, and users' satisfaction toward the designs.

This book is expected to serve professionals, researchers, and students dealing with human factors research and methods applied to the field of business innovation and communication of design. It is organized into four sections:

Creativity, Innovation, Entrepreneurship

Section 1	Creativity, Innovation, Entrepreneurship
Section 2	Creativity and Entrepreneurship in Education
Section 3	Innovative Design

Communication of Design

Section 4 Communication in Design

Each section contains research papers that have been reviewed by members of the International Editorial Board. Our sincere thanks and appreciation to the board members as listed below:

Creativity, Innovation, Entrepreneurship

Communication of Design

We expect this book to provide readers with extensive information of the above-mentioned topics. Intended as a thought-provoking lecture, it is also aimed at inspiring them to contemplate other questions, applications, and potential solutions.

July 2020

Evangelos Markopoulos
Ravindra S. Goonetilleke
Amic G. Ho
Yan Luximon

Contents

Innovative Design

Communication of Design

Creativity, Innovation and Entrepreneurship

Democratizing Innovation. A Geo-Entrepreneurial Analysis and Approach Through the Company Democracy Model

Evangelos Markopoulos[1,2(✉)], George Markopoulos[3], and Hannu Vanharanta[4,5]

[1] School of Management, University College London, Gower Street, Bloomsbury, London WC1E 6BT, UK
e.markopoulos@ucl.ac.uk
[2] Faculty of Engineering and Business, Turku University of Applied Sciences, Joukahaisenkatu 3, ICT-City, C-wing, FI-20520 Turku, Finland
[3] Westminster Kingsway College, 211 Grays Inn Road, London WC1X 8RA, UK
geomark1919@gmail.com
[4] School of Technology and Innovations, University of Vaasa, Wolffintie 34, 65200 Vaasa, Finland
hannu@vanharanta.fi
[5] Poznań University of Technology, Plac Marii Skłodowskiej-Curie 5, 60-965 Poznań, Poland

Abstract. Even that thinking, and innovative thinking, in particular, is supposed to be borderless and unbiased, it seems that most of the innovations globally derive from regions that have built a brand name on it. This limits the opportunities to bright ideas form bright people outside the innovation hubs, resulting in a general loss of intellectual capital for the global economy. This paper aims to democratize innovation by redefining geo-entrepreneurship through a reverse innovation framework that exposes the hidden intellectual capital around the world, evaluates innovation drives and opportunities and empowers the development, commercialization and utilization of innovation. Based on the Company Democracy Model the proposed framework impacts reversely national brain-drain contributes to innovation scouting, strategic partnerships, and redistributes success opportunities. This geo-entrepreneurial approach identifies innovation potential globally, reduces inequalities among all those who can and want to create opportunities regardless of where innovation takes place and by whom.

Keywords: Geopolitics · Geo-entrepreneurship · Geo-economics · Democracy · Reverse innovation · Brain drain · Technology · Innovation · Managemenr · Strategy · Invesments · Startup

E. Markopoulos et al. (Eds.): AHFE 2020, AISC 1218, pp. 3–16, 2020.
https://doi.org/10.1007/978-3-030-51626-0_1

1 Introduction

In the last two decades, the term 'innovation' drives the global economy and development. The rapid and impressive advancements of technology in hardware, software, and communication above all, eliminated distances, created new markets and generated countless types of clients and needs. The continuous quest for the faster, cheaper, better but also smarter products or services created a global entrepreneurial revolution led primarily by students or young professionals and followed massively by the millennials later. Innovation created an entrepreneurial DNA-code with common characteristics and traits on all modern innovators [1].

According to Thomas Friedman and his globalization theory 3.0 [2], the world is indeed flat, and the evolution of innovation makes it not only flat but small as well. However, the way innovators innovate, the cost of innovation, the innovation effectiveness, the areas of innovation, its impact to the society, to the local and the global economies, to the national competitiveness advantage, to the generation of human intellectual capital and to many other dimensions seems to be very different around the world.

Access to technological resources, funding opportunities, market maturity, buying power, exposure and visibility, and partnerships is not equally distributed locally, nationality and internationally.

Such inequalities can be identified in various parts of the world where thinkers from followers are differentiated by political, economic, social, and technical (PEST) factors and incentives. Lack of data, information, knowledge, as well as opportunities and advantages based on political culture, create either brain drain from those who leave, or brain gain from those retained or imported when incentives are given. Such intellectual capital swifts, impact significantly the development and the progress of national economies and markets. Losing brilliant minds, due to lack of opportunities to be developed and rewarded, is the same as importing brilliant minds to recover the inactive ones in a country due to insufficient local or national innovation strategies and culture.

2 Innovation Brands

The value and impact of an innovative idea or even a start-up are usually influenced by the country or city the innovations derive. In the US, \$135,4 billion spend by Venture Capitals and other financial sources the period of 2011–2014 with 60+% in California. Specifically, from the \$ 47.3 billion invested in 2014, 35.6 billion ended up in only three states (New York (4.5), California (27.0), and Massachusetts (4.1)) [3]. Furthermore, 22 of the 27 billion spends in California went to the San Francisco (Bay) area. The same implies to the types of innovations being developed. In 2014 in the US, 41% of the innovations are in technology, 13% in healthcare, and 8% in the renewables (energy). In a similar way the 2nd quarter of 2019, UK gathered € 3.2 billion in tech investments, France 1.8 and Sweden 1.4.

These figures indicate that innovation funding is not spread equally across cities, states or countries but unequally to the ones that have developed an innovation brand name regardless of their population size. This also indicates that the areas or innovation are very limited to what is the trend and not actually what is truly useful. This created a biased approach to what is innovative and what is not and to what shall be funded and what shall not. Terms like Medtech, Cleantech, and especially Fintech monopolized the interest, but life can be much more meaningful without having advanced banks.

A possible explanation of this phenomenon could be that the innovation branded locations attract more investors who wish to invest locally and to trendy innovations. Another explanation could be that investors can be biased towards funding Silicon Valley-based startups than others located in innovation deserts, areas that seem to have a distance from the innovation trends that dominate the public opinion. This thinking, and reality to an extent make innovation funding easier for those being at the innovation branded locations and very difficult of those outsides of it. However, limiting investors to invests in what is withing their region and not consider much what is around or outside of it can be a highly risky logic. Another risk is the return on investment which can be less from the investments done in innovation branded locations but more if done outside of them due to the reverse innovation theory [4]. Investing in less privileged locations reduces the cost of innovation development and increases the return on investments once those innovations are brought to the mainstream markets. To tackle this challenge investors, either private or public, shall approach innovative ideas unbiased from the city, the country or the sector they come from.

It is a need for the development of geo-entrepreneurial thinking by understanding the world and what each place has to offer to an investment at a specified subject, under specific conditions for a specific time. As all people are equal and therefore they can all think, it is for certain that using geo-entrepreneurial thinking the opportunities for both the investors and the innovators can be multiple with greater benefits. The same of course can be applied not only to people but to states of countries that under current or expected political or operational conditions can be ideal for either innovation development or ideal target groups for innovation adaptation.

3 Geo-Entrepreneurship by Example

The term geo-entrepreneurship describes and analyzes the techno-economic entrepreneurial geography. It is the process of identifying the best areas innovation investments can be made at a specific instance, time, subject and target group (market).

An example of geo-entrepreneurship can be given on the innovation area of Cultural Digital Heritage which is a growing market with tremendous opportunities for innovative technologies on the digitization preservations, dissemination and commercialization of natural heritage monuments, and other elements [5]. The European Union invested in this sector 1.46 billion euros from the 78 billion of the Horizon2020 Research and Innovation program budget [6].

In this case and under the geo-entrepreneurship theory, the best possible innovators in this field could be considered the ones deriving primarily from Egypt. The county fulfills many of the geo-entrepreneurship conditions such as original and valid knowledge to be developed, quality of software development, incentives to entrepreneurs, and a wide market for the technology adaptation.

Specifically, since 2000, Egypt invested part of its national development strategy on the digitization and commercialization of its cultural heritage. CULTNAT, the National Center for Cultural and Natural Documentation received world-class international awards, distinctions and recognition for its research on this field and its support to active contributors [7]. Part of the same national strategy has been the development of the Software Engineering Competence Center (SECC) and the Technology Innovation and Entrepreneurship Center (TIEC), among other governmental organizations for the development of the Egyptian innovation. SECC trains, supports and certifies professional software engineers in start-ups, SMEs, but also large organizations, with high impacts software engineering standards and practices such as the CMMi, which is really followed in many developed countries due to their tedious development process on software quality [8]. TIEC has been developed to drive innovation and entrepreneurship for the benefit of the national economy acting as a catalyst for the academia-industry-government triptych [9]. Today TIEC is a leading innovation hub in the region and provides funding schemes, accelerator programs, competitions, and other activities to support the Egyptian start-ups and not only. Besides the maturity of the country on the cultural heritage digitization, digital museums and digital libraries, the quality of software engineering, and then start-ups entrepreneurial support, it must be noted that Egypt has since 1990 a ministry dedicated to the development of ICT. The Egyptian Ministry of Communication and Information Technology is the driving force for the Egyptian technological revolution. Furthermore, the country has endless resources from the 43 universities (20 public with 2 million students and 23 private with 60.000 students) that operate, in which engineering is the dominant course/program of studies.

Taking the geopolitical analysis furthermore, Egypt speaks Arabic, besides English and French. Arabic is a language of 22 countries nearby with a total population of 450 million. Egypt has always been a leading Arabic nation, hosting the Arab League headquarters in Cairo since its establishment, and is one of the most, if not the most, influential country in terms of engineering and technology in the Arab world with highly recognized engineers. These facts, along with the continuous growth of the Egyptian GDP [10] and its 14 technological parks, indicate the wide target group Egyptian innovations can have in adaptation and commercialization.

The given example can be graphically presented in Fig. 1 with the geo-entrepreneurial elements satisfied in this case as an advantage compared to other countries on the areas of Digital Cultural Heritage, and not only.

Fig. 1. Example of geo-entrepreneurial elements

Similarly, to Egypt, there could be cases like Botswana whose technological growth and commitment to software engineering grow significantly after the adaptation of the ITIL by the government. Zambia is also an innovation development hub considered as the country with the best female entrepreneurship results. It is also worth paying attention to small size countries like Malta and Cyprus who were ranked at the 27th and 28th place respectively by the World Intellectual Property Organization in 2019 [11].

4 Geo-Entrepreneurship Criteria

There is a significant difference between geo-entrepreneurship and international business development, even though both involve foreign investments. As of today, the development of established organizations in foreign markets has been driven from economies of scale or cost reductions motives.

Large organizations invest in foreign countries the development of their operations, production, distribution, supply chain, and innovation after analyzing the related costs and target markets. Such initiatives can not be considered as an entrepreneurial strategy since there is no entrepreneurial risk in them. The expansion is well calculated, the costs are proven lower compared to the ones in the organization's country, and the access to the markets is practically justified.

On the other hand, the geo-entrepreneurs are primary start-ups or organizational spin-offs aiming to explore the opportunities a different country offers to develop innovation and operate either in the sphere of influence of that country or globally. Geo-entrepreneurs seek the innovation expertise on a specific sector, subject or service to be delivered best at, and from, a specific country once certain pre and post conditions are fulfilled.

Today international business development and foreign investments are made through several country ranking criteria. The McKinsey Global Institute identifies Azerbaijan, Belarus, Cambodia, Ethiopia, India, Kazakhstan, Laos, Myanmar, Turkmenistan, Uzbekistan, and Vietnam as the less heralded and more geographically diverse outperformers [12]. MSCI (Morgan Stanley Capital International) index on the other hand on its 27 emerging economies includes among others Peru, Colombia, Poland, Czech Republic, Pakistan, Thailand, and even Greece despite its devastating financial crisis and continuous austerity programs of the last decade which still go on. [13]

On the other hand, no such indexes, rankings or investment elements exit today to guide Geo-entrepreneurs on ideas and innovation evaluations from specific countries. The main reason for this gap is the fact the geo-entrepreneurship is innovation-driven, with a great degree of fuzziness, and not conventional product or service driven that have a great degree of certainty due to their existing expertise on the development and distribution process. Tables 1 indicates the main Geo-Entrepreneurship criteria.

Table 1. Main geo-entrepreneurship criteria for investors

Geo-entrepreneurial elements			
Talent Pool	Country size	Country technological growth	Corruption
Intellectual capital	Governmental support on the field	County investment programs	Bureaucracy
Human recourses	Governmental entrepreneurial support	Common Borders/Unions/Leagues	Target Market
University graduates	Country field expertise (innovation area)	Common Cultures	Growing users/clients
Number of patents	Country growth	Common Language	Legislation
Number of startups	Country regional impact	Unemployment	Political Stability

5 Reverse Innovation

Today, working in emerging countries is a strategy and privilege of large-scale organizations by moving their production and operations primarily for cost efficiency. However, such strategies restrict benefits that can be obtained from investing in the intellectual capital the emerging countries have towards achieving reverse innovation. Geo-entrepreneurship challenges this by replacing cost reductions with the utilization of the intellectual capital offered in emerging countries to achieve reverse innovation in two dimensions.

In the first dimension, innovation investors aim to identify ideas, prototype designs, working prototypes and actual innovative products and services with high potential, performance or sales developed while developed with minimal cost. Such innovations can be brought back to the high-end innovation markets, under new branding and marketing strategies applied in the developed countries that can increase the profit margins by selling high and keeping the implementation costs low in the emerging country.

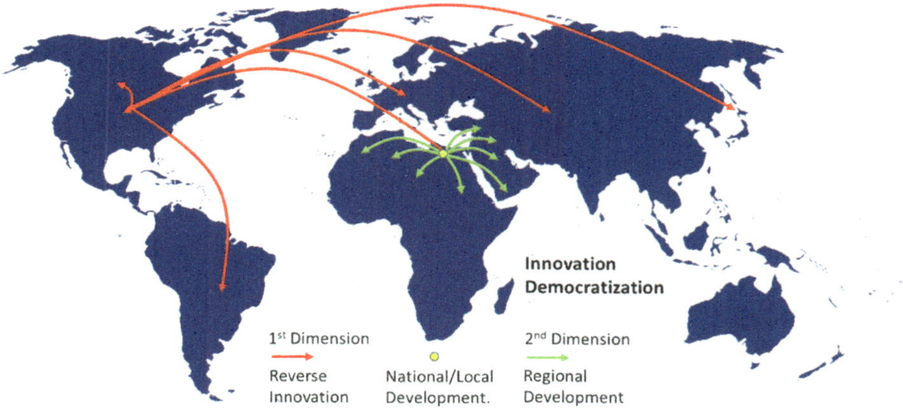

Fig. 2. The geo-entrepreneurial double reverse innovation

In the second dimension, the investment can be direct to the emerging country without necessarily bringing the innovation to the developed country but benefit directly from the massive target groups the innovations are applied. In this case, reverse innovation can be obtained in a reverse way by bringing technology, expertise and ideas from the developed country to enhance the existing technology by using the human recourses of the emerged country, to strengthen its functionality and operations on the markets the innovation exists without indicating or applying change of ownership (Fig. 2).

6 Democratizing Innovation

Geo-entrepreneurship can be achieved with geostrategic thinking, the unbiased thinking needed to clearly and fairly identify opportunities behind people's colors, sex, religions, nationalities, education, lifestyle, social and economic background. Knowledge does not discriminate, and innovative thinking can be achieved by anyone and anywhere who wants to think practically, rationally and effectively. Geo-entrepreneurship can also be defined by innovation democratization.

It is the democratic mentality, thinking and acting that offers the opportunities, circumstances, and infrastructure to those that indicate innovativeness.

Therefore, Innovation opportunities, funding, and support cannot and shall not be a privilege for the ones within the innovation branded locations. It is not only San Francisco, New York, Boston, London, Berlin, Singapore, Beijing or Tokyo that grow the best ideas, in fact, due to the competition and the plethora of ideas the best might get lost in the crowd or never get the chance to be presented. The democratization of innovation via geo-entrepreneurship can reveal the same impressive, or even more impressive ideas and innovations from places not considered as innovation hubs.

The effective adaptation of the democratic geo-entrepreneurship from large scale organizations, investors and even governments can empower the global economy by motivating more startups around the world to keep on innovating knowing that they have a fair opportunity wherever they are.

7 The Democratic Geo-Entrepreneurship Model (DeGeoEM)

The Innovation Democratization for Geo-Entrepreneurship can be implemented through the Democratic Geo-Entrepreneurship Model (DeGeoEM), a derivative of the Company Democracy Model (CDM).

The Company Democracy Model is a knowledge management process framework that transforms organizational knowledge into innovation, competitiveness, and extroversion [14]. The model is composed of six levels evolving from level 1 when knowledge is democratically elicited from anyone in any organization, all the way to

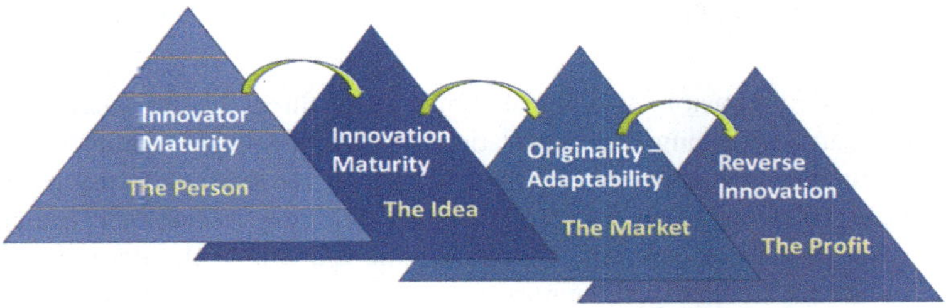

Fig. 3. The democratic geo-entrepreneurship model (DeGeoEM) dimensions

level 6 where the knowledge, and those who contributed, are transformed into organizational assets. Variations of the Company Democracy Model has been used to design frameworks for New Product Development [15], Corporate Entrepreneurship [16], Governmental (Public Sector) Entrepreneurship [17], Green Ocean Strategy [18], Agile Startup Business Planning [19], Teaming [20], and other areas of management.

The Democratic Geo-Entrepreneurship Model (DeGeoEM) maintains the six levels of the Company Democracy Model approached from four different dimensions (Fig. 3) where each one can be considered as a different management methodology that can be implemented separately or jointly with the rest to achieve Geo-Entrepreneurship.

The first dimension indicates the maturity degree of the country the innovation exists or the startup operates. The second dimension indicates the degree of reverse innovation that can be achieved, meaning the success an innovation brings to the investor when it goes to a developed country. The third dimension indicates the credibility of the innovator or the startup that has or runs the innovation. The fourth dimension indicates the innovation strength in originality, adaptability and intellectual property.

8 DeGeoEM Dimension's Levels

Each one of the four DeGeoEM dimensions is composed of six levels in an evolutionary path indicating the completion of one level after the other (Fig. 4).

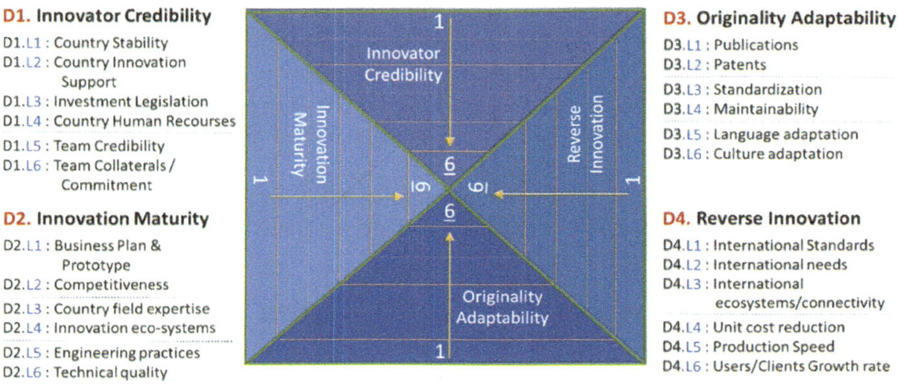

Fig. 4. Democratic geo-entrepreneurship model (DeGeoEM) dimension levels

The levels of the first (1st) dimension (Innovator Credibility) are grouped into two categories. The first four levels analyze the credibility of the innovators' space. Specifically, level 1 deals with the identification of the country stability. Level 2 evaluates the country's innovation support, level 3 analyses the investment legislation and level 4 evaluates the country's human recourses that can be used by the innovator. The last two levels form the second category deals with the innovators' team structure and commitment. Level 5 evaluates the innovators' team structure and coherence, while level 6 verifies the team collaterals and commitment to their innovation.

The levels of the second (2nd) dimension are grouped into three categories. The first category consists of innovation readiness with level 1 to identify the completeness of the business plan and the operations of a working prototype, while level 2 identifies the innovation competitive elements. The second category deals with the innovation infrastructure maturity. Level 3 evaluates the country's expertise in the area of the application and level 4 identifies the existing innovation eco-systems need to empower the development and adaptation of the innovation. The last category deals with technical maturity. Level 5 analyzes the software engineering practices used in the development of the innovation, or other reliable engineering practices, if not technology, and level 6 evaluates the technical quality of the innovation implementation.

The levels of the third (3rd) dimension are also grouped into three categories. The first category deals with the originality of the innovation. This is archived at level 1 with publications, awards and competitions that verify the originality. Level 2 deals with the patents, copyrights and intellectual property right actions applied to the innovation. The second category is related to technical agility. Level 3 evaluates the standardization used on the development of the innovation and level 4 evaluates the degree of maintainability. The third category deals with cultural adaptation elements. Level 5 analyzes the language adaptability and level 6 analyzes the cultural adaptability.

The levels of the fourth (4th) dimension are grouped into two categories. The first category deals with internationalization compliance. Level 1 analyses the international standards the innovation follows in order to be easily adopted in developed countries. Level 2 analysis the relationship of innovation solutions with similar needs in the developed countries. Level 3 analyzes the degree of compatibility with the international ecosystem in terms of functional and technical connectivity (IoT, etc.). The second category is related to the actual reverse innovation drives. Level 4 analyzes the unit cost reduction, level 5 deals with the production speed achieved and level 6 deals with the users/Clients Growth rate.

It must be noted that not all innovations need to reach level six in all dimensions as this is not necessary or feasible based on the stage the innovation is assessed.

9 Geo-Entrepreneurship Controlled Evolution

The four dimensions of the DeGeoEM can be executed either in parallel or in any order suites the investor's strategy. However, the following order of Innovation Credibility, Adaptability and Originality, Innovation Maturity, and Reverse Innovation is a path recommended as the investments are primarily on foreign countries where the credibility of the innovators and the country is the primary concern.

In general Democratization and Democracy can be quite beneficial but risky as well. Providing opportunities to all does not mean that all can utilize the opportunity. Therefore the risk factors on innovation investments in foreign markets and especially in the emerging countries, where many development parameters could be unstable, is quite high but on the same time the degree of profitability once an investment is executed successfully is inversely related to the risk especially when reverse innovation is achieved (Fig. 5).

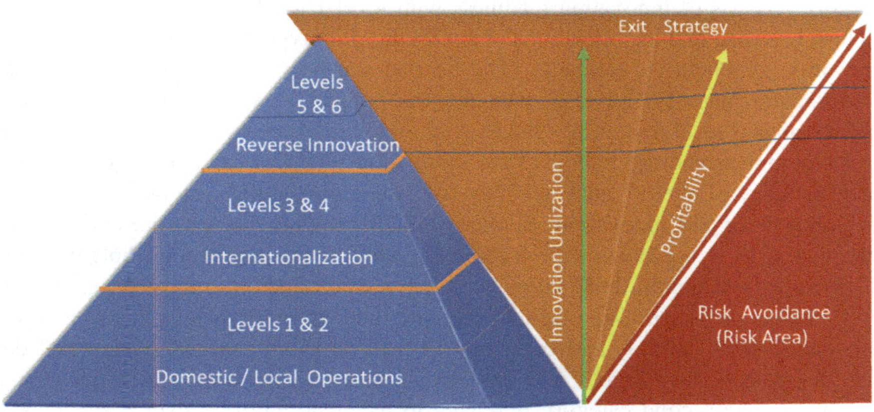

Fig. 5. Innovation maturity, profitability, and risk relationship

The evolution of the maturity and effectiveness of innovation, but also the reduction of risk can be divided into three stages achieved with the completion of two DeGeoEM levels in each. The first (1st) stage is the achievement of success at the domestic or local level. It is the first milestone on which the innovation proves its operations and that the innovators and the startup can stay together after success. Research indicates that 75% of the startups fail after the first success of receiving funding [21]. Thus the risk of achieving the first two levels is quite high. The second (2nd) stage is where the innovation succeeds at international or even at the regional level. This stage proves the internationality of the innovation, its strength on being accepted, able to penetrate and compete in other markets, in other cultures and against local competitors. Lastly, the third (3rd) stage is where reverse innovation is achieved, indicating readiness and maturity to move to developed nations and compete successfully and profitable in markets with high buying power and consumerism.

10 The Democratic Geo-Entrepreneurship Value

The Democratic Geo-Entrepreneurship Model is supported by a scorecard that determines the degree of investment security and startup operations success. The scorecard is based on the values scored on each dimension per case. The values are based on the completion of each level in each dimension.

The maximum net DeGeoEM score in a case is 24 (4 dimensions times 6 levels). Not all levels on each dimeson must necessarily be completed, and the score does not need to be the maximum. In this case, the sum of the score of all dimensions is divided by 24 (maximum score) to indicate the distance from the perfect score (Fig. 6).

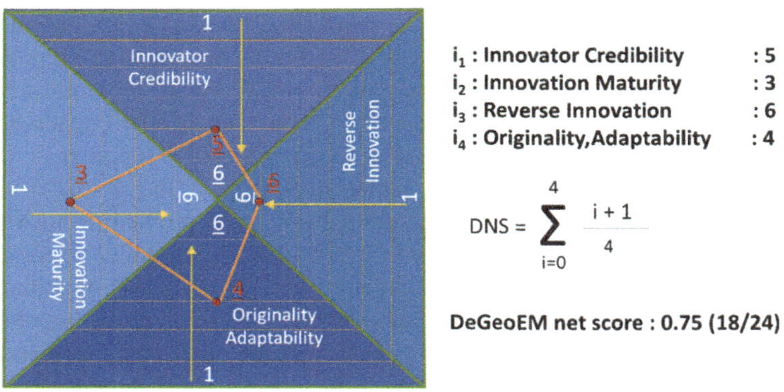

Fig. 6. DeGeoEM net score calculation

However and in order to be precise, the completion value of each level and each dimension is proposed to be calculated with level and dimensions multipliers set up by the investors per case giving the Weighted score.

The Weighted score is more precise as each level receives by the investor a multiplier according to the impact, complexity or criticality this level has for the specific investor. Not all levels have the same weight as some might be more impactful or difficult to be achieved than others. In a similar way, a second multiplier can be applied to the weight of the dimension itself. This indicates the importance of the dimensions based on the strategy of the investor Eq. (1)

$$\mathrm{WNS} = \left[\frac{\sum_{i=o}^{4}\left[(\mathrm{Dim_i} + 1) * \mathrm{Dim\_Multi_i}\right] + \left[\sum_{j=o}^{6}\left(\mathrm{Level_j} + 1\right) * \mathrm{Level\_Multi_j}\right]}{4}\right] \tag{1}$$

The Net and Weighted DeGeoEM scores can be used as a reference and benchmark for other investors in the future forming the DeGeoEM database of intentional startups or project evaluations. Investors seeking a specific type of innovation in India, for example, can search the DeGeoEM score database for similar cases in order to study the score position, level and distance from the reality after the investment has been made.

11 Conclusions

The Democratic Geo-Entrepreneurship Model approaches the democratization of innovation from an economic and social perspective. The economic perspective is related to the opportunities offered to any person who can think and innovate around the world without the need to be within the innovation branded cities and hubs. These opportunities can positively impact the lives of the innovators but also the profits of the investors. The model balances innovation opportunities around the world through the people. On the other hand, this social dimension extends beyond the impact on the lives of the people, to the fairness, equality and diversity the model supports.

The model indirectly supports the social structures of the countries and their development route, buy tackling the devastating brain drain problem from which most of the emerging countries suffer. Providing investment opportunities to innovators to develop their innovations and business while staying, working and operating in their countries, contributes scientifically to the development of the country, the jobs creation in the country and the well-being of the people who prefer to stay, work and grow in their country. The developed countries can still enjoy financial benefits from reverse innovation and secure their investments while reducing the brain gain for them and the brain drain for the emerging markets. Metron Ariston is an Aristotelian principle for success that cannot be achieved in the absence of fairness or the presence of greed.

References

1. Hall, D.: Entrecode: Unlocking the Entrepreneurial DNA. Management Books 2000 (2013)
2. Rugman, A.M.: The theory and regulation of emerging market multinational enterprises. In: Sauvant, K.P., McAllister, G., Maschek, W.A. (eds.) Foreign Direct Investments from Emerging Markets, pp. 75–87. Palgrave Macmillan US, New York (2010). https://doi.org/10.1057/9780230112025_5
3. CB Insignts: The 2014 U.S. Venture Capital Year in Review. Financing & Exit Analysis (2015)
4. Corsi, S., Di Minin, A.: Disruptive innovation … in reverse: adding a geographical dimension to disruptive innovation theory. Creativity Innov. Manage. **23**(1), 76–90 (2014)
5. Markopoulos, E., Markopoulos, P., Liumila, M., Almufti, Y., Romano, C.: Digital cultural strategies within the context of digital humanities economics. In: Ahram, T. (ed.) AHFE 2019. AISC, vol. 973, pp. 283–295. Springer, Cham (2020). https://doi.org/10.1007/978-3-030-20476-1_29
6. European Commission . https://ec.europa.eu/commission/presscorner/detail/en/IP_18_3901
7. Cultnat . https://www.cultnat.org/
8. The Software Engineering Competence Center (SECC). https://www.secc.org.eg/
9. TIEC . https://www.tiec.gov.eg/English/Pages/default.aspx
10. Trading economics. https://tradingeconomics.com/egypt/gdp-growth
11. World Intellectual Property Organization. https://www.wipo.int/pressroom/en/articles/2019/article_0008.html
12. The McKinsey Global Institute. https://www.mckinsey.com/featured-insights/innovation-and-growth/outperformers-high-growth-emerging-economies-and-the-companies-that-propel-them
13. Morgan Stanley Capital International. https://www.msci.com/market-classification
14. Markopoulos, E., Vanharanta, H.: Democratic culture paradigm for organizational management and leadership strategies - the company democracy model. In: Charytonowicz, J. (ed.) Advances in Human Factors and Sustainable Infrastructure. 5th International Conference on Applied Human Factors and Ergonomics, vol 20. pp 190–201 (2014)
15. Markopoulos, E., Gann, E.L., Vanharanta, H.: Democratizing new product development through an industry-society entrepreneurial partnership. In: Ahram, T., Karwowski, W., Pickl, S., Taiar, R. (eds.) IHSED 2019. AISC, vol. 1026, pp. 829–839. Springer, Cham (2020). https://doi.org/10.1007/978-3-030-27928-8_126
16. Markopoulos, E., Aggarwal, V., Vanharanta, H.: Democratization of intrapreneurship and corporate entrepreneurship within the mckinsey's three horizons innovation space. In: Ahram, T., Karwowski, W., Pickl, S., Taiar, R. (eds.) IHSED 2019. AISC, vol. 1026, pp. 1007–1017. Springer, Cham (2020). https://doi.org/10.1007/978-3-030-27928-8_150
17. Markopoulos, E., Vanharanta, H.: Public sector transformation via democratic governmental entrepreneurship and intrapreneurship. In: Ahram, T., Karwowski, W., Pickl, S., Taiar, R. (eds.) IHSED 2019. AISC, vol. 1026, pp. 867–877. Springer, Cham (2020). https://doi.org/10.1007/978-3-030-27928-8_131
18. Markopoulos, E., Kirane, I.S., Piper, C., Vanharanta, H.: Green ocean strategy: democratizing business knowledge for sustainable growth. In: Ahram, T., Karwowski, W., Pickl, S., Taiar, R. (eds.) IHSED 2019. AISC, vol. 1026, pp. 115–125. Springer, Cham (2020). https://doi.org/10.1007/978-3-030-27928-8_19

19. Markopoulos, E., Umar, O., Vanharanta, H.: Agile start-up business planning and lean implementation management on democratic innovation and creativity. In: Ahram, T., Karwowski, W., Pickl, S., Taiar, R. (eds.) IHSED 2019. AISC, vol. 1026, pp. 885–895. Springer, Cham (2020). https://doi.org/10.1007/978-3-030-27928-8_133
20. Markopoulos, E., Vanharanta, H.: Project teaming in a democratic company context. Theor. Issues Ergon. Sci. **19**(6), 673–691 (2018)
21. Entrepreneur Europe. https://www.entrepreneur.com/article/288769

Balancing Innovation and Tangibility Using the Spiraled Agile Design Sprinting Approach

Sylvain Bruni(✉)

Aptima, Inc., 12 Gill Street, Suite 1400, Woburn, MA 01801, USA
sbruni@aptima.com

Abstract. Whether they seek to conduct scientific research or product development, practitioners of applied human factors and cognitive systems engineering must carefully consider how their resources shall be allocated to both yield innovation while ensuring tangibility to the outcomes of their work. For small businesses, academic teams, and consultants, these two objectives are oftentimes antithetic: Expanding resources towards one is done at the expense of the other. To solve this gap, we propose the Spiraled Agile Design Sprinting Approach (SPADES), an activity and schedule-based framework within which all stakeholders collaboratively optimize what they do and how, while ensuring that neither innovation nor tangibility is sacrificed for the other. We applied SPADES to research efforts in critical domains, ranging from aircraft maintenance and intelligence analysis to missile defense analytics and emergency management. We report here on the lessons learned over the course of these efforts.

Keywords: Innovation · Tangibility · Design thinking process · Agile development · Sprints · SAFe · Project and program management · Product design and development

1 Introduction

Practitioners of applied human factors and cognitive systems engineering are required to balance the need to innovate while ensuring tangible outcomes for the scientific and technological activities they perform [1]. From consumer products to highly specialized technology devices, researchers, designers, and engineers seek to combine best practices in their respective areas [2]: human-centered analysis, research, and testing to ensure adherence of the product or service with how humans think or behave; iterative and incremental user experience design to guarantee usability, utility, delight, and trust; and agile prototype development to assure quality, scalability, and resilience of the end product. In turn, the project or program managers must balance antagonistic constraints in allocating resources towards innovating (increased front-end analysis, cycles of design, and iterative assessment) versus building (more back-end implementation, quality assurance testing, and end-user evaluations).

E. Markopoulos et al. (Eds.): AHFE 2020, AISC 1218, pp. 17–23, 2020.
https://doi.org/10.1007/978-3-030-51626-0_2

2 Motivation

While the balance between innovating and building products, systems, and solutions (whether for commercialization or pure research) can be achieved in large, multi-specialized, and well-funded organizations, small businesses and academic institutions must often partially sacrifice innovation for tangibility, or vice-versa.

For example, a typical method for human factors practitioners to understand the users, the context of use, or the job-to-be-done consists in performing a Cognitive Task Analysis (CTA) [3, 4]. However, the cost and logistics of conducting a full CTA are oftentimes beyond reach for small teams on small budgets, leading our community of practitioners to adopt variants on the CTA, such as the applied CTA [5] or the hybrid CTA [6].

More recently, the applied human factors and cognitive systems engineering communities have started adopting and practicing design thinking activities [7, 8] like agile sprints [9] or spark sprints [10] that combine upfront, design-oriented sessions centering the human end-user and their pain points, with iterative prototyping and testing. Although these techniques do not necessarily generate the critical artifacts typical of applied human factors processes, they do shed light on the cognitive aspects of end-users' work and their work environment. But the breadth and depth of the analysis, which, in turn, drive innovation, is by design (so to speak) limited to make way for building activities such as rapid prototyping.

In larger efforts that require additional development (e.g., "runway building," back-end infrastructure, data wrangling and feeding) or that go beyond the scope of scientific research (i.e., products meant to be deployed 'in the real world'), these design activities are subsumed to DevOps [11] or DevSecOps [12]. Intrinsically, these development approaches emphasize continuous development, that is, tangibility, over deep innovation.

The Scalable Agile Framework (SAFe) [13] provides a structured approach to combine most of the aforementioned approaches. But SAFe is prohibitively expensive to many practitioners: There is not sufficient time, money, and personnel to do it all. In critical domains such as healthcare and defense, this conundrum is further complexified with requirements like information security, privacy, or capability maturity [14].

3 SPADES

To resolve this gap, we propose the Spiraled Agile Design Sprinting (SPADES) methodology. SPADES is intended as an activity and schedule-based framework within which all stakeholders collaboratively optimize what they do and how, while ensuring that neither innovation nor tangibility is sacrificed for the other.

SPADES was established by our cognitive systems integration (CSI) team, which is composed of human factors and cognitive systems engineers, cognitive scientists, software and quality assurance (QA) engineers, data and machine learning (ML) scientists, product owners, and scrum masters. Typically, our CSI teams works on Small Business Innovation Research (SBIR) and Small Business Technology Transition (STTR) contracts to research innovative *and* tangible solutions for our government customers. SBIR and STTR efforts are examples of projects where small teams work with small budgets to seek forward-leaning, yet concrete and feasible solutions to urgent and critical problems.

SPADES entails successive spirals of agile design activities that follow the SAFe structure, preceded by a human factors analysis of the domain, data, and users. The concept at the core of SPADES is that of the "*technical nugget*," defined as a key, prioritized capability that solves a specific user pain point and is understood by all team members. Table 1 lists examples of technical nuggets for various critical domains.

Table 1. Examples of SPADES technical nuggets for various domains.

Domain	Example technical nugget
Maintenance	A machine learning-based algorithm to predict part failures in the next couple of weeks, and to alert a maintenance technician
Intelligence analysis	A chat and voice-based interactive modality to search for documents in large databases
Missile defense analytics	A lossless dimensionality reduction algorithm to support human exploration of petabytes of complex modeling and simulation data
Emergency management	A method and interface to learn from human experts how to handle multiple system emergencies in drones

As described in Fig. 1, SPADES is structured in three phases:

1. *An upfront analysis for discovery and understanding*, which leads typically to the production of general design artifacts and the identification of the key technical nuggets of interest;
2. *A series of spirals*, which combine innovation design with tangible development in successive development sprints, each focusing on a single technical nugget; and
3. *A concluding sequence*, typically employed for writing a final report and debriefing customers.

3.1 Upfront Analysis

The pre-spirals analysis of SPADES is calibrated to last for about one-half spiral and to target the identification and specification of the technical nuggets to be developed in the spirals. In this part of the process, the team selects the combination of human factors and design thinking methods they are comfortable with performing *within the time and resource constraints* imposed by the schedule. We have found that a combination of iterative Creative Briefing or spark sprints [10] with the hybrid CTA methodology [6] typically supports a deep and wide analysis for innovation. In so doing, the team not only builds empathy (a key design thinking concept) for the end users, but also gets to understand the cognitive underpinnings of the problems to solve. This upfront analysis supports the creation of high-level use cases and prioritized system requirements. In turn, those help narrowing down which technical nuggets are critical to building a solution.

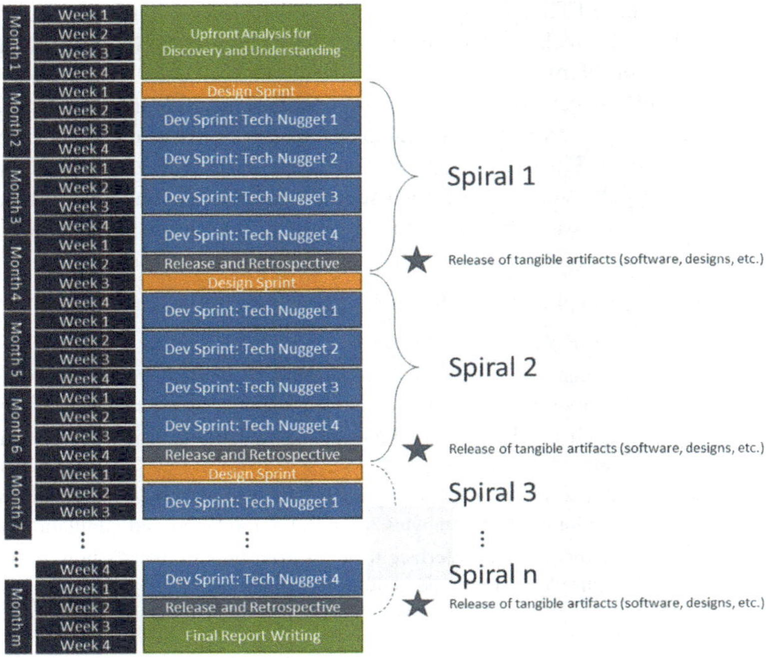

Fig. 1. The Spiraled Agile Design Sprinting (SPADES) approach starts with an upfront analysis to identify technical focus areas, continues with a series of spirals within which development sprints successively focus on each technical nugget, and ends with a closing phase for reporting.

3.2 Spirals

Each spiral in SPADES starts with a design sprint [9] or its scaled-down version, the spark sprint [10], should resources constrain the team, for the purpose of focusing and scoping the spiral *technically* (as opposed to capability-wise, which is done in the upfront analysis). This innovation-focused sprint is constrained to a single week and generates additional or more refined design artifacts as well as a prioritized backlog of tasks and user stories, per the SAFe approach [13]. This sprint constitutes the SPADES equivalent to the SAFe Program Increment Planning sessions.

Each spiral then continues with a series of two-week agile development sprints, one for each technical nugget (as identified in the upfront analysis). Each sprint employs the traditional agile ceremonies (scrums, stand-ups, backlog grooming, retrospectives, and demonstrations). During these sprints, team members adopt key roles such as product owner (who generates and specifies user stories to represent the voice of the user), scrum master (who orchestrates the work and removes logistical impediments), and dev team member (e.g., encompassing front-end designers, software engineers, quality assurance engineers, data scientists and modelers). The driving force for each sprint is the constraint that *one iteration of the technical nugget must be completed within two weeks.* This unique focus shared by the team over a defined period of time ensures agility and productivity.

Finally, each spiral ends with a one-week retrospective analysis, which identifies and prioritizes gaps for subsequent iterations. During that week, the tangible artifacts, whether they are software instantiations, or design artifacts, or artificial intelligence (AI) and ML models, are released and demonstrated to the customer. Demonstration also constitute opportunities for collecting feedback from external stakeholders.

3.3 Concluding Sequence

After the series of spirals is completed, a project is effectively closed with a reporting sequence where all the artifacts designed and prototyped in the spirals (illustrating both innovation and tangibility) are documented per the contractual requirements set by the customer. The duration of this sequence varies from contract to contract, based on how many spirals are feasible within the budget and time constraints under which the team operates.

4 Application and Lessons Learned

Our team has employed SPADES on SBIR and STTR research projects for several services and agencies in the Department of Defense (Table 2). Recognizing that our customers require more than reports at the end of Phase I projects, we applied the SPADES approach to efforts ranging from 6 to 12 months, for the specific purpose of building a tangible proof-of-concept or prototype, without sacrificing the innovative requirements.

Table 2. Examples of innovative and tangible outcomes achieved using SPADES on SBIR and STTR Phase I efforts.

Agency	Domain	Innovation	Tangibility
Army	Intelligence analysis	Context-based multi-horizon relevance algorithms to identify when documents may be useful	Scoped implementation of the algorithms on limited data with front-end instantiation
Army	Mission command	Hybrid methodology (combining human expertise and AI reasoning) for pairing ML approaches to mission command tasks	Ontologies of machine learning approaches and of mission command tasks; instantiation of the methodology for one use case
Navy	Maintenance	7-D technical data representation for supporting distributed asynchronous maintenance teams	Mixed reality prototype of technical data representation automatically ingested from S1000D
Air Force	Intelligence analysis	Mixed modality interactions for collaborative high-volume document analysis	Interactive chat prototype to query a database using natural language inputs
Missile Defense Agency	Simulation analytics	Hybrid lossless and controlled-loss dimensionality reduction for large simulation datasets	Scoped data processing feeding an interactive virtual reality prototype to manipulate data in 3D

Based on the application of SPADES to the efforts listed in Table 2, we share with our fellow practitioners the following lessons learned.

SAFe Ceremonies Are Key to Focus. We have found that adopting all the SAFe ceremonies, which engage all team members, even for as little as 15 min, drastically improves team collaboration, awareness, and understanding. The benefits included delay reduction (e.g., through the immediate sharing of design artifacts to software engineers), less re-do or iterations (e.g., because of a better alignment of middleware analytics models with data feeds and front-end interactions), and better adherence to the scope, thereby reaching a more resilient and tangible outcome faster.

It's Ok to Be Flexible. None of our CSI team member works full time on a single project. Rather, each SPADES practitioner splits their time between several efforts, thereby reducing staff availability and overlap. But we have witnessed that, despite split schedules, our team members were able to operate within the time bounds of each SPADES phase, provided our scrum master built in flexibility in the timing and duration of ceremonies and in the scope of each development sprint within each spiral.

Quality Results from Managed Focus. Between the rigor of SAFe ceremonies and the flexibility required to conform to a small business' staffing model, we realized that managing the focus of the team was the key driver to the quality of innovation and tangibility. Because highly skilled and educated workers consistently strive to 'do more' and dig deeper into the science and novelty of a domain or problem, it is necessary for the product owner and the scrum master to diligently keep the team on task. The initial backlog and use cases of the upfront analysis set the first general boundaries. The scope and focus collected adopted at the start of each spiral refines those boundaries to design and development that is feasible during the spiral. Managing each team member's focus to stay within those refined boundaries is essential to yield quality products.

5 Conclusion and Next Steps

We have presented here a novel method, SPADES, that structures research and development activities in a manner to balance the needs for innovation and tangibility, within the time and resource constraints of small teams working with a limited budget. We exemplified our application of SPADES to Phase I SBIR and STTR contracts and described how our CSI team was able to achieve balance between innovation and tangibility. We finally identified a series of lesson learned, tackling ceremonies, flexibility, and focus management.

We recognize both that (a) this paper is a limited account of our own experience, (b) we have only applied SPADES to one type of contract (although we are now embarking on Phase II efforts which will also leverage SPADES), and (c) that there may exist additional methods or approaches employed by other practitioners in the human factors and cognitive systems engineering communities to reconcile the combined needs for innovation and tangibility.

Therefore, we see our contribution as an invitation to our fellow researchers, scientists, technologists, and product developers, particularly in small businesses and in academia, to share with us their thoughts and feedback on SPADES.

Acknowledgments. The design and implementation of SPADES would not have been possible without the contributions and feedback of many Aptima staff, including Kara Orvis, Adam Fouse, Diane Miller, Chad Weiss, and Mary Freiman. Similarly, the author wishes to thank our Government customers in the Department of Defense who have supported the implementation of SPADES through several SBIR and STTR contracts. The views and conclusions presented in this paper are those of the author and do not represent an official opinion, expressed or implied, of the Department of Defense.

References

1. Skarzynski, P., Gibson, R.: Innovation to the Core: A Blueprint for Transforming the Way Your Company Innovates. Harvard Business School Press, Boston (2008)
2. Fouad, F., Tourabi, A., Lakhnati, G.: The impact of tangible and intangible innovation activities on the new product objectives in three phases of the innovation process: case of the fish industry—Morocco. Technol. Invest. **8**, 11–32 (2017)
3. Klein, G.: Naturalistic decision making: Implications for design. CSERIAC, Wright-Patterson AFB, OH. (Contract No. DLA900-88-0393) (1993)
4. Crandall, B., Klein, G., Hoffman, R.R.: Working Minds: A Practitioner's Guide to Cognitive Task Analysis. MIT Press, Cambridge (2006)
5. Militello, L.G., Hutton, R.J.: Applied cognitive task analysis (ACTA): a practitioner's toolkit for understanding cognitive task demands. Ergonomics **41**(11), 1618–1641 (1998)
6. Nehme, C., Scott, S., Cummings, M., Furusho, C.: Generating requirements for futuristic heterogeneous unmanned systems. In: Proceedings of the Human Factors and Ergonomics Society, San Francisco, pp. 235–239 (2006)
7. Brown, T.: Presentation on innovation through design thinking at MIT World Forum, 16 March 2006. http://mitworld.mit.edu/video/357/
8. Brown, T.: Design thinking. Harv Bus. Rev. **86**(6), 84–92 (2008)
9. Knapp, J., Zeratsky, J., Kowitz, B.: Sprint: How To Solve Big Problems and Test New Ideas in Just Five Days. Simon & Schuster, New York (2016)
10. Banfield, R.: Enterprise design sprints (2019). https://www.designbetter.co/enterprise-design-sprints
11. Dyck, A., Penners, R., Lichter, H.: Towards definitions for release engineering and DevOps. In: IEEE/ACM 3rd International Workshop on Release Engineering, p. 3 (2015)
12. Myrbakken, H., Colomo-Palacios, R.: DevSecOps: a multivocal literature review, pp. 17–29 (2017). https://doi.org/10.1007/978-3-319-67383-7_2
13. Leffingwell, D., Knaster, R., Oren, I., Jemilo, D.: SAFe® 4.5 Reference Guide: Scaled Agile Framework® for Lean Enterprises. Addison-Wesley Professional, Boston (2018)
14. CMMI Institute: CMM development V1.3 quick reference guide (2010). https://cmmiinstitute.com/resource-files/public/marketing/document/cmm-development-v1-3-quick-reference-guide

Social Entrepreneurship in Cross-Cultural Context: Multiple Value Creation

Florentin Popescu(✉), Ingrid Van Rompay-Bartels, Jorcho van Vlijmen, Annet de Lange, and Koos Wagensveld

HAN University of Applied Sciences, Ruitenberglaan 31, 6826 CC Arnhem, The Netherlands
{florentin.popescu, ingrid.bartels, jorcho.vanvlijmen, annet.delange, koos.wagensveld}@han.nl

Abstract. This paper proposes a new model to conduct case comparative studies with respect to social entrepreneurship and sustainable development goals; examining different organizational processes, stakeholder perspectives and taking multiple value-creation into account (social, human, financial, intellectual, natural, and manufactured). This model was developed based on literature research and several pilot case studies of cases of social entrepreneurship in the region of Arnhem and Nijmegen (the Netherlands) with relevant social entrepreneurship in the region of Bangalore (India). Social entrepreneurs should identify, evaluate and prioritize sustainable development issues which maximize outcomes for the six capitals and hence their contribution to the UN Sustainable Development Goals (SDG) targets. To create social value, the activities of social entrepreneurs represented in their business models require inputs, transformation and output in terms of the six capitals. The authors propose a new Social Entrepreneurship Model that frames social entrepreneurship in a cross-cultural context where co-creation is essential by providing stakeholders with knowledge, insights and skills in how they can overcome social problems and improve communities in their local contexts.

Keywords: Social Entrepreneurship · Sustainable Development Goals · Cross-cultural studies · Value co-creation

1 Introduction

Growth is essential for development; however, it is better and more sustainable when it reduces poverty by improving human capital employability and inclusiveness [1]. There is a huge need to adapt and to embrace a more sustainable world that creates more inclusiveness of human capital. Social entrepreneurship is the use of start-up companies and other types of entrepreneurships to develop, fund and implement solutions to social, cultural, or environmental issues. This concept may be applied to a variety of organizations with different sizes, aims, and beliefs.

The difference between social and business entrepreneurship has been clarified by different authors. According to Austin et al. [2], the main difference between social and other types of entrepreneurship lies in the purpose the firm is established; they portray

E. Markopoulos et al. (Eds.): AHFE 2020, AISC 1218, pp. 24–31, 2020.
https://doi.org/10.1007/978-3-030-51626-0_3

social entrepreneurship as an innovative, social, value-creating activity that can occur within or across the non-profit, business, and public sectors [2]. This was supported by Cukier et al. [3], stating that social entrepreneurship focuses primarily on activities with social purposes whereas business entrepreneurship focuses on activities with business profit making purposes.

Entrepreneurs focused on profit typically measure performance using business metrics such as sales, employment, and stock price changes, but social entrepreneurs are either non-profit or combine for-profit objectives with a positive “return to society” and thus use different metrics. Social entrepreneurship typically attempts to further broad social, cultural, and environmental goals often associated with the voluntary sector in areas such as poverty alleviation, healthcare and community development. Profit-making social enterprises may be established to support the social or cultural goals of the organization.

While addressing and facilitating change within the society, social entrepreneurship activities can positively influence the economic growth and social development of the society through reducing poverty and improving large scale economic development [4]. For example, an organization that aims to provide housing and employment to the homeless may operate a restaurant or a paper recycle company, both to raise money and to provide employment for the homeless or human capital not active on the labor market.

Social entrepreneurship has been developed in different cultural contexts and has become an internationally recognized phenomenon. However, we still lack cross-cultural case comparative studies meaning effects of different types of social entrepreneurship (e.g., type of industry or work, type of included human capital) on multiple stakeholders and value creation outcomes with respect to the six capitals (financial, manufactured, intellectual, human, social & relationship and natural) across time (see Fig. 1) [5, 6].

This research has a three folded impact: Academic, Societal and Institutional. The project contributes to knowledge by attempting to quantify how can we create and use social safety nets to enable environmental sustainability and human capital employability through local communities [7, 8] where students’ involvement is essential. Moreover, the idea that society, governments, and new business stakeholders depends on universities to translate their innovative potential into products and services that impact society is relevant and actual to this research as well [9]. Preparing students as professionals for continuous development in their professional practice using different social environments, cultural differences and local social entrepreneurial behavior as frameworks for comparison is paramount for the academic development (use of comparative case studies). Therefore, the impact is obvious.

In a world that is globalizing rapidly, in which the central role of universities in the knowledge economy and in civil society is articulated more strongly and more widely than ever, we do not have a clear sense of what it takes or what it means to understand the relationships between this type of social entrepreneurship and created added values. This project clearly strives to contribute to this discourse.

In a conceptual context, this project is exploring the various tiers of social entrepreneurship and tries to equate them to the overall responses to the talent development of stakeholders from a social and cultural perspective. This project will also

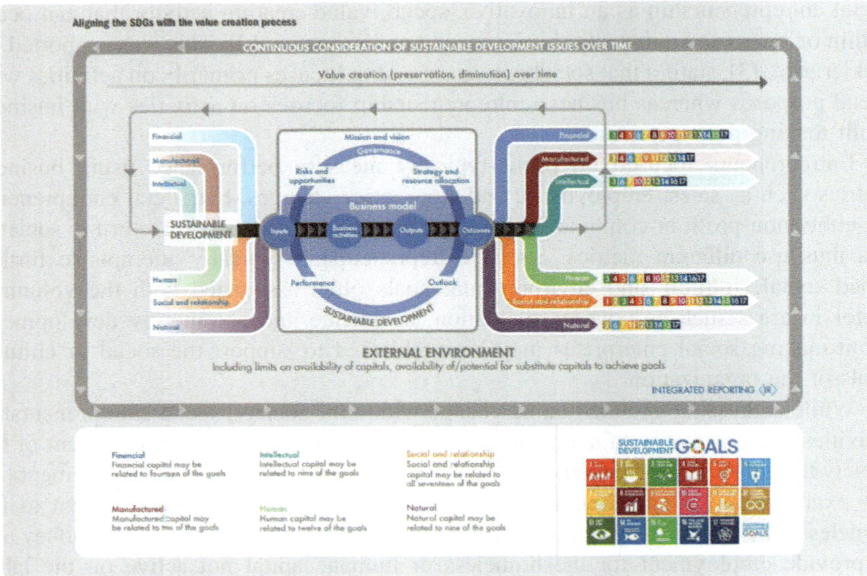

Fig. 1. Integrated Reporting: multiple stakeholder perspective and value-creation in relation to sustainable development goals [5, 6]

benefit other research projects and case studies through the co-creation of knowledge, skills, and insights in lessons learned in local and across borders social entrepreneurship. This new knowledge and skills will be translated into different domains when necessary.

2 Methodology Proposed

This research is based on comparative multiple-case studies with an equal number of cases for the different countries of cases of social entrepreneurship in the region of Arnhem and Nijmegen (The Netherlands) with relevant social entrepreneurship in the region of Bangalore (India). More specifically, the following cases of social entrepreneurship in The Netherlands included: Enspiratie (https://www.enspiratie.nl/), Scalabor (https://www.scalabor.nl/), Intratuin (https://www.intratuin.nl/) versus the following cases: Parivarthana - Christ University (http://csa.christuniversity.in/parivarthana), Beautiful Minds Astitva (http://www.astitvaforspecialneeds.com), Aara Foundation, New Delhi (http://www.aarafoundation.com/) in India. The choice for multiple-case studies is based on the complexity to understand the relations between type of social entrepreneurship and the created values in different national contexts. According to Yin [10] case studies are "an empirical enquiry to investigate a contemporary phenomenon in real-life context, especially when the boundaries between phenomenon and context are not clearly evident"

[10]. Therefore, multiple-case studies allow the researchers to obtain a better understanding of the selected countries in how socio, economic, political and cultural forces may impact and shape the organization of social entrepreneurship in their environments.

The type of case study defined for this research is descriptive in nature. The selection of this type started with the challenge to understand the linkages between social entrepreneurship in the socio-cultural context. A descriptive case study provides opportunities to the researchers to describe the knowledge of social entrepreneurship processes and their linkages with culture (beliefs, practices, norms and values). The multiple-case studies follow a protocol in which participative observations and a set of semi-structured questions is selected for the fieldwork, with a focus on the context at micro level. It means that the beliefs, the practices, the norms and the values of the respondents engaged in social entrepreneurship are in depth questioned and described.

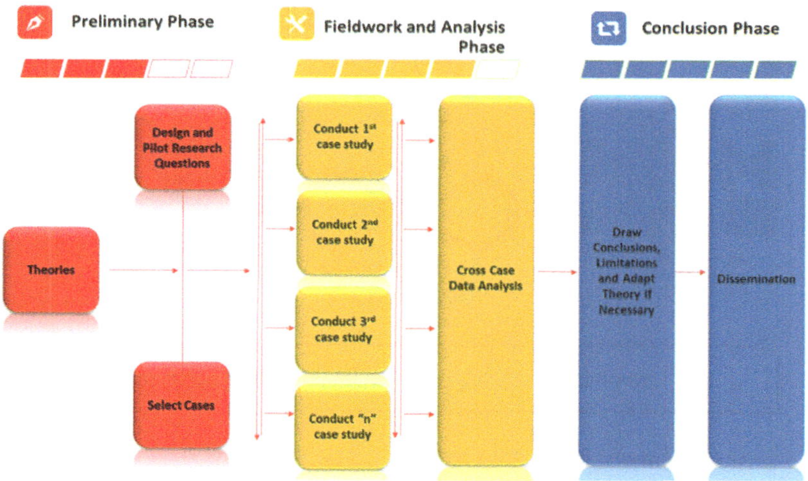

Fig. 2. Social entrepreneurship in cross-cultural context methodology

3 Research Model Design Proposed

Drawn upon the multiple stakeholder perspective and value-creation in relation to sustainable development goals model [5, 6] and the Corporate Social Entrepreneurship model [11] the authors created "Social entrepreneurship in cross-cultural context" Model.

This model is the conceptual model underlying the research project and it functions as the basis for case studies. Central to this model is the notion that organizations do not function in a vacuum, but rather, are connected with and dependent on their surroundings. They rely on their environment for financial, human and physical resources. The internal organizational processes then transform these inputs, into outputs such as services, products, knowledge and other gains. In the case that the context regards the transformation of inputs to contribute to economic growth and/or social development the organization has created added value and it will remain possible to extract resources [4]. In the case that there is no perceived added value, the organization will lose its legitimacy. Organizations, therefore, are organismic entities [12]. This organismic view on organizing is consistent with the idea that social entrepreneurship is closely related to society and its challenges and problems, and cannot be understood without understanding the relations between the organization and its context. Hence the model consists of three major parts: Input, Transformation and Output, where various dynamics are interlinked, studied and analyzed to add value for multiple value-creation (social, human, financial, intellectual, natural, and manufactured) with respect to social entrepreneurship (Fig. 3):

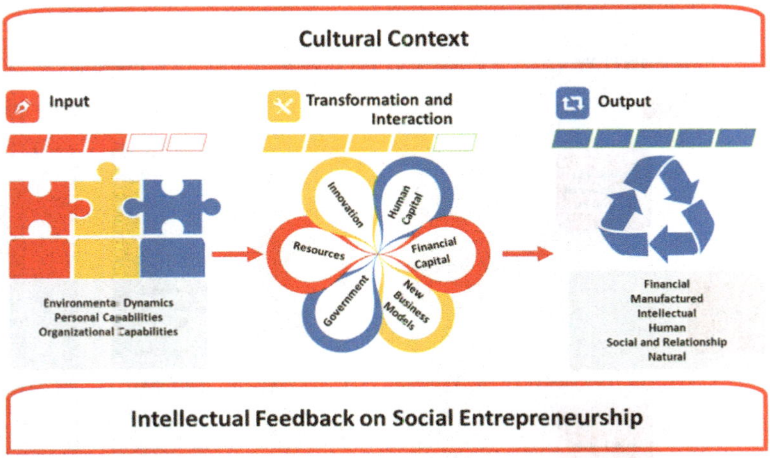

Fig. 3. Social entrepreneurship in cross-cultural context

3.1 Input

What does the social entrepreneurial organization need or use in order to function? Three dimensions are distinguished. Environmental dynamics concerns the organization's external context. Demographical, economic, social, technological, ecological and political (DESTEP) factors may be useful in comprehending environmental dynamics of social entrepreneurship. Agrahawal en Sahasranamam [11] refer to the relevance of socio-economic conditions, problems and relevant regulations. Different aspects such as the prevalence of social problems, government regulation, religion, etc. are also taken into account.

The second dimension in the first part of the model is the personal capabilities of the entrepreneur (entrepreneurial skills, personal growth, moral and religious motivation and personal resources - skills, financial, network etc.). Van den Dool [13] describes how social innovation starts with an entrepreneur that identifies a (social) problem, then starts to relate to it, and eventually decides to do something about it. In doing this the entrepreneur makes use of his/her talents, skills, and resources.

The third dimension is the organization's internal environment where the organization's history, capabilities, resources, inclusiveness and talent development within the organization are relevant in supporting social entrepreneurship approach in doing business.

3.2 Transformation and Interaction

Just like 'regular' enterprises, social enterprises transform input into output. This part of the model focuses on how these 'normal' and social enterprises are similar or different: how they, in their own way, manage and balance their processes from the perspective of the different goals the organization may have. These transformational processes may include resource commitments, renewal and improvement of products, services and processes within smart industries, social innovation and creation of a new business model, developing collaborations with NGO's, local entrepreneurs and government, balancing between social and organizational goals.

3.3 Output

According to the model of multiple value creation enterprises can create value in multiple ways (new financing and revenue models, new market creation, new customer acquisition, new business models, human capital development while building trust and goodwill within the community, legitimacy and new business collaboration, etc.). The model differentiates between financial, manufactured, intellectual, human, social and relational, and natural value. We have added two more concepts to the model. The model enables us to investigate social entrepreneurship in different cultural contexts throughout case studies. Therefore the cultural context is added. Social enterprises develop knowledge and skills concerning social entrepreneurship, and can be considered as intellectual value. Since the aim of this applied research project is to create knowledge on the concept of social entrepreneurship the organization's feedback loop was added to the model.

4 Conclusions and Further Research

The topic of social entrepreneurship in relation to the sustainable development goals is becoming more and more urgent considering complex societal issues like climate change, growing poverty etc. To date, no conceptual model was available to further examine this topic in an applied research setting. As a result, we have developed a new conceptual model based on earlier research and new case studies highlighting the importance of input, transformation and output processes in relation to the six capital values. This research project is looking for an applied research perspective in implementing results and findings, where conclusions and recommendations are intended to address different as well as interlinked actors: higher education institutions involved, companies and NGO's participating, etc.

The results will form (conform the knowledge transfer triangle: Education, Research, Professional Field) a basis of transferrable international social skills and knowledge for the stakeholders involved; various business models on how to renew and improve products, services and processes within smart industries involved in the project and beyond; motivate and social responsibility awareness to all the stakeholders.

As future research agenda, we would like to call for new research to further test and validate the current conceptual model (see Fig. 2) in different contexts (small, medium sized companies versus multinationals, different sectors etc.) to develop interventions to facilitate and develop social entrepreneurship on a greater global scale. Developing more integrative research studies examining the whole process of input, transformation and output and their contribution to the UN Sustainable Development Goals (SDG) targets is inherent to further proposed research projects.

Acknowledgments. We would like to thank all of our stakeholders Enspiratie (https://www.enspiratie.nl/), Scalabor (https://www.scalabor.nl/), Intratuin (https://www.intratuin.nl/), Parivarthana (http://csa.christuniversity.in/parivarthana), Beautiful Minds Astitva (http://www.astitvaforspecialneeds.com), Aara Foundation, New Delhi (http://www.aarafoundation.com/) who made this research possible, especially Arnhem Business School, The Netherlands (students assistants Hoda Maksoud and Alex Benz) and Christ University, Bangalore, India.

References

1. United Nations. The sustainable development goals report 2018. https://unstats.un.org/sdgs/files/report/2018/TheSustainableDevelopmentGoalsReport2018-EN.pdf
2. Austin, J.E., Stevenson, H., Wei-Skillern, J.: Social and commercial entrepreneurship: same, different, or both? Entrepreneurship Theory Pract. **30**(1), 1–22 (2006)
3. Cukier, W., Trenholm, S., Carl, D., Gekas, G.: Social entrepreneurship: a content analysis. J. Strateg. Innov. Sustain. **7**(1), 99–119 (2011)
4. Zahra, S.A., Gedajlovic, E., Neubaum, D.O., Shulman, J.M.: A typology of social entrepreneurs: motives, search processes and ethical challenges. J. Bus. Ventur. **24**(5), 519–532 (2009)
5. Adams, C.A.: The international integrated reporting council: a call to action. Crit. Perspect. Acc. **27**, 23–28 (2015)
6. Adams, C.A.: The sustainable development goals, integrated thinking and the integrated report. https://integratedreporting.org/wp-content/uploads/2017/09/SDGs-and-the-integrated-report_full17.pdf
7. Detaille, S., de Lange, A.H.: Fit for the Future. Futureproof HR Policies. Vakmedianet, Alphen aan den Rijn (2018)
8. De Lange, A.H., Kooij, D., van der Heijden, B.I.J.M.: Human resource management and sustainability at work across the lifespan: an integrative perspective. In: Finkelstein, L.M., Truxillo, D., Fraccaroli, F., Kanfer, R. (eds.) Facing the Challenges of a Multi-Age Workforce: A Use-Inspired Approach, pp. 50–79. Routledge, London (2015)
9. Leloux, M., Popescu, F., Koops, A.: New skills for entrepreneurial researchers. In: Kantola, J., Barath, T., Nazir, S., Andre, T. (eds.) Advances in Human Factors, Business Management, Training and Education. Advances in Intelligent Systems and Computing, vol. 498, pp. 1251–1263. Springer, Cham (2017)

10. Yin, R.K.: Case Study Research: Design and Methods, 3rd edn. Sage, Thousand Oaks (2003)
11. Agrawal, A., Sahasranamam, S.: Corporate social entrepreneurship in India. South Asian J. Global Bus. Res. **5**(2), 214–233 (2016)
12. Morgan, G.: Images of Organization, Executive edn. Berrett Koehler Publishers/Sage Publications, San Francisco/Thousand Oaks (1998)
13. Van den Dool, E.C.: Spiritual dynamics in social innovation. An organizational context, lived spirituality and a school of spirituality (2017)

Tool Development in Support of Story-Planning for Design Pitching

Qiong Peng(✉) and Jean-Bernard Martens

Eindhoven University of Technology, Eindhoven, The Netherlands
{q.peng,J.B.O.S.Martens}@tue.nl

Abstract. Industrial design students generally have difficulty in making good design pitching to convincing the audiences to subscribe a design concept. Despite the values of stories for design communication have been greatly proposed by literature, in students' practices there is usually no story or no complete story in their pitching. As what often been complained by the students that making stories takes effort and time, appropriate tools are promising to help. This paper presents an exploration of such kind of tools with the aim to support story-planning for design pitching. A tool called StoryPrea is introduced and a user test of this tool is reported. The results shown the positive role of the tool and also indicated the directions for design iteration in future work.

Keywords: Story · Story-planning · Design pitching · Tool development

1 Introduction

Incorporating stories into the design process has been promoted by many researchers and practitioners. While its multiple benefits for design are obviously highlighted, literature actually doesn't tell much about how to practically involve stories into design pitching. Meanwhile, the adoption of stories in design pitching by ID (Industrial design) students seems poor as observed in a previous study that they were prone to plan design pitching initially. On one hand, there was no story or no complete story in their pitching. ID students often focused on introducing what a design concept it is and its features. Some essential elements and details which are important to convince the audiences were neglected. On the other hand, it is seemed that students actually didn't accept stories. Despite they are often encouraged or even required to use stories, they primarily stressed the problem in ease of using stories by complaining it as an activity both effort-requiring and time-consuming. Given to this situation, appropriate tools are thus promising for help. This paper is actually an exploration of tools which can support ID students for planning stories for their design pitching. In this paper, a tool called StoryPrea is introduced and a user test for this tool is presented as well. Insights were collected from students' feedback.

There are two contributions in this paper: First, a tool called StoryPrea is developed as a feasible tool with the aim to support story-planning for design pitching. It is also an attempt to fill in the gap in literature which lacks enough empirical studies. Second, the exploration also provided evidence for our assumption that ID students' acceptance of stories could be enhanced with the support of this tool. The paper is structured in three

E. Markopoulos et al. (Eds.): AHFE 2020, AISC 1218, pp. 32–38, 2020.
https://doi.org/10.1007/978-3-030-51626-0_4

sections. We started with a broad review of literature about story/storytelling and the related tools. Next, the tool StoryPrea is introduced and a user test with its results are reported and discussed. We finally concluded by identifying the main insights obtained from the test and proposing the future work.

2 Related Work

Stories and storytelling have been early highlighted in presentations [e.g. 1–5] or pitches [e.g. 6, 7]. Richard Maxwell proposes use storytelling to pitch better [8] because the secret of selling lies in stories which are how we organize our minds. Weissman stresses to tell stories in presentation to win [2] as stories are value laden [9]. Jerry, W's suggestions of seven steps for crafting the content into a comprehensive and meaningful story [10] provides a general strategy for practitioners to build a story when preparing a prestation. Nancy, D highlights using storytelling for resonating and persuading the audience [5, 11]. Particularly, storytelling can facilitate design pitching as it owns advantages in creating a sense of empathy [12, 13], fueling imagination [14], building understanding [15], introducing change [16], and communicating a design solution in a human-centered way [17, 18]. Hence, the audience could easily understand the information like design background, user needs, using context, design impact of a design idea. However, despite designer are viewed as storytellers [19], there is not sufficient work and empirical studies on using stories in the context of design pitching. The storytelling approach-impact framework is a representative study which explores using storytelling in design pitching [20, 21], but they didn't offer a concrete method or tools that students can use for making a story for pitching. StoryPly [22] is a method for crafting stories, but it more fits for discussing user experience in conceptual design.

As to the typical tools, most of them are specifically supportive for presentations, such as PowerPoint-related tools in support of slides building and visual material management [e.g. 23]. Some are the ones related to presentation rehearsal-support [e.g. 24]. In recent years, online platform or cloud-based tools for presentation emerge greatly, such as Prezi [25], Visme, Microsoft Events, Haiku Deck, Pitcherific, Canva, SiteCamp, Powtoon, etc. These tools are developed by making use of the benefits of Internet and interactive technologies to support presentations by providing different kinds of support for information preparation, rehearsal and feedback. None of all these tools can actually work for the purpose of story-planning for design pitching. It seems necessary to explore and develop alternatives to fill this gap.

3 Introduction of the StoryPrea Toolkit

The StoryPrea tool is developed based on the similar elements in story and product design (shown in Table 1). The tested version of the StoryPrea tool is consisted of 11 ready-made templates and guidelines. The working scenario (shown in Fig. 1) is proposed to first put the templates on the wall or laid flat on a big table, and then fill in the blanks on the templates with post-it notes with the support of the guidelines.

Table 1. The similar elements in story and product design

Story elements	5W1H in product design	Design elements in details
Character	Who	Users, stakeholders
Setting	When, where	Time, place, environment
Activity	What, how	Behavior, activity
Conflict	Why	Design problem
Resolution	What, how	Design solution
Ending	Why	Impact/influence

Fig. 1. The working scenario of the StoryPrea tool

The 11 templates (examples are shown in Fig. 2) are categories into two groups. Template 1 to 5 in the first group support to think about what the design concept it is now. Template 6 to 11 in the second group support to plan a story for pitching. All the templates are in A3 size with titles, sub-titles, 8 blank squares, logo and page numbers. The titles explain the related design elements of a concept including who, when and where, what happens, design problem and design solution. Subtitles indicate five basic story elements including character, setting, activity, problem and solution. There are 8 blank squares with dot outlines on each template with the purpose to encourage users to fill in the information. Users can use post-it notes to fill in the blanks by writing text or sketching. Users also enjoy freedom to add, delete, and reposition the post-it notes on the templates for editing. The guidelines are proposed in the form of provocative questions with the aim to support users to fill in the blanks.

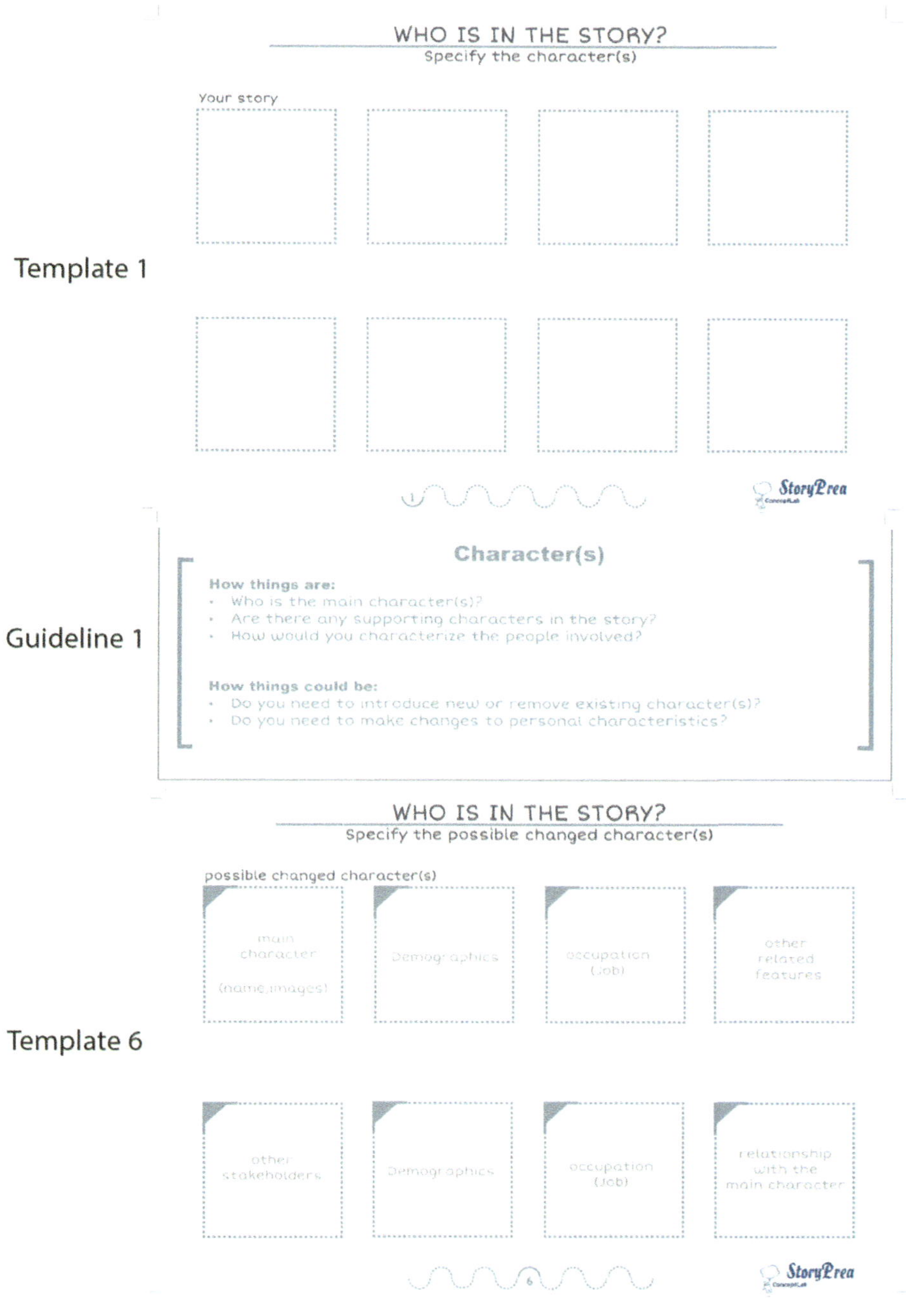

Fig. 2. Examples of the StoryPrea tool

4 Results

The StoryPrea was tested by 30 students with the background of Industrial design. Both the undergraduate students (55.6%) and master students (44.4%) participated in the test individually (16 students) or in groups (14 students, each group with 2 or 3 students for a same design project) to plan stories in order to pitch their design concept that they had finished. All the participants were involved in interviews after they used the tool, and the interviews were audio recorded and analyzed by clustering the 143 quotes into 5 categories (shown in Table 2).

Table 2.

Category	Quote no.	Quote example
[1] useful for story-planning	42	"It's useful to plan a story with the elements provided by the tool." "It makes create a story for pitching quite easy"
[2] ease of use	35	"It's easy to fill in the templates." "The tool is easy to use as we can only put them on the table and fill in the templates"
[3] problems	28	"I don't understand some of the guideline questions." "It's time-consuming. We spent more than 30 min to fill in all the templates"
[4] suggestions for improvement	25	"Maybe you can give more explanation or examples so that we can fully understand it." "Improve the visual design, or make it digital. I don't want to take many templates every time"
[5] enhanced awareness of stories	13	"Yes, by using this tool, I pay special attention to stories when planning a pitch." "We all know the importance of telling stories but we often neglect it. The tool not only reminds us of the story but also help us to create a story with the elements"

5 Discussion

To make a successful design pitch is never easy for ID students. As literature indicated, planning stories for pitching could increase the probability of success. With the help of appropriate tools, the story-planning would be feasible for most of the students, and much easier and effective to be applied. We thus developed the StoryPrea tool with this specific goal to support planning stories for design pitching. The overall opinions based on the feedback from interviews of the students demonstrated its usefulness in planning stories. The templates provided an environment in which ID students were pushed to re-look at their design concepts and think about the related aspects. The StoryPrea tool provided opportunities in which students can check the missing information and add essential details. For instance, the detailed information about the people that they designed for, such as their expectation and/or their behavior, made the character rich and vivid. Through the structured process of reflecting upon the design concepts and analyzing the story elements, to pitch a design concept by planning a concrete story becomes easy to be a success.

Two aspects of limitation are also reflected as follows. Firstly, we have to acknowledge that there is a space for optimization for the current version of the StoryPrea tool, both in its visual design and usability, as tools are supposed to support our work with less effort and time. Secondly, to make a new tool applicable and acceptable, a huge sample of the participants should be involved in a user test [15]. In our study, there were limited number of participants. Many other related aspects such as different understanding of story-planning in different cultural backgrounds, or different education environment for product design, for instance, students from an art college and students from a technology university may show different opinions to story and story-planning for design pitching, might need to be taken into consideration. Finally, to plan a story cannot absolutely ensure the success of design pitching. We didn't assess the final results of pitching. The conclusions were made only based on the feedback of the student users and by the analysis of qualitative data.

In the future, we will primarily focus on improving the problems that the participants encountered during their tool-using. Meanwhile, other aspects such as enhancing users' engagement and user experience, etc., will also be considered.

6 Conclusion

This paper presents the development and assessment of a tool called StoryPrea which is aimed to support planning stories for design pitching. The results have shown its potentials to support plan a story through analyzing the story elements. Meanwhile, ID students' awareness of using stories for design pitching seemed being enhanced through the process of tool-using. Despite the positive feedback from the students, we found some usability problems such as time-consuming, vague understanding of the guidelines, etc. still need to be solved. The direction for the future work has been also identified.

Acknowledgments. The authors would like to thank all the participants involved in the study.

References

1. Abela, A.: Advanced Presentations by Design: Creating Communication that Drives Action. Wiley, Hoboken (2008)
2. Weissman, J.: Presenting to Win: The Art of Telling Your Story. FT Press, Upper Saddle River (2008)
3. Reynolds, G.: Presentation Zen: Simple Ideas on Presentation Design and Delivery. New Riders, Berkley (2011)
4. Reynolds, G.: Presentation Zen Design: Simple Design Principles and Techniques to Enhance Your Presentations. New Riders, Berkley (2013)
5. Duarte, N.: Slide:ology: The Art and Science of Creating Great Presentations, vol. 1. O'Reilly Media, Sebastopol (2008)
6. Hankins, G.: The Power of the Pitch: Transform Yourself Into A Persuasive Presenter and Win More Business. Dearborn Trade Publishing, Chicago (2005)

7. Scott, K.: Story Pitch: The How to Guide for Using A Pitch to Create Your Story. CreateSpace Independent Publishing Platform, Scotts Valley (2017)
8. Maxwell, R., Dickman, R.: The Elements of Persuasion. HarperCollins, New York (2007)
9. Barker, R.T., Gower, K.: Strategic application of storytelling in organizations: toward effective communication in a diverse world. J. Bus. Commun. **47**(3), 295–312 (1973)
10. Wilson, J.S., Blumenthal, I.: Managing brand you: 7 steps to creating your most successful self. AMACOM DivAmerican Mgmt Assn (2008)
11. Duarte, N.: HBR Guide to Persuasive Presentations. Harvard Business Press, Boston (2012)
12. Lämsä, A.M., Sintonen, T.: A narrative approach for organizational learning in a diverse organisation. J. Workplace Learn. **18**(2), 106–120 (2006)
13. Buskermolen, D.Ö., Terken, J.: Co-constructing new concept stories with users. In: Collaboration in Creative Design, pp. 233–249. Springer, Cham (2016)
14. Tassinari, V., Piredda, F., Bertolotti, E.: Storytelling in design for social innovation and politics: a reading through the lenses of Hannah Arendt. Des. J. **20**(sup1), 3486–3495 (2017)
15. Lawson, B.: How Designers Think. Routledge, Abington (2006)
16. Boje, D.M.: Consulting and change in the storytelling organization. J. Organ. Change Manag. **4**(3), 7–17 (1991)
17. Erickson, T.: Design as storytelling. Interactions **3**(4), 30–35 (1996)
18. Quesenbery, W., Brooks, K.: Storytelling for User Experience: Crafting Stories for Better Design. Rosenfeld Media, New York (2010)
19. Bertolotti, E., Daam, H., Piredda, F., Tassinari, V.: The Pearl Diver. The Designer as Storyteller. DESIS Philosophy Talks-Dipartimento di Design, Politecnico di Milano (2016)
20. Parkinson, D.: Developing an Understanding of Storytelling at the Design Pitch: Relating Approach to Impact. Doctoral dissertation, Northumbria University (2014)
21. Parkinson, D., Warwick, L.: Stimulating thinking at the design pitch. Storytelling approach and impact. Des. J. **20**(suppl1), 4509–4518 (2017)
22. Atasoy, B., Martens, J. B.: STORYPLY: designing for user experiences using storycraft. In: Collaboration in Creative Design, pp. 181–210. Springer, Cham (2016)
23. Winn, J.: Avoiding death by PowerPoint. J. Prof. Issues Eng. Educ. Pract. **129**(3), 115–118 (2003)
24. Trinh, H., Yatani, K., Edge, D.: PitchPerfect: integrated rehearsal environment for structured presentation preparation. In: Proceedings of the SIGCHI Conference on Human Factors in Computing Systems, pp. 1571–1580. ACM (2014)
25. Perron, B., Stearns, A.: A review of a presentation technology: Prezi (2010)

Pink Ocean Strategy: Democratizing Business Knowledge for Social Growth and Innovation

Evangelos Markopoulos[1,2(✉)], Maria Barbara Ramonda[3], Lisa Marie Carolin Winter[4], Haseena Al Katheeri[5], and Hannu Vanharanta[6,7]

[1] School of Management, University College London, Gower Street, Bloomsbury, London WC1E 6BT, UK
e.markopoulos@ucl.ac.uk
[2] Faculty of Engineering and Business, Turku University of Applied Sciences, Joukahaisenkatu 3, ICT-City, C-Wing, FI-20520 Turku, Finland
[3] Centro Motor Toyota SA, Av Colón 5077, X5000 Córdoba, Argentina
bramonda@centromotorsa.com.ar
[4] UVEX Winter Holding GmbH & Co. KG, Wuerzburger Street 181, 90766 Fuerth, Germany
lisa.winter@uvex.de
[5] College of Technological Innovation, Zayed University, P.O. Box 144534 Abu Dhabi, UAE
Haseena.AlKatheeri@zu.ac.ae
[6] School of Technology and Innovations, University of Vaasa, Wolffintie 34, 65200 Vaasa, Finland
hannu@vanharanta.fi
[7] Poznań University of Technology, Plac Marii Skłodowskiej-Curie 5, 60-965 Poznań, Poland

Abstract. In a constantly and rapidly changing world, innovation and organizational sustainability keep on being redefined. The integration of Blue and Green Ocean strategies can deliver social innovation and impact directly the lives of millions of people globally while assuring the needed organizational sustainability and profitability. This approach introduces the Pink Ocean Strategy concept, which cares about human life, integrity and dignity to booster innovation efforts. The new Pink Ocean Strategy is executed through the Company Democracy Model, the model for people, to embrace anyone who has and can share data, information, knowledge as well as wisdom creation for the good of the society, community and the humanity. The paper presents the six stages of the Company Democracy Model adjusted to navigate organizations into Pink Oceans, the overall strategic management framework, the pre and post conditions of each stage and the expected results per stage, enables this navigation to Pink Oceans.

Keywords: Pink Ocean · Green Ocean · Blue Ocean · Growth · Social · Sustainable Development Goals · Innovation · Democracy · Strategy · Management · UN · Shared value · Social innovation

E. Markopoulos et al. (Eds.): AHFE 2020, AISC 1218, pp. 39–51, 2020.
https://doi.org/10.1007/978-3-030-51626-0_5

1 Introduction

The continuous and rapid evolution of technology, as well as the global entrepreneurship dragged by the .com bubble and the start-up revolution, created a new entrepreneurial reality driven by new business ethics and values. The desire of the investors, to utilize bright young minds with innovative thinking, together with the dream and faith of the young entrepreneurs to succeed, in fame and wealth, as their start-up role models, shifted innovation directly to full capitalism where fast and profitable ideas dominated the attraction of interest, investors and funding.

This overall revolution created new markets where profitable innovation stands out of the competition in uncontested markets spaces as W. Chan Kim and Renee Mauborgne described it in the Blue Ocean Strategy [1]. Blue Oceans can be considered more of a metaphor or synonym to successful innovation and less than a strategy, as innovations that do not stand out in their market or outside of it, cannot be considered innovations. Therefore, innovation strategy is genuinely the Blue Ocean Strategy if innovation is effective and efficient. Building on that and following the significant global concerns related to global environmental and sustainable challenges, Evangelos Markopoulos, Ines Kirane, Clarissa Pipper and Hannu Vanharanta, extended the Blue Ocean Strategy into the Green Ocean Strategy for the development of innovative, sustainable and environmental technology-driven solutions in the sphere of Clean Tech [2].

The Green Ocean identifies the Blue Ocean's innovative elements and combines them with the knowledge needed to further innovate through the Company Democracy Model for democratic knowledge elicitation. The roadmap towards Green Oceans is purely strategic as decisions, dilemmas, risks, and opportunities need to be analyzed prior to entering the very niche green market. Furthermore, the Green Ocean Strategy comes in a time that is not only necessary but is also aligned with the maturity of the world on going green [3]. The public opinion towards a carbon-free word and environmentally sustained is stronger than ever. This situation strengthens innovations that in other periods could have been questioned much on their business sustainability, originality, profitability, and scalability. However, there are significant societal issues that are not being addressed with the capitalistic Blue Ocean Strategy and the environmentalist Green Ocean Strategy. Above the creation of wealth and a carbon-free world, it is essential to help people live their lives decently in order to enjoy the carbon-free world with economic prosperity. Most of the 17 United Nations Sustainable Development Goals are directly related to the efforts towards the development of social behavior to address the significant human needs. These goals, among the rest, have care, love, and support as a common nominator to achieve and sustain human dignity above all.

2 Innovation Evolution

The term 'Innovation' is constantly identified as a fundamental phenomenon of social and economic business advancements [4]. Firms advocate for innovation and its ability to develop their standing, ideally respective to their competitors, within their market. The conversion of ideas to the invention and their commercialization in products or services creates value for their customers and shareholders. Innovations, however, should be replicable at an economical cost that can satisfy a specific corporate need and target group. This process comprises the intentional application of information, imagination, and initiative to derive values from established resources to spawn or adapt useful products or services [5].

Innovation can be primarily categorized into Radical, Incremental, and Lateral Innovation areas. In Radical Innovation inventions eradicate existing business models, exterminate existing systems or processes and substitutes them with something completely new. Incremental Innovation improves existing systems, processes, methods, products or services, while Lateral Innovation replicates existing innovations in the industry.

Within these categories, different types of innovation exist, focusing on Technology, Processes, Business Models, Positioning, and others, ranging from Disruptive Innovation to Frugal Innovation to Service Innovation or Shared Value Innovation. Figure 1 presents a classification of the innovations types with the three main categories.

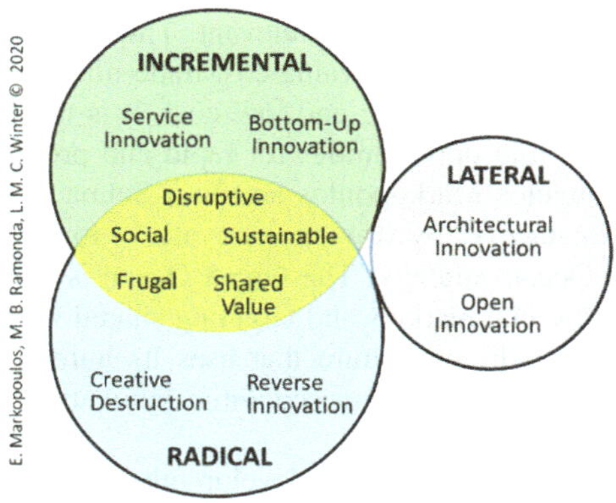

Fig. 1. Innovation categorization

In brief, Disruptive Innovation builds unique and new markets and reshapes existing ones [6] while Frugal Innovation discovers new business models, reconfigures value chains, supplies functional solutions with fewer resources and redesigns products towards affordability scalable and sustainable manners [7]. Bottom-Up Innovation focuses on implementing inventions by employees, in contrast to top-down innovation [8]. Social Innovation targets social needs through the implementation of new strategies and active collaboration with constitutions across governments, businesses and the non-profit world [9]. Open Innovation uses multiple external sources to drive innovation instead of relying on their internal knowledge and resources. Reverse Innovation applied to the inventions and production of ideas, products or services in emerging economies prior to their commercialization in developed economies [10]. Service Innovation changes the processes of how customers are served to increase value, while Architectural Innovation takes technologies, processes or ideas and implementing them into new markets Sustainable Innovation implements sustainable development solutions for the environment, society, and the economy. Lastly, Shared Value Innovation meets societal needs profitably, by leveraging resources and innovations from the private sector; creating new solutions for society's most demanding problems [11].

3 Innovation Oceans. From the Red to the Blue, to the Green

In 2004 the Blue Ocean Strategy revolutionized the business sector by urging organizations to set aside competition between them and give room to the expansion of markets through innovative ideas [1]. The term Red Ocean was an invention to describe the saturated 'bloody' market-driven competition in existing market spaces, continue exploitation for demand on the same target, value-cost trade-off groups and price differentiation. The majority of the organizations today continue to compete defined and predictable markets, with known competitors and with traditional management structure. The fear of the unknown, the cost of innovation, and the long term results in many cases keep organizations in Red Oceans to keep on trying to make the most of the existing markets [1]. Alternatively, Blue Oceans mobilize companies to generate new market spaces and make the competition irrelevant. Thus, it creates and captures new demand through innovation, breaks the value-cost trade-off and aligns the activities of the organization towards differentiating and low cost. It is those who thrive to create innovative ideas that will find opportunities for rapid and profitable growth [1].

Recently, Prof. Evangelos Markopoulos and Ines Selma Kirane created an extension of the Blue Ocean theory, by driving innovations towards environmental sustainability with Green Ocean Strategy. The Green Ocean is an evolution of the Blue Ocean by capturing new social markets and targeting shared value competitiveness. Its essence is of a democratic driven culture that uses its human intellectual capital to develop environmentally sustainable and economic innovation by the society for the society [2].

Attempts have been made in the past to develop other oceans emphasizing strategic alliances towards going Blue but did not seem to have the expected acceptance. An example of this is the Purple Ocean Strategy by Dr. Kittichok Nithisathian and Prof. John Walsh, which integrated the Red and Blue Oceans, based on the logic that Blue Oceans do not guarantee long term organizational success and sooner or later they will turn Red again. Furthermore, it considers organizational operations is both the contested and uncontested markets and suggests beating the competition by focusing on the customer and emphasizing the value of "free", zero cost [12]. The Purple Ocean was positioned between the Red and the Blue Oceans, indicating that it is wise to stand in the middle without being conservative and compete in a saturated market but also without seeking the new big idea.

The constant development of new strategies however, under the concept of the 'ocean' as new market space, highlights an innovation gap that can be fulfilled with more specialized and targeted strategies.

4 Mind the Innovation Gap

The emergence of entrepreneurs and innovators who intend to combine financial sustainability with social impact is a great novelty. There are many types of social enterprise but the essential aspect is based on having pure intentionality for change that is more powerful and impactful than profit.

While it is true that the engine driving social innovation is made up of collaboration, empathy, sensitivity to dilemmas, as well as the desire to strengthen the environment of societies, this entrepreneurial passion does not seem to be enough. It is important for them to utilize the best any strategy and model available regarding social innovation for their efforts to be sustainable and simultaneously philanthropic.

Despite the evolution of the various Ocean Strategies, from Red to Blue and Blue to Green, innovation continues to grow rapidly. Therefore, there is yet a need for a strategy to covers socio-economic trends and needs.

The Blue Ocean Strategy focuses on finding unexplored markets through innovation, whilst the Green Ocean Strategy focuses on extending innovation towards environmental sustainability and viability. By observing the evolution of the innovation oceans, it seems that one ocean comes to complete the other.

The Red Ocean Strategy created a gap of innovation that was covered by the Blue Ocean with its pure capitalist nature but created a gap of environmental and sustainability innovation as Green markets were smaller with more expensive products and demanding costumes. The Green Ocean Strategy covered this gap by utilizing the global environmental movement and customer maturity. However, not all needs have been covered and it is not possible to be covered by only one strategy. The need for humanitarian innovations is a major one. Blue Ocean does not address it directly and Green Oceans leans on it satisfying the environmental dimension. This innovation gap indicates that traditional approaches must be complemented by the exploration of change, and new approaches need constant modernization.

5 Social Innovation Challenge

In a world where the expansion of wealth goes above all for many, and the battle for a carbon-free world is taking over, the value of people's dignity must not be forgotten. This social dimension must be supported by providing people help to live their lives honorably and satisfactorily in order for them to savor the Green carbon-free world with the Blue economic prosperity.

The 2030 Agenda for Sustainable Development which was adopted by all the United Nations Member States in 2015 proposed 17 Sustainable Development Goals (SDGs). The goals provide a blueprint for peace and prosperity for people and the planet, today and in the future. SDGs embody a crucial call for action by all countries in a global partnership. They honor that ending poverty and other deprivations go in close association with strategies that can improve health and education, reduce inequalities and drive economic growth, all while confronting climate change and working to preserve life on land and below water [13].

Most of the 17 SDGs such as goal 1 (No Poverty), 2 (Zero Hunger), 3 (Good Health and Well-Being), 4 (Quality Education), 5 (Gender Equality), 6 (Clean Water and Sanitation), 8 (Decent Work and Economic Growth), 10 (Reduce Inequalities), 11 (Sustainable Cities and Communities), and 17 (Partnerships), align the efforts towards developing social behaviour to address the major human needs of the society [14].

The common denominator characteristics of the UN SDGs are care, love, and support, which are needed to achieve and sustain human dignity above all. It is vital to focus on complying with these SDGs to achieve prosperity in the business world, and reach economic growth while protecting the environment and empowering societies.

6 The Pink Ocean Strategy - Thinking Beyond Profit

The process of integrating the UN SDGs in a strategy interlinked with the current Blue and Green Ocean strategy leads to the creation of the Pink Ocean Strategy (Fig. 2). The Pink Ocean Strategy can directly impact the lives of millions globally while at the same time assuring the needed sustainability and viability in the environment and society, with profitability growth and scale-out business operations development.

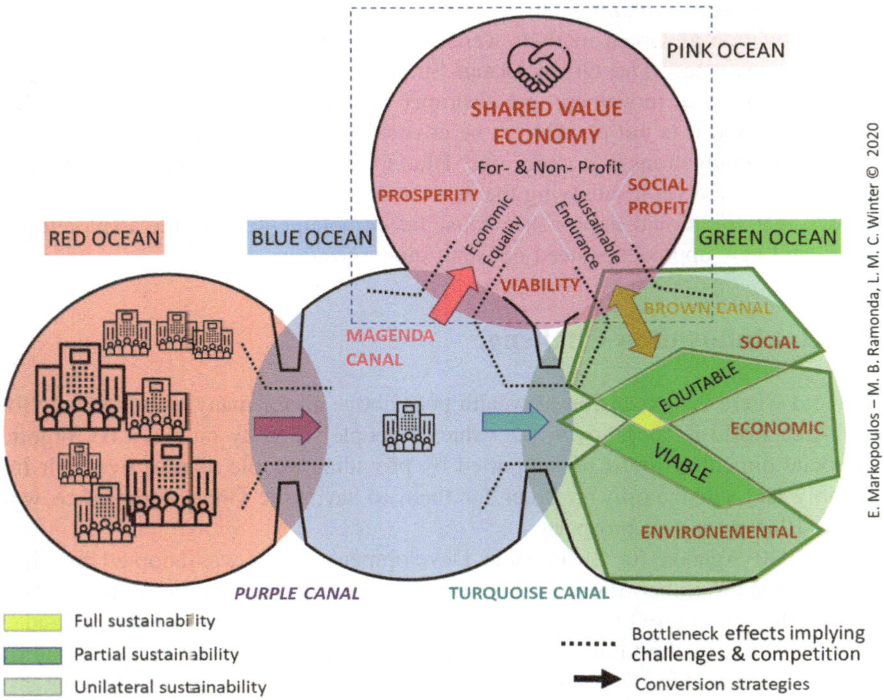

Fig. 2. The Pink Ocean Strategy

The Pink Ocean Strategy is introduced at a time the world seeks new alternatives of innovative thinking to rising problems such as climate change and inequality. The old paradigms on the paternalistic assistance of the state or the companies to address these problems are, indeed, long gone. In fact, the world has evidenced a revaluation of the role of innovation in society and the enormous impact it can have on the sustainable growth of nations. Therefore, in order to contribute to the development of existing innovation tools the Pink Ocean Strategy aims to create a space for growth which allows contribution of ideas towards the construction of a social innovation ecosystem, but also an environments to test ideas, approaches, dynamics and models, that are likely to become useful solutions to global and regional social concerns.

To navigate to the Pink Ocean, an organization has to go through the Blue Ocean since innovation remains a precondition for progress beyond the competition. Achieving social innovation requires innovation, and innovation is achieved through Blue Oceans.

However, the thinking of utilizing an innovative idea, process or product is what differentiates the path an innovation can follow or stay into the pure capitalism and profit-driven Blue Ocean. There is nothing wrong with innovations to stay in Blue Ocean as this is a significant achievement on its own, but extending beyond the Blue, there must be a drive for it and a fair market for the innovation to be sustainable.

The innovation transition process from Blue to Green or Pink is achieved through a process canal. Green Ocean is reached through the turquoise canal, a process that transforms innovation into environmental social and sustainable. Pink Ocean, however, is reached through the magenta canal, a path that emphasizes more into sustainability by sharing and caring for the innovation benefits. It is the economic equality that leads organizations from the Blue to the Pink Ocean where a shared value-driven culture to sustain human dignity overrules any other success goal.

Reaching Pink Oceans opens the pathways to extent into Green as both oceans are characterized by social and sustainable values and goals. The success an innovation can reach in social communities and assure sustainable endurance can extend the social good and contribution to the society by transforming parts of its operations or the entire innovation to address a more sustainable goal, enlarging this way its target group.

7 Pink Ocean Strategy Pathways

Organizations can implement Pink Ocean Strategy either through various paths connected with the Blue and Green Ocean Strategies (Fig. 3). The main characteristics of the Pink Ocean Strategy are compassion and care driven for those who need help, opportunity and an environment to create, develop and prosper. Achieving prosperity by offering to people motivation, fairness, and care, it is certain that success will return sooner or later to the investments made on these people.

The Pink Ocean shapes a compassionate market space and not competitive. It is a long term people based and not profit based investment. The Pink Ocean coincides with the competition as opportunities are offered to those who seek them. This approach creates and shifts the demand for opportunities to the less privileged ones. As a shared value strategy, Pink Ocean shares the value-cost trade-off and aligns the overall system

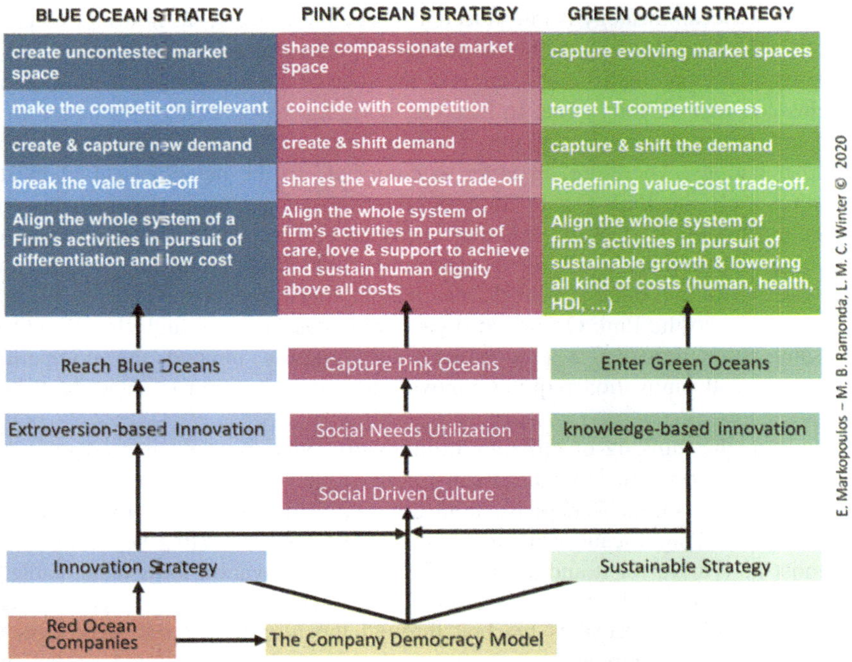

Fig. 3. Pink Ocean Strategies classifications and paths for achievement

of the firm's activities in pursuit of care and support towards achieving and sustaining human dignity above all costs.

Unlike Red Oceans, the Pink Ocean requires a democratic organizational culture that can enable shared and care value thinking and support innovation efforts based on that. It is for certain that not all organizations, and even more not all people feel or believe that share and care value innovations can achieve a return on investments. To tackle this issue employees in an organization must democratically be selected, and not obliged, to think and work towards Pink Ocean innovations.

Corporate democratization can either be the base for the organizational innovation strategies, or derive from the Red Ocean strategy as an alternative part to the more expected, accepted, and profitable Blue Oceans. The same path can be followed for Green Ocean strategies as both Pink and Green Oceans differentiate the profitability strategy form the Blue Oceans which is the next and logical evolution.

8 The Pink Ocean Strategy Processes

The element of democratization in an organizational culture is crucial for reaching Pink Oceans. Share and care value cannot be achieved by imposing and mandating people to share or care. It can only be possible if those who do care and share are identified in the company and supports by turning their ideas into effective innovations in products and services. The Company Democracy Model by Prof. Evangelos Markopoulos and Prof. Hannu Vanharanta forms the base on which the Pink Oceans process evolves. The mode is structured into an evolutionary six-level pyramid in which knowledge is identified at level 1 and transformed into a global competitive advantage in level 6 [15]. The Company Democracy Model has been primarily developed for the creation of Blue Oceans but over time variations of it have been created to address various organizational goals and needs, with one of them to be the creation of shared value. This challenge has been approach by emphasizing on the identification of the human intellectual capital and the progression of the individual, and not only the idea or the innovation, over the six evolutionary levels, with the innovation benefits to be shared between the employee and the organization [16]

The six levels of the Pink Ocean strategy start with the identifications of the share and care value thinkers and ends with the institutionalization of the innovation and its adaptation on a global scale (Fig. 4).

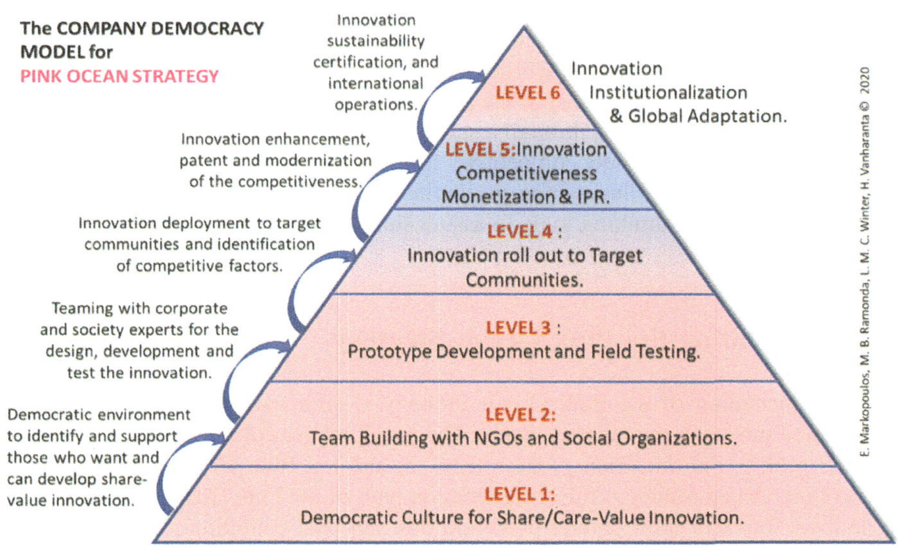

Fig. 4. The Pink Ocean Strategy levels

Level 1 on the Pink Ocean Strategy implementation model deals with the creation of an organizational culture open to share and care value ideas, knowledge and contributions. It is the level that allows all who feel and need to share their knowledge for a common good, to express themselves and propose their solutions.

Level 2 is based on the creations of teams between those who have shared value ideas with the experts needed to implement them, but also with external parties such as NGOs and social organizations related to the innovation concept and needed to provide their experience and expertise.

Level 3 is based on the implementation of the shared value idea into a working prototype that can be tested by the NGO or the social organization that joined the team at level 2. This is the level where the actual knowledge and idea is practically implemented into a product or service and tested as well.

Level 4 extends the implementation of the innovation achieved in level 3 by rolling it out to the members and user of a community to identify areas of improvement, usage concerns, receive valuable feedback and actually let the innovation mature for some time while the users/clients enjoy the innovation services with limited charges or free.

Level 5 measures the success of level 4 and identifies the most competitive and rewarding elements of the innovation that can be monetized for the innovation to generate the needed income to maintain operations and build on it. The selection and the monetization of such elements are related to the value they return to the user/client.

Level 6 moves the successful innovation (from level 4), and sustainable as well (from level 5) to comply with international sustainable goals, standards, protocols, and treaties to become institutionalized and roll out its application and distribution globally. This is a double target/benefit level. The institutionalization of the innovation can help its financial sustainability but also its recognition and adaptations by various organizations, NGOs and governments worldwide, which will support it financially as well.

It must be noted that the colors of the pyramid change from pink to blue and back to pink again as the process goes through a stage where innovation monetization practices are applied to the innovation competitive features in order to obtain financial sustainability needed for the innovation to continue services the people. These practices follow more of a Blue thinking logic to assure suitability than the Pink thinking caring logic.

9 Implementing the Pink Ocean Strategy

The implementation of the Pink Ocean Strategy is archived by the effectiveness an organization moves in the magenta canal. The canal, whose color is the combination of blue and pink is the transition process from the needed Blue Oceans Strategy to the desired Pink Ocean Strategy. It is the execution path of the Pink Ocean Strategy levels. To archive, this transition several preconditions and post-condition much be satisfied but also the timing the organizations decides to go Pink (Fig. 5).

The preconditions are related primarily to the needed innovation culture the organization shall have. Regardless of the type of innovation or the innovations strategy, the organization must empower innovative thinking, promote collectiveness, and reward contributions that improve its operations either as simple processes or advanced innovations on revolutionary products and services. Such preconditions can be achieved with the adaptation of the Company Democracy Model prior to moving into the Pink Ocean Strategy. This can also be an alternative path for the organizations to reach the needed Blue Oceans required to ignite the initiative towards Pink Oceans.

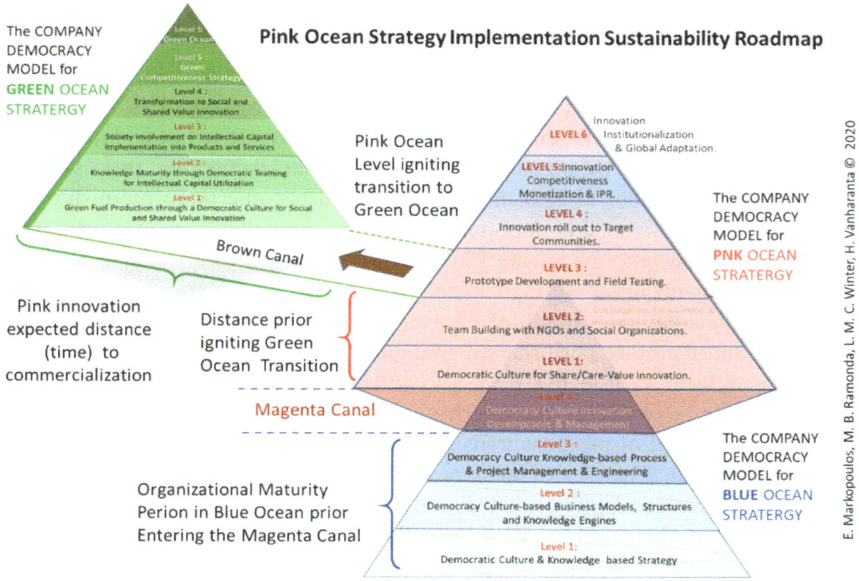

Fig. 5. Pink Ocean Strategy implementation and sustainability roadmap

The pre-condition of Blue Oceans is important for one more reason which is sustainable for the Pink Ocean initiatives until they become sustainable themselves. Pink Ocean has driven initiatives to achieve financial sustainability at levels 4 or 5. This means that in more than three-quarters of the Pink Ocean process, all the effort is being funded only by the organization itself, without any revenue.

It is therefore recommended for an organization to enter the magenta canal once the Blue Oceans have profitability excess. This is usually achieved when the organization reaches level 4 in the Company Democracy Model, meaning that the Blue Ocean innovation passed level 3, which is the market accept level, and it is at global operations by utilizing its competitive advantages. However, financially stable organizations can enter the magenta canal at the early stages of the Blue Ocean journey (level 2 or 3 of the Company Democracy Model).

The postconditions of the Pink Ocean Strategy Implementation is its financial sustainability. Shared value and social innovation can lack enough financial sustainability even if they can massively attract and sustain innovation users, followers or clients in the best case. Therefore, it is important for the innovation to extend into spaces where similar social concerns are shared, where the population of these innovation areas is more keen pay, maintain subscriptions or become loyal to the offered products. One of these spaces can be environmental sustainability which also touches upon several UN SDGs that form the target groups of the Pink Ocean. Therefore the transition from Pink to Green in either selective innovation elements holistically is a postcondition to be determined by the degree of the innovation success along the Pink Ocean.

10 Areas of Further Research

The Pink Ocean Strategy model presented in this paper is a first attempt to model a new innovation-based organizational management strategy. The model is based on the concept of organizational democratic cultures which makes sense but its practically not easy to be achieved even by organizations proud of their democratic processes.

In order for the model to gain more strength and become more promising on the results excepted to deliver, it is important to develop a number of metrics and indexes that can monitor the progress of the Pink Ocean Strategy execution and the benefits gained by the organizations through them. Of courses, this metrics-driven execution process contradicts the ideas that Pink Oceans are generated to bring innovation to the ones in need and therefore performance measurements cannot be applied in this case. This valid statement is true but it is also true that Pink Oceans achieve financial sustainability late, as well are late return on investment. Until then the organizations need to track any kind of progress translated into any kind of benefits which does not need to be financial. This set of metrics that will extend this work emphasizes the identification of various benefits that directly or indirectly will positively impact the organizations to continue seeking Pink Oceans. Such benefits can be related to the time an organization stays in the magenta canal, to employee satisfaction and loyalty, to the impact of the Pink Ocean activities on the organizational reputation on ethical governance, to social corporate responsibility, to sustainability with the UN SDGs and others targets. The sum of these measurable impacts will provide the Pink Ocean Strategy a quantified progress picture and the support needed to reach its goal.

11 Conclusions

According to Heraclitus ‘change is constant’. The evolution of management, especially the last fifty years with the internationalization and the globalization eras, has moved into multidisciplinary dimensions through which organizations constantly seek the transformation that will not only progress them but above all will sustain them.

The Pink Ocean Strategy presented in this paper provides a new strategic management and leadership thinking organizations can explore building reputation benefits besides the financial ones. The UN Sustainable Development Goals were created to provide help and relief to the people in need and the planet, but at the same time, they opened massive target market sensitive to the societies, the SDG movement and the people as well. Approaching effectively this new distribution of priorities is as effective as approaching new target groups aligned with the pubic feeling and concerns.

The Pink Ocean Strategy aims to extend the term of innovation from the old meaning served by the Blue Ocean concept to a new social meaning. It is a strategy that places the humans and the societies first and above any profitable target. Profitability will be achieved after all by the determination organizations have to truly sail on Pink Oceans.

References

1. Kim, W.C., Mauborgne, R.A.: Blue Ocean Strategy: How to Create Uncontested Market Space and Make the Competition Irrelevant. Harvard Business Review Press, Brighton (2015)
2. Markopoulos, E., Kirane, I.S., Piper, C., Vanharanta, H.: Green ocean strategy: democratizing business knowledge for sustainable growth. In: Ahram, T., Karwowski, T., Taiar, W. (eds.) IHSED 2019. AISC, vol. 1026, pp. 115–125. Springer Nature Switzerland AG, Basel (2020)
3. Jacobs M.: Green growth: economic theory and political discourse. Grantham Research Institute on Climate Change and the Environment. Working papers 108 and 92 (2012)
4. Andersen, E.S., Martin, B.R., Fagerberg, J.: Innovation Studies: Evolution and Future Challenges. Oxford University Press, Oxford (2013)
5. Yaacob, A.S., Fazal, S.A., Kumar, N., Mohd Zawawi, N.F., Al Mamun, A.: Defining the concept of innovation and firm innovativeness: a critical analysis from resource-based view perspective. Int. J. Bus. Manag. **11**, 87–94 (2016). Canadian Center of Science and Education
6. Harvard Business Review. https://hbr.org/2008/10/how-ti-spot-disruptive-innovat
7. Sammut-Bonnici, T., McGee, J.: Frugal innovation. In: 12th Wiley Blackwell Encyclopedia of Management of Strategic Management. Wiley, Hoboken (2014)
8. Park, S., Kim, J.N., Krishna, A.: Bottom-up building of an innovative organization: motivating employee intrapreneurship and scouting and their strategic value. Manag. Commun. Q. **28**, 531–560 (2014)
9. Stanford Graduate School of Business. https://www.gsb.stanford.edu/faculty-research/centers-initiatives/csi/defining-social-innovation
10. Seke Mboungou Mouyabi, J.: Reverse innovation: a revolution of the present and future of the emerging markets. Int. J. Eng. Res. Manag. **04**, 6–14 (2017)
11. Porter, M.E., Kramer, M.R.: Creating shared value. Harv. Bus. Rev. **89**, 62–77 (2011)
12. Nithisathian, K., Walsh, J.C.: Purple ocean strategy concept paper: content analysis from interviews and literature review. In: SIU International Conference, Bangkok (2013)
13. Sustainable Development Goals, https://sustainabledevelopment.un.org/?menu=1300
14. About the sustainable development goals. https://www.un.org/sustainabledevelopment/sustainable-development-goals/
15. Markopoulos, E., Vanharanta, H.: Democratic culture paradigm for organizational management and leadership strategies - the company democracy model. In: Proceedings of the 5th International Conference on Applied Human Factors and Ergonomics, Poland (2014)
16. Markopoulos, E., Vanharanta, H.: Company democracy model for development of shared value. Procedia Manuf. **3**, 603–610 (2015)

Analysis of the Cognitive Processes Underlying Discussions in Complex Problem Solving

Yingting Chen[1(✉)], Taro Kanno[1], and Kazuo Furuta[2]

[1] Department of Systems Innovation, School of Engineering, The University of Tokyo, Hongo, Bunkyo-Ku, Tokyo 113-8656, Japan
yingtingsherry@cse.t.u-tokyo, kanno@sys.t.u-tokyo.ac.jp

[2] Resilience Engineering Center, School of Engineering, The University of Tokyo, Tokyo, Japan
furuta@rerc.t.u-tokyo.ac.jp

Abstract. The efficiency of solving complex problems in groups determines the productivity of a society. Existing guidelines for these collaborations are action-focused, and the few cognitive-oriented ones require time and training to be executed accurately. This research aims to propose intuitive and light-weighted recommendations for Complex-Problem-Solving (CPS) collaborations. The underlying cognitive process in CPS discussion was explored, especially the explicit knowledge used. It was found that episodic memory functions better at expanding the conversation scope, while semantic memory appeared to be a more straightforward foundation to initiate new ideas. Since the episodic memory serves as an outstanding primer in the conversation, the results could imply that the better episodic memory is communicated, the more fluence the discussion could be.

Keywords: Cognitive process · Episodic memory · Semantic memory · Problem complexity

1 Introduction

A complex problem is a problem that has no clear definition, nor a clearly defined goal. The means of moving towards the diffusely described goal state are also unclear. Many social challenges can be seen as complex problems: choosing a career, finding a life-partner, climate change, population growth, etc. [1]. A meeting has been identified as a means of Complex-Problem-Solving (CPS) in collaboration, however, also a source that leads to losses in productivity [2]. In Japan, the enterprises spend 67000 h on non-productive meetings every year [3]. A survey from Harvard Business Review shows that 129 out of 182 senior managers in a range of industries said meetings are unproductive and inefficient [4].

The difficulty of holding a productive problem-solving meeting brought an increased attention to the Design Thinking (DT) Workshop, a methodology for creative problem-solving. The operations employed by DT originate from the cognitive process that designers engaged in solving complex problems [5]. Given that knowledge and

E. Markopoulos et al. (Eds.): AHFE 2020, AISC 1218, pp. 52–59, 2020.
https://doi.org/10.1007/978-3-030-51626-0_6

experience have been regarded as critical components of thinking processes aimed at the creation of the new [6, 7], DT helps the non-designers to leverage their knowledge and experience in generating novel ideas. Although the DT workshops are now widely employed in real-world problem-solving, they still cannot replace meetings. This is because these workshops (1) require significant execution time, usually 1 to 4 days, to fit the entire analytic and iterative design process; (2) need experienced facilitators to lead the thinking process. Since the issue is only partially solved, an intuitive and light-weighted collaborative problem-solving method is still needed.

Nevertheless, not all meetings are ineffective. It is curious that some people are capable of generating more quality solutions than others, given the same environmental settings.

This research aims at proposing recommendations for CPS discussion. In order to do so, the underlying cognitive process in CPS discussion needs to be explored, especially the knowledge and memories used. This objective can be divided into two parts - (i) To understand the ability of problem-solving regarding the complexity; (ii) To identify the pattern of memory and knowledge used in a discussion.

This paper covers the results of the current work in progress: for objective (i), the methodology of problem complexity evaluation has been developed. For objective (ii), one preliminary discussion was conducted to test the experimental formats and data analysis techniques.

2 Literature Review

2.1 To Understand the Ability of Problem-Solving Regarding the Complexity

In early works, the ability to solve complex problems is typically measured via dynamic systems that contain several interrelated variables that participants need to alter. Researchers used simulation scenarios with different degrees of fidelities to measure the ability to solve complex problems of individuals [1, 8, 9]. However, there were difficulties in transferring the research results into real-world scenarios for the researches were using simulated scenarios and toy questions.

In this research, instead of using variables set by researchers, the complexity was identified by individuals and measured using the score of conflicts among the set of solution strategies.

2.2 To Identify the Pattern of Memory and Knowledge Used in a Discussion

There were many researches for hosting productive meetings from behavior science perspectives [10, 11]—many of them concerning the ‘actions’ (for example, sending the agenda beforehand) rather than the cognitive process in solving complex problems.

CPS requires creative combinations of knowledge and a broad set of strategies [1]. Recently, the use of explicit memory, especially episodic memory, has been actively studied in problem-solving creativeness.

Episodic Specificity Induction (ESI) has been shown to selectively enhance performance on divergent thinking tasks by boosting the fluency and flexibility of ideas [12]. It could be assumed that the more effective the participants communicate episodic memories (personal-related information), the more effective the discussion could be.

On the other hand, semantic memory (universe-related information) is thought to support creative thinking by presenting a knowledge base of facts that can be coupled to solve creative problems and generate novel ideas [13].

3 Methodology

3.1 Experimental Preparation

An intuitive base topic was chosen for the experiment: 'improve presentation skills.' The topic was selected because it is a typical scene for the potential participants. The complexity was controlled by adding various constraints to the base topic. Shown in Table 1, five topics were created with 5 different constraints. The OR constraints extends the problem space, while AND constraints limit the problem space.

Table 1. Decomposition of topics' structure after adding constrains

Topics	Constrains
To help improving one's presentation skills in a foreign language	~OR improve one's foreign language
To help improving one's presentation skills for those with poor memory	~OR improve one's memory
To help improving one's presentation skills using online tools	~AND using online tools
To help improving one's presentation skills during the commute	~AND using commuting time
To help improving one's presentation skills in 4 weeks	~AND using 4 weeks

There were three steps in the experiment: two questionnaires and one experimental discussion. The objectives, data to-be-collected, and sample size of each step are listed in Table 2. The detailed of each step are explained in Sect. 3.2, 3.3 and 3.4 correspondingly.

Table 2. Experimental process

Steps	Questionnaire 1: personal relevancy	Questionnaire 2: problem complexity	Experimental discussion: problem solving process
Objectives	Select participant with relevant experience and motivation	Evaluate the subjective problem complexity	Identify the memories used in the solution
Data to-be-collected	Rating of: Personal interest Personal involvement	Solutions to a topic Rating of: - Difficulty of each solution - Conflict among the solutions	30-min-discussion of 2-person group: - Audio data - Video data - Writings during the discussion
Sample size	22–55 people	22–55 people	4–5 two-person groups

3.2 Questionnaire of Personal Relevancy

In this research, personal relevancy to a problem was measured from two perspectives: interest and involvement in problem-solving. 'Interest' indicates the participant's motivation of solution creation. 'Involvement' shows the relative level of episodic memory one possesses.

The participants were asked to provide ratings using a five-point scale, to four relevancy factors regarding their interests and involvement in a set of topics in Table 1: R_1 = I am interested in this topic; R_2 = I am interested in creating solutions for this topic; R_3 = I am/was involved in this topic; R_4 = I am/was involved in creating solutions for this topic. The personal relevancy will be calculated as the average value of the captioned factors.

3.3 Questionnaire of Problem Complexity

In the questionnaire of problem complexity, participants were asked to create two to five solutions to the selected topics based on the results in Questionnaire of Personal Relevancy. After that, the participants were asked to evaluate the necessity and difficulty of executing each solution, moreover, the potential conflicts among the created solutions.

The problem complexity is calculated using the mathematical definition that follows the structure of COSYSMO parametric estimator shown in Eq. 1 [14]. It contains additive factors when the variable has local effects; multiplicative factors when the effect is global; and exponential factors when the variable has global and emergent effects depending on the size of the variable [14].

$$\text{Complexity} = \left(\sum\nolimits_{i=1}^{n} \text{sol.difficulty}_i \cdot \text{sol.necessity}_i\right) \cdot \prod\nolimits_{j=1}^{m} \text{H}_j \qquad (1)$$

where:
n = number of essential solutions in a solution-set
m = number of applicable types of conflict within the solution set
H = number of solutions having conflict type j.

There are four types of conflicts, which are based on heuristics to identify conflicting components in a solution: phases of matter, resource, laws of physics, and logic [14]. The complexity is defined by the unavoidable conflicts within the solution set, if a participant believes there is no conflict, the problem is not complex to this participant.

3.4 Experimental Discussion: Problem-Solving Process

The participants were asked to conduct two discussions in a two-person group, using their native language. The participants with similar relevancy and problem complexity ratings were arranged into the same group. This is to ensure the balance of knowledge contribution in their discussion. Discussions were held on separate days to minimize the carried-over solutions from one topic to another. Participants could use a white-board to assist their discussion. The fundamental data to be analyzed is the transcript of the conversation. The writings on the whiteboard, together with the video recording, serve as references.

The discussion was coded into episodic memories, which related to participant's experiences; semantic memories, which related to the knowledge participants learnt through text; initial idea, a new concept of the solution; developed idea, a concept with more details added into the initial idea; finally, interpreter, the change of representation of the problem. The coding scheme will be adjusted on an evolving base.

4 Results

4.1 Participant's Information

At the current stage, 51 responses for Questionnaire of Personal Relevancy were collected. The participants aged from 24 to 33, with a majority of Chinese and Japanese native speakers (Chinese Mandarin: 26, Japanese: 16, English: 3, Others: 6). There were 12 responses obtained for problem complexity (tentatively). One preliminary experiment of problem-solving process was conducted by two Japanese-speaking participants.

Relevancy is calculated as the mean of the scores of the four sub-questions (interest, interest in solving, involve, involve in solving). Topic 1 has the highest relevancy (3.84 out of 5), followed by topic 5 with relevancy 3.10. Therefore, only topic 1 and 5 can be considered as relevant to the participants.

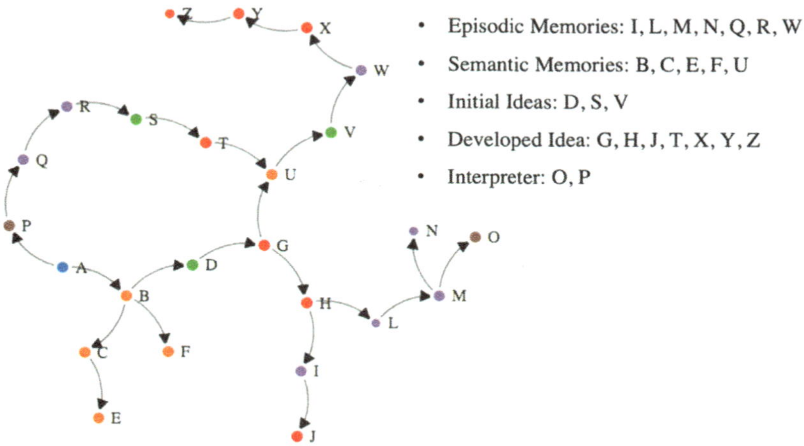

Fig. 1. The two participants conducted a discussion about Topic 1. The cognitive process of discussion coded based on 25 identified utterances. During this discussion, the two participants used more episodic memory (7 out of 25) compared to semantic ones (5 out of 25) to develop their solution. They produced twice as many developed ideas (7 out of 25) from the ones initially proposed (3 out of 25). There was one successful problem reinterpretation, which generated one initial and one developed idea.

4.2 To Identify the Pattern of Memory and Knowledge Used in a Discussion

The two participants suffered from impasses of resolving the conflicts after they kept brainstorming for 15 min. Figure 1 was plotted based on the discussion of the 15 min productive discussion. The nodes represent the coded utterances in the discussion. The letters indicate the sequence of these utterances in the discussion.

4.3 To Understand the Ability of Problem-Solving Regarding the Complexity

After comparing the solutions of topic 1 and 5, a negative correlation between the complexity difference and sematic similarity of the solution was found. Semantic similarity is obtained by comparing the solution descriptions. From the 12 responses, there was a moderately negative correlation (−0.362) between problem complexity difference and semantic solution novelty.

5 Discussion

Participants tend to generate ideas based on their semantic memory rather than episodic ones. One of the possible explanations is generating ideas by communicating, and analyzing episodic memory requires more working memory. However, since the episodic memory serves as an outstanding primer in the conversation, the better they are communicated, the more productive the discussion could be.

Developing an existing idea appears to be easier than generating a new idea in this discussion. However, since this process only recorded the cognitive process before the impasse happened, it is unclear whether this trend is a temporary or general nature in longer discussions.

In the analysis of the results of the discussion, fragility of the data processing method was found. The cognitive process analysis relies heavily on verbal analysis. Due to the difficulty of utterance extraction and data cleaning, the amount of information preserved, and finally analyzed is hard to be evaluated, which makes it difficult to replicate the experiment.

More data is needed to validate the observation in Sect. 4. If the negative correlation could still be observed, it could be suggested the solution to one problem could be used as semantic priming to another. Whereas the two problems shall share the same general settings, and possesses similar complexities.

6 Conclusion

This research aims at proposing recommendations for CPS discussion. The underlying cognitive process in CPS discussion was explored, especially the knowledge and memories used. This objective was divided into two parts - (i) To understand the ability of problem-solving regarding the complexity; (ii) To identify the pattern of memory and knowledge used in a discussion.

To achieve objective (i), methods for evaluating personal relevancy and perspective toward problem complexity was developed. The participants are found to be sensitive toward the specific scenarios, which created a spectrum of complexity and relevancy for questions under the same general background.

To achieve objective (ii), this research also implicated a preliminary analysis of the cognitive process of problem-solving discussion. It was found that the episodic memory functions better at expanding the conversation scope, while semantic memory appeared to be a more straightforward foundation to initiate new ideas.

As a next step, a more robust analysis method needs to be developed to capture the information in the discussion. Besides, more conversation factors needed to be incorporated in order to analyze the collaborative effects.

References

1. Dörner, D., Funke, J.: Complex problem solving: what it is and what it is not. Front. Psychol. **8**(JUL), 1–11 (2017). https://doi.org/10.3389/fpsyg.2017.01153
2. Rogelberg, S., Scott, C., Kello, J.: The science and fiction of meetings. MIT Sloan Manag. Rev. **48**, 18–21 (2007)
3. Maruyama, H.: Enterprises spend 67000 hours, 1.5 billion yen on non-productive meetings every year. Asahi News Digital, 21 April 2019
4. Perlow, L.A., Hadley, C.N., Eun, E.: Stop the meeting madness. Harv. Bus. Rev. (2017)

5. Razzouk, R., Shute, V.: What is design thinking and why is it important? Rev. Educ. Res. **82** (3), 330–348 (2012). https://doi.org/10.3102/0034654312457429
6. Ward, T.B.: What's old about new ideas? In: Smith, S., Ward, T.B., Finke, R.A. (eds.) The Creative Cognition Approach, pp. 157–178. The MIT Press, Cambridge (1995)
7. Kim, E.J., Kim, K.M.: Cognitive styles in design problem solving: insights from network-based cognitive maps. Des. Stud. **40**, 1–38 (2015). https://doi.org/10.1016/j.destud.2015.05.002
8. Buchner, A., Funke, J.: Finite-state automata: dynamic task environments in problem-solving research. Q. J. Exp. Psychol. **46**, 83–118 (1993). https://doi.org/10.1080/14640749308401068
9. Gray, W.D.: Simulated task environments: the role of high-fidelity simulations, scaled worlds, synthetic environments, and laboratory tasks in basic and applied. Cogn. Sci. Q. **2**, 205–227 (2002). citeulike-article-id:8050468
10. Serrat, O.: Knowledge Solutions: Tools, Methods, and Approaches to Drive Organizational Performance, pp. 1–1140 (2017). https://doi.org/10.1007/978-981-10-0983-9
11. LeBlanc, L.A., Nosik, M.R.: planning and leading effective meetings. Beh. Anal. Pract. **12** (3), 696–708 (2019). https://doi.org/10.1007/s40617-019-00330-z
12. Madore, K.P., Addis, D.R., Schacter, D.L.: Creativity and memory: effects of an episodic-specificity induction on divergent thinking. Psychol. Sci. **26**(9), 1461–1468 (2015). https://doi.org/10.1177/0956797615591863
13. Kenett, Y.N., Faust, M.: A semantic network cartography of the creative mind. Trends Cogn. Sci. **23**(4), 271–274 (2019). https://doi.org/10.1016/j.tics.2019.01.007
14. Salado, A., Nilchiani, R.: The concept of problem complexity. Procedia Comput. Sci. **28** (Cser), 539–546 (2014). https://doi.org/10.1016/j.procs.2014.03.066

Virtual Reality (VR) Safety Education for Ship Engine Training on Maintenance and Safety (ShipSEVR)

Evangelos Markopoulos[1,2(✉)], Mika Luimula[2], Pasi Porramo[3], Tayfun Pisirici[4], and Aleksi Kirjonen[4]

[1] School of Management, University College London, Gower Street, Bloomsbury, London WC1E 6BT, UK
e.markopoulos@ucl.ac.uk

[2] Faculty of Engineering and Business, Turku University of Applied Sciences, Joukahaisenkatu 3, ICT-City, C-Wing, FI-20520 Turku, Finland
mika.luimula@turkuamk.fi

[3] Ade Ltd., 144 Hämeen Valtatie, 20540 Turku, Finland
pasi@ade.fi

[4] Wärtsilä Land and Sea Academy, Stålarminkatu 45, 20810 Turku, Finland
{tayfun.pisirici,aleksi.kirjonen}@wartsila.com

Abstract. The training and educational skills expected today from maritime engineers surpass their ability to follow them consistently. Such knowledge gap can impact significantly ship safety and operations. On this reality, ship engineers face multiple challenges related with the time required to be educated on each engine type, time needed to spend off the ship to acquire this training, the continuous training cost, and the easy access to complete engineering documentation and schematics. This paper attempts resolve these challenges with ShipSEVR (Ship Safety Education with Virtual Reality). The technology delivered has been achieved with the contribution of the smart technology company Wärtsilä and Ade with the contribution of VR development. TUAS, Ade, and Wärtsilä work close on research to optimize this engine room virtual reality training environment and impact ship and maritime safety. The paper demonstrates the industry challenge, technology architecture, working prototype, implementation methods, and adaptation techniques.

Keywords: Virtual Reality · Game design · Engine room · Engine · Maritime · Ship · Safety · Management · Technology · Learning environment · Wärtsilä

1 Introduction

The continuous growth of the maritime industry along with the advancements of engineering creates a tremendous technical complexity on operating and maintaining engines and engine rooms safely and effectively. Lack of training time and training costs are critical factors that impact safety in the engine rooms. Due to the nature of maritime, it is challenging to facilitate training which would require both classroom and field work activities. To achieve or maintain professional level of engineering skills

E. Markopoulos et al. (Eds.): AHFE 2020, AISC 1218, pp. 60–72, 2020.
https://doi.org/10.1007/978-3-030-51626-0_7

requires abilities to utilize engineering documentation just in time efficiently. Even that safety cannot be categorized by the degree of industry importance or areas of application, there are specific sectors, such as the maritime, where the need for absolute effective safety procedures must be followed precisely either proactively, or even reactively. A vessel sailing in the middle of an ocean is totality on its own when an accident happens. The response time for help to be provided can be unpredictable as the vessel might not be able approachable, or the help might be insufficient. According to the Japan Transport and Safety Board, 199 marine incident cases were recorded from January 1st to November 30th 2019 (Fig. 1). From those, 84%, or 167, were due to engineering and engine room incidents such as machinery and propulsion failure, or insufficient fuel, lube, oil or cooling water [1].

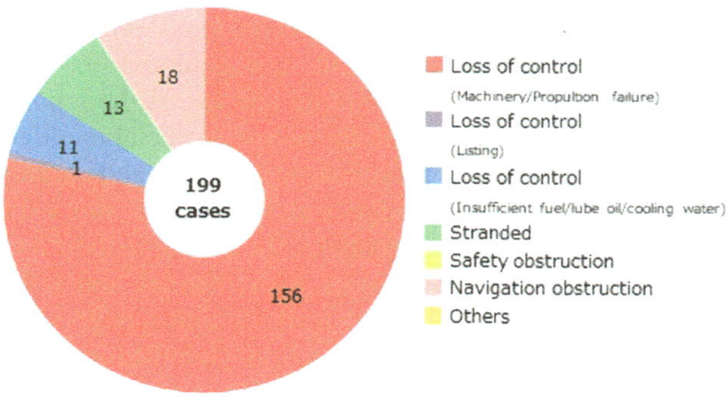

Fig. 1. Number of marine incidents in 2019 [1]

Similarly, the Maritime Injury Guide indicates that equipment failure, lack of maintenance, lack of proper training/human error and collisions are the three main categories of maritime accidents [2]. Specifically, lack of constant and proper maintenance for equipment failure prevention, failure to monitor a ship's systems and lack of the necessary protective parts or equipment are among the major factors that contribute to maritime accidents.

2 On-Ship Safety Training

The maritime industry is one with many dangers and risks as accidents and injuries can happen out of nowhere, leaving behind devastating results. Accidents don't pick sides when they happen. It can be the ship owners' fault for not providing proper tools and training to the crew but also the crew's fault for not keeping up with needed training. Both cases have the cost as common denominators. Effective and professional training is a significant cost for the shipping companies and requires substantial time and effort from the crew to go through it successfully. However, there is one more factor that contributes to the poor or lack of maritime training and this is the time period a ship can lose for the completion of a training program. This impacts the operations costs of the shipping company as the more a ship is on the move the more profit it makes. Therefore, the need to move to on-board training is important and critical [3].

As of today, the on-board training is delivered mainly with either on-line courses or physical training in classrooms or in designated work areas. The first type has limited effectiveness as the seafarers study on their own while the second type cannot be delivered on all ships and can be very expensive if all the training costs are counted (Planning, Programme development and design, Delivery and Evaluation) [4].

3 Futuristic Technologies on Ship Safety Training

The transition from the on-shore training to the on-board training in the maritime sector is not only a need but also a new area of application for advanced and futuristic technologies [5]. The maritime sector has been considered quite conservative to keep up with the technology needed to implement such solutions [6] but this change is happening, and the technology is quite tempting and cannot be ignored.

On board training can be achieved with the use of Artificial intelligence (AI) expert systems, decision support systems, blockchain technology and machine learning. AI constantly transforms data into knowledge and knowledge into expertise. Orca AI systems develops AI-backed navigation tools for the maritime industry to help prevent costly collisions [7]. The technology works with thermal and low light cameras, onboard sensors and an AI-powered vessel tracking and navigation system through which the crew can avoid collision by ship detection. Gamification has been also being used in maritime training as the technology contributes much beyond training. Gamified training provides information on the ways a game is played. This generates information from the player's habits, behaviors, signs, images, sounds, gestures, etc., while a game is played, based on the way the contents is structured and delivered [8].

Virtual Reality, on the other hand, extends the maritime simulation-based training by bringing the simulation through the VR glasses to wherever the trainee is. Virtual Reality reduces the time and cost needed for simulation-based training while increases the accessibility and opportunity to on-board maritime training [9]. MarSEVR is a technology developed at TUAS to handle unlimited maritime safety applications and scenarios. The first MarSEVR application was a decision making and situation awareness case by replicating a ship bridge environment and creating various training scenarios in a close cooperation with maritime specialist from Aboa Mare with difficulty levels. The application emphasizes on watch change between two navigation officers on a cargo ships bridge and on the events that happen after the watch change [10].

Lastly holograms seem to enter the maritime industry as well. The Japanese maritime services company, JRCS, imagines the future of maritime to be with self-navigating ships and digital captains. This can be achieved with the use of mixed reality, artificial intelligence and holograms [11]. While developing MarSEVR and now ShipSEVR one of our intention is to design a learning environment which can be integrated later in remote controlled and at the end when autonomous vessels are reality remote monitored systems as an advanced user interface.

4 The ShipSEVR Project

In an attempt to keep up with the maritime technology trends, research was conducted to examine and develop ShipSEVR (Ship Engine Safety Education Virtual Reality), a VR Training next generation learning environment focused specifically on ship engines and engine rooms safety procedures. The technology has been designed and developed at AVR Turku Innovation and Competence Factory (AVR Turku) for Wärtsilä Land and Sea Academy (WLSA) utilizing latest VR technologies. AVR Turku is a joint innovation environment owned by TUAS where researchers, students and companies are working closely together. In this project, TUAS and its industrial partners Wärtsilä and Ade focus on VR development utilizing successfully the AVR Turku innovation strategy by starting the innovation process with students and researchers and finishing it with the commercialization of the first ShipSEVR learning episode.

The delivered learning episode consists of a 3D ship engine room space where trainees are expected to find certain devices and equipment by utilizing the available technical drawings. The technology was tested intensively by WLSA at their Turku premises and the promising results helped the project continue.

The development process had been split in three phases, (a) reviewing the state-of-the-art solutions globally, (b) prototyping with teams five students latest UI solutions from the gaming industry, and (c) finalizing the first commercial training episode. The reviewing phase conducted in early 2018 revealed some bottlenecks current VR training solutions typically have especially in usability, and user experience.

The initial research team, composed from five ICT students studying game technology at TUAS, was involved in autumn 2018 and worked with unique system check lists provided by Wärtsilä LSA. Their work was coached by TUAS academics, lecturers and engineers in a close cooperation with VR specialists from Ade Ltd. Wärtsilä LSA training specialists were frequently involved to provide valuable feedback and receive ideas from gaming engineers on how state-of-the-art in VR gaming could be utilized outside the gaming industry and particularly in shipping, where users are not experienced gamers but maritime engineers with limited computer skills.

5 The ShipSEVR Research Process

This project became part of the TUAS Advanced Game Technologies course in autumn 2018. Wärtsilä shared one 3D model of Wärtsilä LNG-carrier ship design (Fig. 2) through ftp server, and documentation for five system check lists on Fuel System, Cooling System, Lubricating Oil System, and Start Air System, and Exhaust Gas.

Fig. 2. Wärtsilä LNG-carrier 3D ship design

Each check list was supported with the relevant documentation and technical diagrams.

The research students worked in groups on the checklists under specific instructions and challenges. For example, the group that focused on Cooling Water System received commissioning instructions for checking for all engine and auxiliary equipment connections, leaks on running pumps, and the position of a HT thermostatic valve. They were supported with the diagram of the external cooling water system for in-line engines (Fig. 3), with photos of an installation and with general training slides that have been used in traditional marine commissioning trainings.

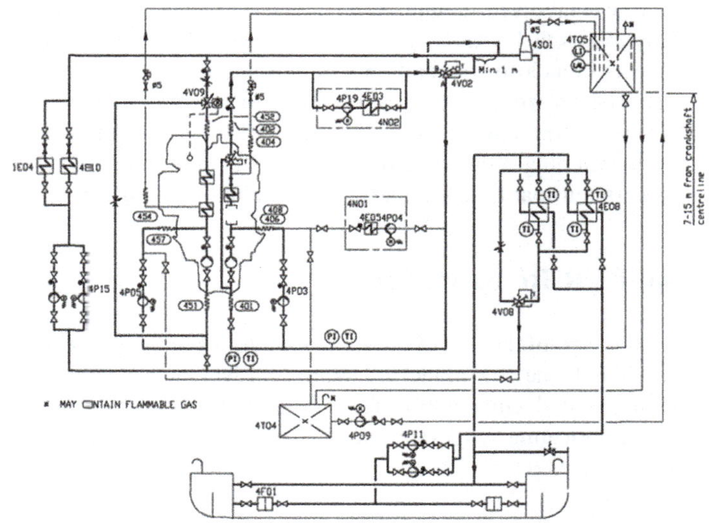

Fig. 3. External cooling water system, in-line engines

However even that 3D design of these types of systems are available, challenges in usability, user experience and complex user interface designs have made the usability of these 3D assets very complex and less insufficient for training purposes.

Through this project Wärtsilä aimed to explore the kind of innovations game developers could bring on board due to their gaming background. The game industry, through serious games, can tackle the challenges Wärtsilä had in usability, user experience, and UI design. In order to achieve better results Wärtsilä provided a detailed 3D model of Wärtsilä 34DF engine generating set (Fig. 4) which got integrated in the technology.

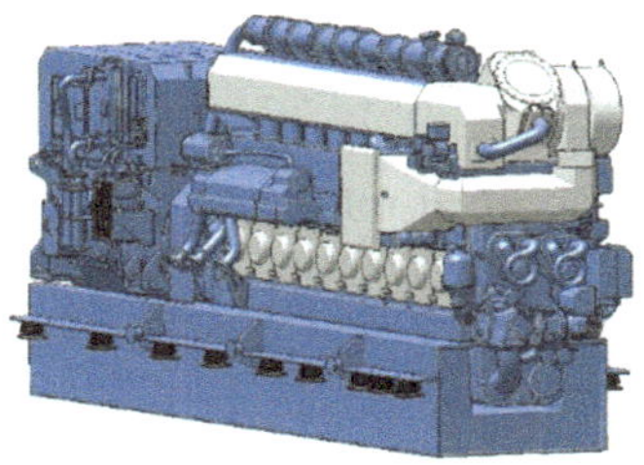

Fig. 4. 3D model of Wärtsilä 34DF engine generating set

6 Initial Demo Results of ShipSEVR

The first deliverables and demos of the technology from this research were presented in November 2019 as first demos in the Allied ICT Finland seminar in Aboa Mare maritime academy and training center receiving significant and constructive feedback. The Cooling Water System team developed a well-designed VR tutorial to tackle usability and user experience challenges (Fig. 5). This was an important Wärtsilä requirement. The tutorial was located at the entrance of the ship engine room. The interaction methods and the UI were explained in detail to the users before entering the room (Fig. 6).

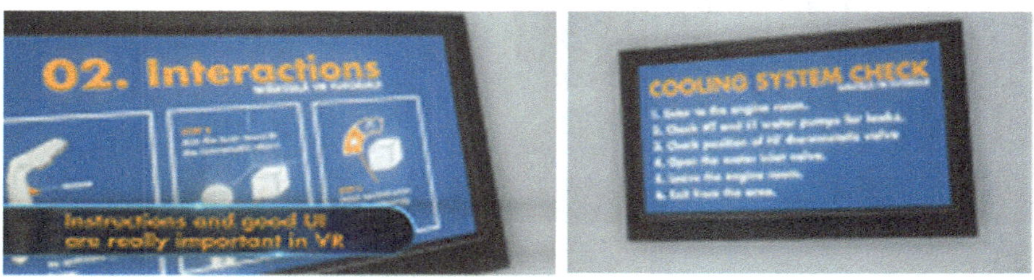

Fig. 5. VR tutorial

Fig. 6. Engine room entrance with the tutorial infront of it

To increase the easiness, a virtual robot guide was invented to instruct the users during the training session. In practice, instructions of the task and the content of the system check list were available when ever needed (Fig. 7).

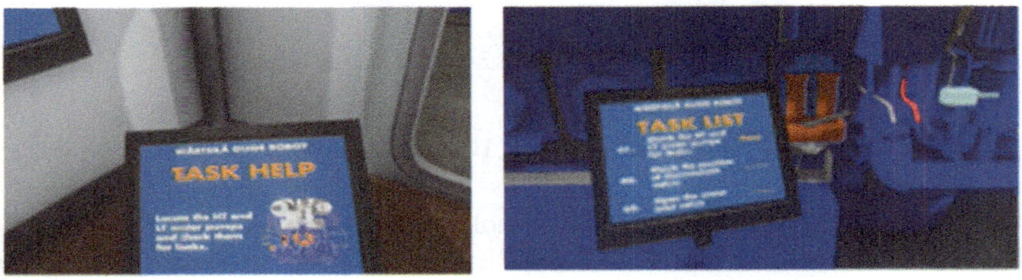

Fig. 7. Cooling water system instructions and task list.

To avoid any misunderstandings, feedback was given just in time. The feedback including highlighting of the active objects by utilizing both informative UI menu elements and concrete animations (Fig. 8).

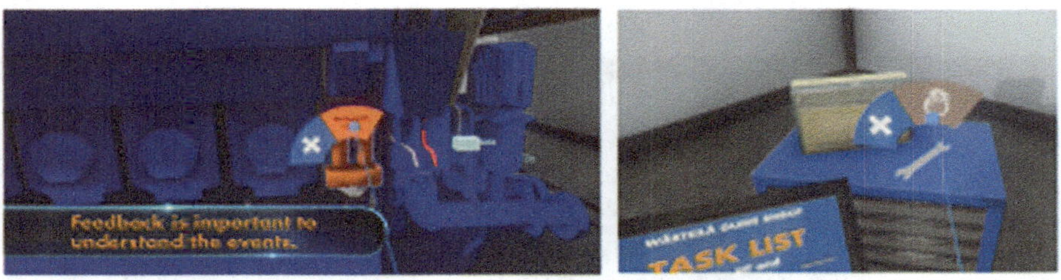

Fig. 8. Animated feedback.

7 Results and ShipSERV Commercialization

The first complete version of the ShipSEVR Technology was completed on December 2018 with the final presentation tasking place at the TUAS Labs with the participation of Wärtsilä's and Ade's representatives. The results were impressive, Wärtsilä was quite satisfied and decided to apply innovative gaming elements of five student groups in a commercial version.

This commercialization phase started early January 2019. Engineers from TUAS together with Ade's VR specialists worked together in a close cooperation with Wärtsilä for the enhanced version of the technology. In addition to the system check lists, ShipSEVR now includes an extra application at the introduction phase before starting the actual training system checks. This VR solution uses the "exact" ship's 3D model of Wärtsilä's WSD Platform Supply Vessel designed by Ship Design unit at the Stord office in Norway (Fig. 9).

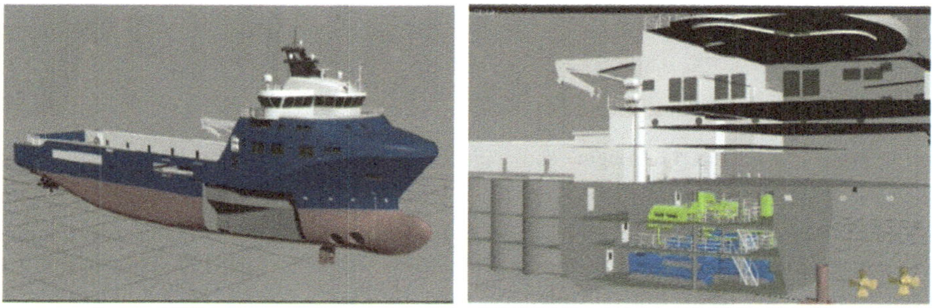

Fig. 9. 3D model of Wärtsilä's WSD platform supply vessel

Ship Technical specifications: Length: over all 92.50 m, Deadweight: 5900 tonnes, Deck area: 1090 m^2, Speed: 14.7 knots, ERN: 99.99.99.9

Design highlights: Dual Fuel. LNG/MDO, DP, ERN 99.99.99.99 Compact high capacity cargo area.

In ShipSEVR, the ship was equipped with following main generator sets (Fig. 10):

2× Wärtsilä 6L20DF 1014 kW and 2× Wärtsilä 6L34DF 2510 kW. Thus, it is a Diesel-electric propulsion (running mainly on gas) with Wärtsilä LNGPac, engine specific gas valve units (GVU) and engine auxiliary equipment.

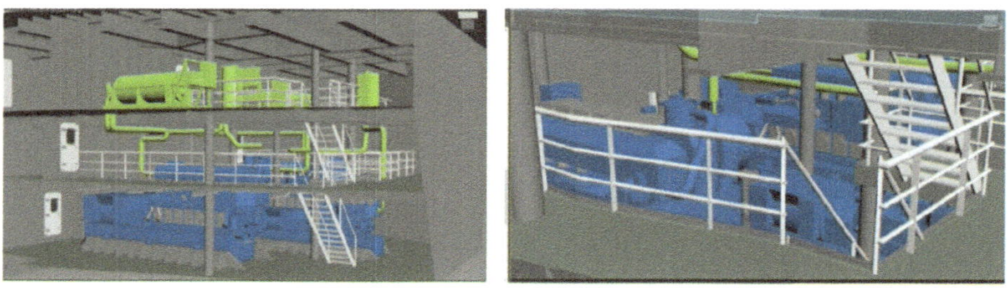

Fig. 10. 3D models of ShipSEVR, ship generator sets

The generator models helped ShipSEVR to develop a realistic virtual environment of the ship's engine and engine room. The user is able to move freely around the engine room by using teleport moving system (Fig. 11).

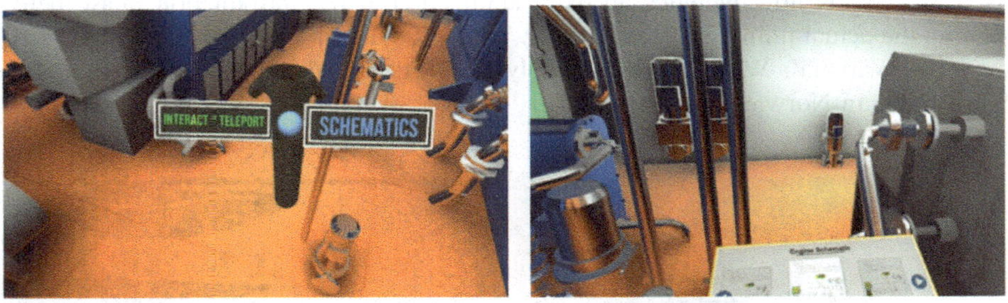

Fig. 11. Engine and engine room teleport navigation

During the tests, the users were asked to find certain devices by using real, detailed and office engine schematics documentation from Wärtsilä (Fig. 12). This solution is designed for engineers but can also be used in TUAS engineering education. For engineers, this solution brings just in time all required documentation visually and easily available. This approach is totally unique 1) compared to traditional way of learning first in classroom without understanding of real-world challenges and context and 2) compared to field work training where especially novices have challenges to read and understand documentation efficiently in harsh ship engine room conditions.

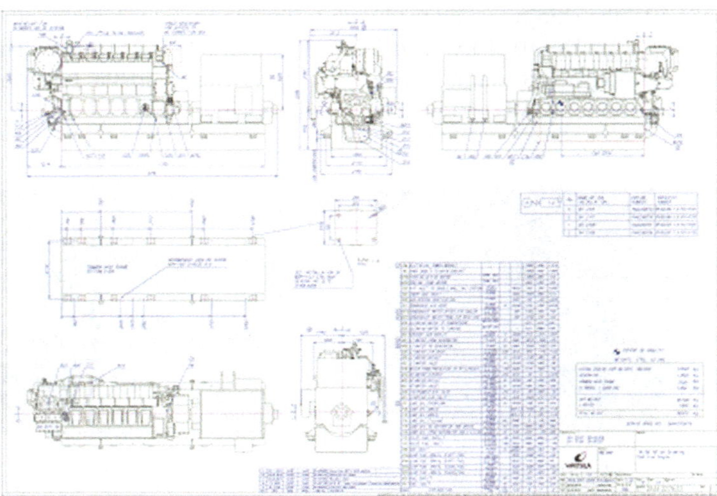

Fig. 12. Engine installation drawing

Consequently, at the end of this commercialization phase, TUAS and Ade provided a solution where trainees can open easily all engine schematics whenever needed. In practice, schematics could be browsed by using VR controllers' trackpad buttons (left/right). In addition, these schematics could be zoomed in/out.

8 VR Extends Training and Education to Behavioral Analysis

ShipSEVR extends to a very important factor in training, often neglected, which is the utilization of data analytics related to the trainee behavior during one or more training sessions. When such data are not collected training remains at formal educational goals, in which knowledge and skills are transmitted to the trainee. However, the use of technology can provide behavioral and psychology analysis of the trainee. Virtual reality is a powerful tool for human behavior research. The ability the technology offers to create 3D visual scenes and to measure the responses to the visual stimuli, allows behavioral researchers to test hypotheses in a manner and scale unfeasible previously [12].

VR provides a unique advantage for such analysis as the eyes of the trainee are recorded and tracked at every instance. Hardware wise, ShipSEVR intends to integrate one of the most promising new VR technologies which is the Varjo's VR-1 glasses with human-eye resolution. This type of glasses can offer, especially in maritime, opportunities to focus on solutions where detailed textual or other visual information is a crucial part of learning aspects in maritime education. Also Leap Motion and other finger recognition technologies will offer not only better usability experiences but also provide new tools for analyzing trainees' behavior. If for example the trainees hesitate between making a choice on how to approach a machine, what tool to use or what metric instrument to consult, these thoughts are recoded and identified with sight traction. Staring at a wrong tool for example is a mistake even if the trainee does not decide to use the wrong tool and finally select the right one. The time to response can also measure by the time the sight of the trainee remains still, at the wrong place, or is lost in the environment.

The combination of such behavioral metrics can be analyzed to identify the readiness and the emotional and the psychological status of the trainee at the time of emergency situations or at routine maintenance work in the engine room.

9 ShipSEVR Impact on the Shipping Industry

ShipSEVR was developed as a research project to identify the degree of safety training effectiveness in engine rooms and its impact to the shipping industry. As the shipping industry indicates high rotation on the on-board crew, the need for continuous training on the new crew to obtain the needed engagement towards achieving safety performance is critical and important [13]. However, the ability to deliver physical and hands on training on the engines and inside the engine rooms while a ship sails is not only possible but also very dangerous.

ShipSEVR can transform any location on the ship, even a cabin, into a virtual space with a precise simulation of the real engine room conditions, space layout and engines. The technology provides accessibility to engine and engineer room schematics, technical diagrams and documentation at the trainee eyes, as a portable digital technical library. The easiness to access, read, analyze the details (zooming), search and store technical documentation is a significant contribution to shipping engine maintenance as most of the physical technical documentation is not easily accessible when needed and not in good conditions due to their usage. Unlimited training scenarios can be delivered in ShipSEVR as various challenges can be combined on type of work to delivered, physical challenges in the engine room and emergency or critical situations.

ShipSEVR was developed for safety training and education in engine rooms but its operations aims to extend the safety scope with the integration of behavioral analytics that predict not only the ability of the engineer to perform safety operations effectively but also on recognizing the emotional and physical state of the engineer to deliver a technical safety challenge.

Transforming this functionality and contribution into financial terms the benefits for both the shipping companies, the shipyards and the engine manufacturers are highly considerable in both financial and reputational costs. The effectives of ShipSEVR reduces the training costs, identifies capable crew to trust the engine room, reduces the possibility to have any type of accident, increases the safety standards and practices, and ranks rewardingly the well-trained crew though its pontification system.

Furthermore, ShipSEVR increases the shipping companies reputation for their efficient and safe operations but also the engine manufactures' reputation for their adaptation and alignment with advances technologies as supportive tools to their customers.

10 Areas of Further Research

The development on ShipSEVR is expected to continue towards the integration of behavioral and cognitive science technologies.

The research will be directed on the identification and the development of user response metrics that will be scored based on the readiness, the behavior and the correct answers given, or acts executed.

The technology intends to prioritize each engine room challenge based on its criticality and complexity, and to provides response time intervals and accepted angels of view, as behavioral characteristics that can indicate the concentration of the trainee and the practical deliver of the knowledge gained in the theoretical training. This behavioral analysis can be supported with behavioral biometrics as every person has a unique way to express his/her self with individual behavioral characteristics such as body movement, coordination and usage [14]. Behavioral biometrics can reveal the psychology of the trainee when executing a task. VR and psychology have come closer with the commercial availability and accessibly of the technology [15]. The phycological analysis of maritime engines, and in general of any member of a vessel's crew is quite critical taking into consideration the operational conditions of a vessel.

Other technologies and scientific disciplines that have been studied and can be integrated in ShipSEVR over the time are computational intelligence for the identification of the training scoring process, affective neuroscience and learning, embodied agents, and emotional assessment using electroencephalography (EEG), for the emotional status recognition of the trainee. In our previous studies, we have cooperated with neuroscientists and developed a virtual reality evaluation toolkit for driving inspection [16]. The data gathered can be supported with machine learning and artificial intelligence for the optimization of the trainee's profile, the development of the appropriate training program to follow, and for the identification of his/her suitability on the execution of specific tasks.

Lastly, a technological area that has equal importance is the digital holography and the avatar technology. The two technologies can contribute significantly in the interactive training of the trainee and on the trainee's safety respectively. Holographic communication can transform the way feedback is given, how a subject is explained, and improve the way teaching, mentoring and coaching is delivered. Likewise, avatar technology can transform the VR experience by coordinating the avatar of a trainee on a critical and risky challenge is the digital engine room, avoiding physical exposure to risks.

11 Conclusions

Maritime training and education on safety issues has always been a prime concern for the industry. However, to maintain safety at the expected desired and requires standards it is important to reduce the seafarers' turnover. The increasing ship-to-shore mobility on the seafarers (officers and unlicensed mariners) requires attention and the retention of the experienced ones, taking into consideration the years spend at sea. Research indicates that retention issues in the shipping industry are related mostly with the employee and employer satisfaction on the job, the opportunities for career progression and the good working conditions [17].

ShipSEVR has been designed to address these issues by proving innovative workspaces that improve working conditions, increase job satisfaction, contribute to career development and retain the experienced seafarers. The concept behind the technology is not only the delivery of safety training on engine rooms and engines, but the execution of this delivery in pleasant, easy and rewarding way for all the parties involved. ShipSEVR revolutionizes the tools for safety training delivered by both the shipping company and the engine manufactures, offering developed opportunities to the seafarers to demonstrate their safety skills and career commitment. The commercialization of the technology has begun and will continue following the research progress.

Acknowledgments. The authors would like to thank all five student teams, and staff members from Wärtsilä, Ade, and Turku University of Applied Sciences who have participated in the development of this training VR technology. This work was supported by Business Finland, several companies, and the Turku University of Applied Sciences.

References

1. Japan Transport Safety Board. https://www.mlit.go.jp/jtsb/statistics_mar.html
2. Maritime Injury Guide. https://www.maritimeinjuryguide.org/blog/top-reasons-maritime-accidents/
3. Sampson, H., Tang, L.: Strange things happen at sea: training and new technology in a multi-billion global industry. J. Educ. Work **29**(8), 980–994 (2016)
4. Videotel. https://videotel.com/counting-the-costs-of-maritime-training/
5. Sanchez-Gonzalez, P.-L., Díaz-Gutiérrez, D., Leo, T., Núñez-Rivas, L.: Toward digitalization of maritime transport? MDPI Sens. **19**(4), 926 (2019)
6. Lukas, U.: Virtual and augmented reality for the maritime sector – applications and requirements. IFAC Proc. Vol. **43**(20), 196–200 (2010)
7. Maritime Journal. https://www.maritimejournal.com/news101/navaids/ai-navigation-tools-help-prevent-collisions
8. Nikitakos, N., Sirris, I., Dalaklis, D., et al.: Game-based learning for maritime education and training: the case of trader of the world. WMU J. Marit. Affairs **16**, 265–291 (2017)
9. Cao, F.: A ship driving teaching system based on multi-level virtual reality technology. Int. J. Emerg. Technol. Learn. (iJET) **11**(11), 26–31 (2016)
10. Markopoulos, E., Lauronen, J., Luimula, M., Letho, P., Laukkanen, S.: Virtual training technology in maritime safety. In: Proceedings of the 10th IEEE International Conference on Cognitive InfoCommunications, CogInfoCom, 23–25 October 2019, Naples, Italy, vol. 1, pp. 283–288 (2019)
11. Microsoft. https://news.microsoft.com/apac/features/technology-and-the-sea-autonomous-ships-and-digital-captains/
12. Brookes, J., Warburton, M., Alghadier, M., et al.: Studying human behavior with virtual reality: the unity experiment framework. Behav. Res. Methods **52**, 455–463 (2019)
13. Bhattacharya, Y.: Employee engagement as a predictor of seafarer retention: a study among Indian officers. Asian J. Shipp. Logistics. **31**(2), 295–318 (2015)
14. Pfeuffer, K.J., Geiger, M.J., Prange, S., Mecke, L., Buschek, D.: Behavioral biometrics in VR: identifying people from body motion and relations in virtual reality. In: Proceedings of the CHI 2019: CHI Conference on Human Factors in Computing Systems, May 2019 Paper no. 110, pp. 1–12 (2019)
15. Pan, X., Hamilton, A.: Why and how to use virtual reality to study human social interaction: the challenges of exploring a new research landscape. Br. J. Psychol. **109**, 395–417 (2018)
16. Luimula, M., Besz, A., Pitkäkangas, P., Suominen, T., Smed, J., Izullah, F.R., Hämäläinen, H.: Virtual evaluation tool in driving inspection and training. In: Proceedings of the 5th IEEE Conference on Cognitive Infocommunications, Gyor, Hungary, pp. 57–60 (2015)
17. Caesar, L.D., Cahoon, S., Fei, J.: Exploring the range of retention issues for seafarers in global shipping: opportunities for further research. WMU J. Marit. Affairs **14**, 141–157 (2015)

Contribution of Open Source Software to the Management of Inclusive Entrepreneurships

Esteban Crespo-Martinez[1,2,3(✉)], Diego S. Suarez-Briones[1,2], Juan Pablo Carvallo[3], Fabian Carvajal[3], Pedro Mogrovejo[1], and Rosalva Vintimilla[1,3]

[1] Escuela de Ingeniería de la Producción y Operaciones, Universidad del Azuay, Cuenca, Ecuador
{ecrespo, ssuarez, rvintimilla}@uazuay.edu.ec, pedromogrovejo@es.uazuay.edu.ec

[2] Departamento de Vinculación con la Comunidad, Universidad del Azuay, Cuenca, Ecuador

[3] LIDI, Universidad del Azuay, Cuenca, Ecuador
{jpcarvallo, fabianc}@uazuay.edu.ec

Abstract. People with disabilities, in Ecuador, are part of population that has been marginalized and discriminated against through processes of exclusion or restriction that have hindered the enjoyment of rights under equal conditions as the rest of the inhabitants. The Support Found "Messengers of Peace" together with the University of Azuay manages the social soap enterprise Aquamarinna, whose objective is to strengthen family and social capacities to identify, support and include people with disabilities in community living, allowing them to make informed decisions and respecting their diversity. As a support strategy, it was considered to adopt OSS, according to CMMI2 procedures, in the design and modeling of business processes by a BPM software solution. As a result, the adoption of this technology increased productivity, accelerate decision-making, response to market changes, and simplify daily tasks.

Keywords: BPM · Open Source Software · Social entrepreneurships · Vulnerable social groups · Disability

1 Introduction

People with disabilities, identified as a part of the main vulnerable groups in Ecuador, include all human beings who have long-term physical, mental, intellectual or sensory impairments that, by interacting with various barriers, prevent their full and effective participation in society, regardless of gender, ethnics or socioeconomic status. This population, broadly speaking, has been marginalized and discriminated against through social policies and processes of exclusion or restriction that have hindered and prevented in many cases, the enjoyment of rights on equal terms as the rest of society, a situation that has been affirmed by the World Health Organization (WHO) in its World report on Disability. In the Republic of Ecuador, on February 2, 2015, the Government

E. Markopoulos et al. (Eds.): AHFE 2020, AISC 1218, pp. 73–79, 2020.
https://doi.org/10.1007/978-3-030-51626-0_8

established the Participatory Inclusion Policy, which is deployed through a community-based methodology.

In an effort made by the Support Found "Messengers of Peace" in addition with the Production and Operations Engineering School of the University of Azuay, the Aquamarinna project is managed, a kind of social enterprise related to the manufacture of handmade soaps, whose objective is to strengthen the capacities of the family group and the community with the purpose of identify, give support and include 13 people with disabilities in community life, allowing them to make informed decisions and respecting their status and diversity.

Supporting the strategic direction of this emerging SME, it was considered to adopt FLOSS (Free, Free and Open Source Software), according to CMMI2 procedures, in the design and modeling of business processes using Bonita BPM, a business process management solution, with the purpose of structuring Aquamarinna's internal processes to support its business strategy by contemplating the four planes of the Balanced Scored Card suggested by Kaplan and Norton, which will subsequently increase productivity in the elaboration of products, acceleration in decision making, improvement in efficiency, increase in agility in response to market changes, and simplify the tasks that employees perform as part of their activities daily.

Carvallo et al. [5], in their article "Systematic Literature Review: Success, Failure, Risks, Benefits and Barriers Factors in the Adoption of Open Source Software", they expose some interesting characteristics that have served as a starting point to consider the adoption of this solution based on FLOSS, with the intention of achieving the stated objectives. Thus, the methodology is developed considering these variables to which the procedures established in the CMMI2 are added as part of the implementation of a computer solution, respecting the processes and corporate culture. Alvarado [1], in her research work, describes the management of business processes as "the corporate methodology that aims to improve the efficiency of organizations through the management of business processes, which must be modeled, organized, documented and optimized continuously".

The following paper are exposed in seven sections: i) state of art, where we analyze the related works with this research, ii) the methodology that was used, both in the modeling of the enterprises processes and the integration of the IT into the business processes, iii) the results obtained after the adoption of this software, iv) the evaluation and comparation of the achieved goals, v) the general conclusions of this work and the furthers works to be developed, and vii) the bibliographical sources that were consulted.

2 State of Art

The CMMI (by the acronym Capability Maturity Model Integration) is a model for the improvement and evaluation of development processes, maintenance and operation of software systems (Institute CMMI). The latest version was published in 2018. COBIT is a framework that includes all the best practices exposed for IT Government (ISACA). ECU@Risk is a methodology for the information risk management focused in MSMEs, developed in Ecuador and based on ISO 27001, 27002, 27005, 31000,

37000; CRAMM, Octave-S, MS Risk y Magerit methodologies; IT government frameworks COBIT and COSO III, also the national and international laws and rules [4].

Bonita Soft, is an open-source platform (low-code) of workflow and business process management (BPM) applications developed in 2001. Alvarado [1] adds that it is a set of office applications for process management Business, downloadable under the GPL V2 license, is multiplatform according to Taco and Rojas [10]. Moreno [8] notes that it has two types of editions: a community and a commercial one. Unlike an ERP, a BPM can design processes from start to finish, and has faster change management, even in real time. A BPM, according to Farrance [6], is a tool that allows to modeling business processes graphically, simulating the processes that are part of the business value chain.

According to Alvarado [1], the French organization APEC uses Bonita Software to optimize its registration processes on the portal, which allows it to obtain greater capacity in data transfer and execution on a custom-made platform. It also cites the success story of BBVA, where Bonita helps to evaluate the process of analyzing customer invoices, with the purpose of validating their data in relation to the service provider, facilitating the understanding and determination of debt. On the other hand, DirectTV adopted this BPM solution to create process flows and catalogs in 10 Latin American countries, interacting with other third-party solutions. By the other hand, Gartner indicates that applications made with Bonita are much faster than other iBPMS more complex. In the comparison they emphasize the support provided, the ease of installation and deployment, integration with other APIs and standard tools, as well as the general capabilities of the product, which generates a high business value for the users.

3 Method

The CMMI v2 practice areas were applied for the deployment of the solution. According to the CMMI Institute, this model helps a company to understand its current level of capacity and performance through practices that increase and optimize these aspects. This obliges to fulfill three values proposed in the model: i) the implementation of an effective and sustainable improvement program; ii) a performance-based approach; and iii) management's commitment and support to visibly and actively support the effort to improve. These processes were explained in the Fig. 1.

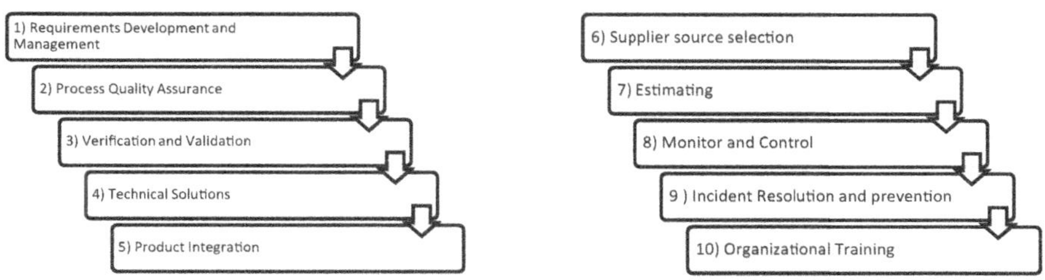

Fig. 1. CMMI based processes used in the FOSS adoption

To support the evaluation of the context suggested by the first process, we relied on tools suggested by COBIT 5 and others recommended by ECU@Risk methodology [4] to identify the business context, such as the type of structure, style management, processes, regulations, etc., with the purpose of feedback and detailing the information that will serve as input for the other processes. The "Incident Resolution and Prevention" process was complemented with the threat and countermeasure detection processes for the information assurance suggested by ECU@Risk.

In order to verify and enable the improvement of the quality of the processes that were contemplated in the modeling through the use of the BPM tool and the resulting products, the "Process Quality Assurance" process was applied at level 2, with the objective of developing and maintain an updated plan based on historical quality data, in addition to communicating quality problems and non-conformities, and recording and using the results of quality assurance activities. For the development and implementation of this project, the use of Scrum was considered [11] a process that allows collaborative work to obtain the best result of a project where the requirements may vary, and where innovation, flexibility, and productivity are fundamental and also associating with the research results of "On the Use of Non-Technical Requirements for the Evaluation of FOSS Software Components" wrote by (Carvallo, Crespo, Carvajal, Mendez, Torres and Vintimilla).

For the validation of the processes designed for people with disabilities, students and teachers of Production and Operations Engineering School accompanied through work and training, because their learning is by imitation, as [4] assures, and contributing with "Consejo Nacional para la Igualdad de discapacidades" (by it acronyms in Spanish CONADIS).

4 Results

Conduct evaluations with objectivity for the quality assurance of the process as suggested by the quality assurance process established by the CMMI were relevant to achieve the success of the project. This application showed a more consistent process implementation, offered a better visibility of the results and problems that were presented during its deployment, as well as demonstrated greater visibility in the project performance, which, as COBIT government processes suggest it, they were communicated to their sponsors.

For synchronization tasks of the project members, with a 15 min meeting per day applying SCRUM, the goals of each iteration were raised on a blackboard. In this way, the team identified the tasks performed by each of those responsible, involving the dependencies between tasks, the progress achieved to the objective and the obstacles that could impede their achievement. Each team member answered three specific questions: i) What have I done since the last meeting to support my team's goals?; ii) What will I do now specific to help my team achieve the objectives set?; iii) What do I need, or what does my work team need to achieve the objectives?

Together with the interested parties, as for the documentation of the processes, a record of the phases was kept during the execution of the project, adapting the strategies that had been proposed and with it the adjustments and improvements of the procedure. In coordination with the School of Production and Operations Engineering, the facilitator, through its management, allowed overcoming obstacles and controversies that were beyond the scope of the team.

The adoption of Bonita software v.7.9.3, contemplating the processes suggested by CMMI v2, contributed to modeling the internal management processes and connectors with greater ease thanks to the graphical interface and APIs that allowed a better integration with the others technological solutions that this organization currently maintains.

This Open Source Software allowed to design more effectively the soap making processes and other activities that enable the incorporation of people with disabilities to it. Six beneficiaries have been included in the production process with more objective tasks, who provide support in manual cleaning, packaging and transport to the warehouse activities.

5 Discussion

The CMMI Institute mentions a reality that can be verified in professional practice: i) 50% of organizations do not have standard processes, process assets or work aids; ii) 42% of them do not have established standard planning processes; iii) 41% of organizations do not adequately prepare people for the future, by developing skills for the generation of organizational capacities; iv) 33% of organizations consider that controlling plans and making adjustments becomes a difficult task, and; v) 54% of them do not measure what really matters. The following Fig. 2 represents the exposed values:

Fig. 2. Organization reality. Source: CMMI Institute

Because the CMMI V2 model does not imply any specific methodology, nor does it consist of a set of processes that must be implemented, it is clear to know that it must be applied according to the organizational context. That is why it was important to consider the administrative management tools suggested by the COBIT Government framework and ECU@Risk methodology.

Many OpenSource-based BPM projects are available on the GitHub site (reference BPMN Tool Matrix), several of them designed to work on Windows platforms. In this work it was possible to verify the result obtained in the adoption of the community version of BonitaSoft, coinciding with Rodriguez, Bazan and Dias [9] who also affirm that the community version provides the opportunity to extend the components, connectors or functionality of the same, thanks to the solid support of the community, such as the ability to update the instances of the processes in execution and synchronization with an LDAP server. In the application of the Bonita tool in the Mobilization System of Universidad Politécnica de Cuenca, Angamarca [2] shares its result indicating that "the implementation of a BPM solution can significantly improve the control and efficiency of the processes, to the once it collapses in the reduction of loss of resources and setbacks when executing business processes in institutions", where the result of its application was feasible, limiting that the processes raised and embodied in diagrams, can be automated through the adoption of the solution.

Quality assurance during the design and implementation of the BPM solution proved to be an important factor, since it verified and ensured compliance in the identification of incorrect and incomplete requirements, as well as detecting unresolved defects and release and implementation planning. Consequently, the validation allows to ensure that the product works as expected in its design environment. Project management based on Scrum has been remarkable when the applied social projects, especially in the case of persons with disabilities where team morale is relatively low.

6 Conclusions and Further Works

The success in the implementation of a solution based on FLOSS consists in the thorough identification of the environment where it will be implemented, as well as the application of standards to monitor the quality of the project throughout its life cycle, to which the successful incorporation is added of agile methodologies during its approach, execution and closure. Agile methodologies allow to maintain a harmony between quality and the objectives of adopting a technological solution that is adapted according to organizational requirements. The adoption of BPM software into the organization allowed to identify the flow of internal processes with greater objectivity, considering the four business management scenarios suggested by Kaplan and Norton in the Balanced Scored Card. With the experience acquired in this project, a work to be carried out would encompass a broader objective, which consists in the implementation of the BPM in the processes of the manufacturing and supply chain.

Acknowledgments. This work has been partially financed with ADOPTADOSS and Aquamarinna research project funds of Universidad del Azuay.

References

1. Alvarado, P.: BONITA SOFT: Gestor de procesos de negocio BPM. Universidad Nacional de Colombia (2011)
2. Angamarca, P.: Implementación del sistema de movilización en la Universidad Politécnica Salesiana de Cuenca utilizando el Business Process Management (BPM) BonitaSoft Open Solutions. [En línea] (tesis de pregrado). Universidad Politécnica Salesiana, Quito, Ecuador (2015)
3. BPMN Tool Matrix. GitHub project. https://bpmnmatrix.github.io/
4. Crespo, E.: ECU@Risk, Una metodología para la gestión de Riesgos aplicada a las MPYMEs. In: INCISCOS 2017, Enfoque UTE, Quito, pp. 107–121 (2017)
5. Carvallo, J., Crespo, E., Carvajal, F., Vintimilla, R.: Systematic literature review: success, failure, risks, benefits and barriers factors in the adoption of open source software. In: 2018 International Conference on Information Systems and Computer Science (INCISCOS), pp. 328–336. IEEE Press, Quito (2018)
6. Farrance, M.: What is "BPM?" Business Process Management. In: 10th IEEE International Symposium on High Performance Distributed Computing, pp. 1–6. IEEE Press, New York (2001)
7. Gartner. Reviews for Intelligent Business Process Management Suites (iBPMS), https://www.gartner.com/reviews/market/intelligent-business-process-management-suites
8. Moreno, M.: Automatización del proceso de gestión de reparación de la EP EMAPAR utilizando el Business Process Management (BPM) "Bonitasoft Open Solutions". Escuela Politécnica del Chimborazo, Riobamba (2017)
9. Rodríguez, A.S., Bazán, P., Díaz, F.J.:: Características funcionales avanzadas de los BPMS: análisis comparativo de herramientas. In: XXI Congreso Argentino de Ciencias de la Computación, Junín 2015
10. Aguirre, R., Aracely, X., Paucar, T., Vladimir, M.: Levantamiento, diseño y automatización del proceso de gestión de incidentes para Magmasoft, utilizando la suite de BPM Open Source Bonitasoft e integración con Alfresco como repositorio documental, mediante la utilización de la plataforma Java Enterprise. Escuela Politécnica Del Ejército, Quito, Ecuador (2014)
11. Schwaber, K., Sutherland, J.: La Guía de Scrum (2013). http://creativecommons.org/licenses/by-sa/4.0/
12. Carvallo, J.P., Carvajal, F., Crespo, E., Mendez, L., Torres, M.J., Vintimilla, R.: On the use of non-technical requirements for the evaluation of FOSS software components. In: Piattini, M., Rupino da Cunha, P., García Rodríguez de Guzmán, I., Pérez-Castillo, R. (eds.) Quality of Information and Communications Technology. QUATIC 2019. Communications in Computer and Information Science, vol. 1010. Springer, Cham (2019)
13. Consejo Nacional para la Igualdad de Discapacidades: Valores, misión, visión (2020). https://www.consejodiscapacidades.gob.ec/valores-mision-vision/
14. Abela, J., Ortega, J., Pérez, A.: Sociología de la discapacidad Exclusión e inclusión social de los discapacitados, Revista del Ministerio de Trabajo y Asuntos Sociales (2003)

Blockchain and Organizational Characteristics: Towards Business Model Innovation

Khuram Shahzad(✉)

School of Technology and Innovations, Production/Innovation and Entrepreneurship InnoLab, University of Vaasa, Yliopistonranta 10, 65200 Vaasa, Finland
khuram.shahzad@univaasa.fi

Abstract. Blockchain seems to challenge the current business models by providing opportunities for new value creation. However, several research gaps remain in literature in evaluating how firms can leverage new approaches to innovation management and opportunities created by blockchain. Supporting organizational characteristics affecting digital innovation management process also need attention in order to challenge the traditional theories while developing unique fundamental assumptions between innovation processes and outcomes. Thus, blockchain and organizational characteristics need to be understood as an encompassing, overarching and interrelated ecosystem in digital innovation management. Grounding on digitalization and innovation management, this research conceptualizes how blockchain technology and supporting organizational characteristics (i.e., R&D investment, strategic alignment, cultural support, top management knowledge and involvement, insights from customers and end-users) can be integrated for business model innovation. This research develops a conceptual framework involving multi-disciplinary collaborative actions that strengthen and empower business model innovation.

Keywords: Blockchain technology · Organizational characteristics · Digitalization · Business model innovation · Industrial ecosystem · Innovation management process · Digital innovation

1 Introduction

Grounding on digitalization and innovation management, this research conceptually determines how blockchain technology and supporting organizational characteristics (i.e., R&D investment, strategic alignment, cultural support, top management knowledge and involvement, insights from customers and end-users) can be integrated in a smarter way to develop business model innovation that is a key factor in enabling innovation management process. This research helps industrial firms to find ways in decreasing silos of disruptive technological challenges and promotes organizational harmony in implementation of such technologies. Scholars such as [1] define digitalization as *"use of digital technologies to innovate a business model and provide new*

E. Markopoulos et al. (Eds.): AHFE 2020, AISC 1218, pp. 80–86, 2020.
https://doi.org/10.1007/978-3-030-51626-0_9

revenue streams and value-producing opportunities in industrial ecosystems" which holds that it is more than just the application of numerous digital technologies. Based on this definition, this research includes the perspectives of digital technologies as well as the value-producing opportunities of developing business model innovation. Similarly, the concept of blockchain has increasingly gained ground in the literature on strategy, innovation management, entrepreneurship and digital economy. Scholars have developed a set of definitions and concepts in a variety of contexts, employing blockchain with different labels and, in some cases, with different meanings and purposes e.g., digital chain, smart contracts etc. Scholars such as [2] defined blockchain as *"a digital, decentralized and distributed ledger in which transactions are logged and added in chronological order with the goal of creating permanent and tamper-proof records".*

As a decentralized technology, blockchain eliminates the maneuvering and somehow exploiting role of middleman as well as the transactions costs and time. The entire process of authorizing transactions and adding blocks in a public blockchain is completely distributed in a way that no single regulatory authority exists. It provides encrypted transactions data with more security and verified platforms on a network using cryptographic algorithms [3, 4]. Thus, blockchain seems to challenge current business models and provide opportunities for new value creation and a significant role in developing business model innovation, however, prior literature neglected it [5]. Several research gaps remain in literature in evaluating how firms can leverage new approaches to innovation management and opportunities created by blockchain. It can be grounded on digital innovation management to transform their business models in order to attain sustainability benefits. Yet, the prior literature remains unclear in developing an understanding of value-creation opportunities derived from blockchain [1] and how does this technology can contribute to digital innovation management. The prior discussion of blockchain remained with very few high profile examples, which necessitate conducting a rigorous action research in order to develop rich understanding of how firms adopt blockchain and begin to integrate it into their operations [4]. Furthermore, supporting organizational characteristics that affect successful digital innovation management process [6] also need attention in order to challenge the traditional theories on innovation management while developing unique fundamental assumptions. Thus, blockchain and organizational characteristics need to be understood as an encompassing, overarching and interrelated ecosystem in digital innovation management. There remains an immense need to align and integrate the blockchain with organizational characteristics to ensure the successful innovation management of firms through developing business model innovation.

Against this backdrop and the crucial need to develop successful innovation management process, efforts should be made to build research-based knowledge and novel theorizing on digital innovation management that draws on the rich and rapidly emerging research on blockchain. More attention should be given to the organizational characteristics that support in the concrete, viable and scalable implementation of blockchain developing business model innovation. This research develops a conceptual framework of next generation blockchain-based digital innovation management that involves multi-disciplinary collaborative actions strengthening and empowering business model innovation.

2 Significance of the Research in Relation to Current Knowledge

Digitalization has become a major enabling and transformational force in rapid technological changes and emergence of disruptive technologies [1, 7]. In current digitalized era, firms are under immense pressure requiring revolutionizing their innovation management processes. The growing importance of this notion is driving firms to overwhelmingly reinvent the way their operations and processes are conducted in industrial value chains through using several current technologies such as IIoT, blockchain, cloud computing, advanced algorithms, AI, automation, big data and analytics etc. [1, 8]. As building blocks of fourth industrial revolution – known as Industry 4.0 [9], these digital technologies possess such capabilities that can exponentially transform the spectrum of innovation management and competences by offering significant potential for value-creation – thriving in a digital world. This next generation "digital disruption" [10] in innovation management is challenging existing theories thus requiring firms to capitalize on digital transformation of operating models [11, 12]. The concept of "Factories of the Future 2020" delivers European added value by helping firms to make the necessary adaptations in terms of innovative and digitalized excellence and flexibility, including skills development, to capitalize on the increasing demand for more customized and higher-quality products. Open innovation perspective drive firms to extend their product development and innovation capabilities with external partners providing them with valuable competence and knowledge. This rise of global knowledge economy has intensified a critical need for research-based knowledge and novel theorizing on digital innovation management that draws on the rich and rapidly emerging digital technologies.

Digitalization is compelling every phase of the innovation process – from gathering and evaluating customer insights, to producing and vetting ideas, to designing and manufacturing new products, and, finally, to tracking the triumph of products once launched. Scholars [13] have segmented firms into three different strategic innovation models (i.e., need seekers, market readers and technology drivers). First category i.e. need seekers usually involve customers in generating new ideas to develop products and services, while market readers as fast followers create ideas by carefully observing their markets, customers, and competitors, concentrating mainly on generating value through incremental innovations to existing products. Technology drivers rely mainly on their in-house technological capability to cultivate new products and services, generating both breakthrough innovation and incremental change to meet known and unknown needs of their customers via new technology. However, each category of firm in this innovation models require new approaches to innovation management and opportunities created by digitalization. Similarly, the Forbes' top ten World's Most Innovative Companies (from "ServiceNow" at No. 1 to "Facebook" at No. 10) have a distinctly digital feel, meaning that their innovation management models are mainly based on digitalization of their operations.

Digital innovation management including these disruptive technologies does not only contribute to automation and optimization of the processes (increasing the efficiency by 15–20%) but also provide value-producing opportunities and profitability

(20% of revenue generation) by reducing transaction costs [14] and minimizing errors [1]. Technological applications however are not enough for innovation excellence as several incumbent firms are ill prepared; it necessitates transformation to business model innovation [15], based on digital technologies such as blockchain. Evidently, several firms have had substantial triumph and business development opportunities with the full range of digital technologies. However, there might be some challenges of implementation and their proper usage. According to recent research, only 1% of CIOs reported blockchain implementation and only 8% of respondents get involved in short-term planning and pilot planning [16]. The report based on [17] states that 74% of their respondents see a convincing business case using blockchain technology while 34% respondents stated that their firms have initiated blockchain adoption in some way or the other.

There is a limited attention on blockchain in literature and a lack of consensus remains around the value and limitation of this technology, reflecting an immature state of the literature [1, 4]. Digital technologies and business model innovation also face challenges of continuous improvement in keeping up with their competitors and offering long-term value to their customers [18]. Digital technologies are not as simple as to "plug and play" thus necessitates to ensure developing an understanding across teams in a way that everyone involved in using these technologies are trained enough and using them with the same pace [6]. This requires firms to develop such an organizational culture that support in integration of these digital technologies.

2.1 Blockchain and Business Model Innovation

Blockchains is an emerging technology which has drawn significant attention from energy supply firms, start-ups, technology developers, financial/non-financial institutions, national governments and the academic community [5, 19]. Blockchain has the potential to transform applications and redefine the digital economy [20]. The capabilities of blockchain run beyond bitcoin facilitating current technology applications to be massively upgraded [21]. It is known as a distributed ledger technology/not a single technology but rather functions on a pile of technologies that include the internet as an infrastructure and utilizes the blockchain practice for keeping transaction records and establishing consensus rules [22]. Blockchain is currently revolutionizing industrial applications by providing solutions while driving economic revolution on a worldwide scale and making the whole operational process more transparent as it is unchallengeable and redefines trust that allow secure, quick, reliable, and transparent public/private results. The longer the chain, the harder it is to make modifications in previous blocks, and thus the higher the level of trust [3–5, 20]. For example, in high-value businesses of diamond trading, this technology can transform the whole supply chain transparent and trustworthy.

The potential role of blockchain could actually transform the business model and process of innovation management by making the whole process transparent and trustworthy. This will enable firms to reconsider their ways of conducting businesses. Blockchain is emerging in several fields, however, it's significant role in the "core" of business – business model innovation and innovation process need to be defined at first hand, as it would help firms to rethink about their purpose of being in digital era.

Blockchain technology provides several prospects to cultivate businesses model innovation. [5] called for further empirical research on how blockchain technology effects business model of creating, delivering and capturing value consisting nine building blocks; 1) customer segments, 2) value proposition, 3) channels, 4) customer relationships, 5) revenue streams, 6) key resources, 7) key activities, 8) key partnerships, and 9) cost structure. They also have argued that consortia-led blockchain projects public blockchain projects can engender architectural and radical innovations, which can be interesting to find out in empirical settings. Similarly, [1] stressed that future research could focus on the business model innovation of value creation, value delivery and value capturing opportunities from emerging blockchain technology. They argued that blockchain and smart contracts could play an intriguing role in changing the value-capture component of business model as well as in future company competitiveness by enabling increased transparency among multiple actors.

2.2 Organizational Characteristics Supporting Innovation Management Process

The ability to innovate and to be innovative can be regarded very important for firms to be able to secure their survival and success. The importance of being bold characterizes firms and their inclination towards digital innovation management. The firms that practice the supporting organizational characteristics (e.g., R&D spending, strategic alignment, cultural support, top-management involvement, and involving customers and end-users) can outperform their competitors by building an integrative network in innovation management process and digitalization [13]. Spending on internal and external R&D have a significant role in process innovativeness by offering the most relevant knowledge [23] and facilitating the adoption and implementation of technologies such as blockchain. R&D spending among the "Global Innovation 1000" overall increased 11.4% in 2018, to a record high of $782 billion, reflecting R&D spending increases in all regions and nearly all industries [13]. Furthermore, the firms who consider commencing blockchain initiatives need to consider strategic alignment between their innovation strategy and business strategy. It means the innovation strategy (introducing novel components) must be aligned with business strategy and value creation in making the transition to advanced service business model [24]. Involving customers directly to generate new ideas help them to approach market first [5, 25].

Rapid transformation in technological innovation also requires firms to undergo an internal cultural shift that nurtures innovation management process for sustainable development amidst global competition [1]. Top management knowledge significantly facilitate successful innovation management process and business model innovation as top management involvement in R&D investments and strategy create business opportunities by aligning their capabilities with innovation programs. Moreover, top-management's role is crucial in the management of innovativeness, especially as the driver and communicator of change [13, 26]. Lastly, the insights from customers and end-users can lead to better decisions that will resonate throughout the life cycle of product development, necessitating customers and end-users' involvement in creating, capturing and delivering value in industrial ecosystem [6, 13].

3 Concluding Remarks

This research develops a conceptual framework involving multi-disciplinary collaborative actions that strengthen and empower business model innovation. However, blockchain is a rather new concept and it seems it is finding its applications in different fields. Blockchain and organizational characteristics need to be understood as an encompassing, overarching and interrelated ecosystem in digital innovation management. Thus, a holistic understanding of blockchain, its processes, implementation, and practices is required for further research in order to produce a clear understanding for both research and practice. It necessitates exploring and mapping the usage of blockchain technology by firms to automate transactions as well as revising and renewing their business model in order to achieve successful innovation management process. In addition, current developments in digital innovation management research require further investigation on exploring and integrating the organizational characteristics that support in implementing blockchain technology in their value chain. Exploiting blockchain technology and supporting organizational characteristics go hand-in-hand with business model innovation and innovation management processes, which requires an integrated framework and process that define how value can be created, delivered and captured among value chain actors [1].

References

1. Parida, V., Sjödin, D., Reim, W.: Reviewing literature on digitalization, business model innovation, and sustainable industry: past achievements and future promises. Sustainability **11**, 391 (2019)
2. Treiblmaier, H.: The impact of the blockchain on the supply chain: a theory-based research framework and a call for action. Supply Chain Manag. Int. J. **23**(6), 545–559 (2018)
3. Reyna, A., Martín, C., Chen, J., Soler, E., Díaz, M.: On blockchain and its integration with IoT. Challenges and opportunities. Future Gener. Comput. Syst. **88**, 173–190 (2018)
4. Cole, R., Stevenson, M., Aitken, J.: Blockchain technology: implications for operations and supply chain management. Supply Chain Manag.: Int. J. **24**(4), 469–483 (2019)
5. Morkunas, V.J., Paschen, J., Boon, E.: How blockchain technologies impact your business model. Bus. Horiz. **62**(3), 295–306 (2019)
6. Jaruzelski, B., Loehr, J., Holman, R.: The global innovation 1000: navigating the digital future. Strategy+Business (73) (2013)
7. Gauthier, C., Bastianutti, J., Haggège, M.: Managerial capabilities to address digital business models: the case of digital health. Strateg. Change **27**(2), 173–180 (2018)
8. Bharadwaj, A., El Sawy, O.A., Pavlou, P.A., Venkatraman, N.: Digital business strategy: toward a next generation of insights. MIS Q. **37**, 471–482 (2013)
9. Gilchrist, A.: Industry 4.0: The Industrial Internet of Things. Apress, Berkeley (2016)
10. Weill, P., Woerner, S.L.: Thriving in an increasingly digital ecosystem. MIT Sloan Manag. Rev. **56**(4), 27 (2015)
11. Lerch, C., Gotsch, M.: Digitalized product-service systems in manufacturing firms: a case study analysis. Res.-Technol. Manag. **58**(5), 45–52 (2015)

12. Parida, V., Sjödin, D.R., Lenka, S., Wincent, J.: Developing global service innovation capabilities: how global manufacturers address the challenges of market heterogeneity. Res.-Technol. Manag. **58**(5), 35–44 (2015)
13. Jaruzelski, B., Chwalik, R., Goehle, B.: What the top innovators get right. Strategy+Business (93) (2018)
14. Shahzad, K., Ali, T., Takala, J., Helo, P., Zaefarian, G.: The varying roles of governance mechanisms on ex-post transaction costs and relationship commitment in buyer-supplier relationships. Ind. Mark. Manag. **71**, 135–146 (2018)
15. Porter, M.E., Heppelmann, J.E.: How smart, connected products are transforming companies. Harv. Bus. Rev. **93**(10), 96–114 (2015)
16. Gartner.: Blockchain Status 2018. Market adoption reality (2018). https://www.gartner.com/doc/3869693/blockchain-status–market-adoption
17. Deloitte.: 2018 Global blockchain survey (2018). https://www2.deloitte.com/insights/us/en/topics/understanding-blockchain-potential/global-blockchain-survey.html
18. Sjödin, D.R., Parida, V., Lindström, J.: Barriers and conditions of open operation: a customer perspective on value co-creation for integrated product-service solutions. Int. J. Technol. Mark. **12**(1), 90–111 (2017)
19. Andoni, M., Robu, V., Flynn, D., Abram, S., Geach, D., Jenkins, D., Mccallum, P., Peacock, A.: Blockchain technology in the energy sector a systematic review of challenges and opportunities. Renew. Sustain. Energy Rev. **100**, 143–174 (2019)
20. Helo, P., Hao, Y.: Blockchains in operations and supply chains: a model and reference implementation. Comput. Ind. Eng. **136**, 242–251 (2019)
21. Underwood, S.: Blockchain beyond bitcoin. Commun. ACM **59**(11), 15–17 (2016)
22. Smith, T.D.: The blockchain litmus test. In: 2017 IEEE International Conference on Big Data (Big Data), pp. 2299–2308. IEEE Press (2017)
23. Rodriguez, J.A., Wiengarten, F.: The- role of process innovativeness in the development of environmental innovativeness capability. J. Clean. Prod. **142**, 2423–2434 (2017)
24. Foss, N.J., Saebi, T.: Fifteen years of research on business model innovation: how far have we come, and where should we go? J. Manag. **43**(1), 200–227 (2017)
25. Adams, P., Freitas, I.M.B., Fontana, R.: Strategic orientation, innovation performance and the moderating influence of marketing management. J. Bus. Res. **97**, 129–140 (2019)
26. Singh, S.K., Gupta, S., Busso, D., Kamboj, S.: Top management knowledge value, knowledge sharing practices, open innovation and organizational performance. J. Bus. Res. (2019, in press)

Design for Social Innovation: Redefining the Concept

Meg Schwellnus(✉)

School of Industrial Design, Faculty of Engineering and Design, Carleton University, Ottawa, Canada
megschwellnus@cmail.carleton.ca

Abstract. Social innovation is currently an ill-defined concept, which hinders understanding of it. This paper addresses the issues with current definitions and proposes a framework within which to rewrite the definitions to enhance comprehension of the term 'social innovation'.

Keywords: Social innovation · Concept · Definitions · Design process

1 Introduction

Design for Social Innovation is a growing field of practice in design with exploration on how designers can contribute to social innovation. However, Social Innovation (SI) is a concept with many definitions and ideas associated with it, and none are agreed upon universally [1]. Discussion of SI can thus be complicated by misunderstandings, should the members debating it have varied perceptions of the term 'social innovation'. Furthermore, during the design process, a set of criteria is required to keep the design outcome in line with its goal. In designing for SI, designers and innovators face the challenge of choosing which of these many definitions and ideas they should use as a guide to create for, or as a set of dimensions to create within. Without a universal decision, each definition can be considered opinions, allowing debates to continue.

The determination of a single definition for social innovation is difficult due to the term 'social', and the very many contexts and cultural views it can occur in. The definition must account for all these differences and determine restrictions on what can be considered as a social innovation. The current core concept of social innovation universally agreed upon is that SI changes society [1–5]. This concept is too broad to determine what is and what is not a social innovation - according to it every single man-made thing that exists in our world can be considered a social innovation, as when created they changed society in some way. Thus, the current definition of social innovation is rendered useless as it provides no boundaries to include and exclude one thing from another.

To assess the suitability of existing definitions for the field of design, a review of literature with focuses on defining social innovation was performed. A total of 15 definitions were found, in addition to the use of a systematic review that included 252 definitions by Edwards-Schachter and Wallace [3]. The elements and content of the

E. Markopoulos et al. (Eds.): AHFE 2020, AISC 1218, pp. 87–93, 2020.
https://doi.org/10.1007/978-3-030-51626-0_10

definitions were assessed for differentiation and applicability to the field of design and associated fields.

2 The Concept of Social Innovation

There are over 260 definitions in use to describe the meaning of the term 'social innovation'. The three elements of the definitions for SI are context, focus, and end users. These elements were determined based on their prevalence in the definitions, demonstrated below. These elements vary within themselves – one definition might use one context, the next will use another, and the definition following will use yet another context.

2.1 Prevalence of Elements

These elements were found often in combination in a definition, but sometimes found alone. For example, in the following definitions: "*a combination of existing and known elements of culture, material and/or non-material, or a modification of one (combination) to form a new one*" [1], and "*an idea that works for the public good*" [4], there is the use of culture as the context, and no other elements found. Definitions with only a focus can be exampled by the following: "*fundamental transformation of the social system and the structures that support it*" [1], "*new combination of social practice*" [1], and "*inventions and initiatives that are imitated*" [1], where the focuses are on a fundamental transformation of the social system, new combination of social practice, and imitation of inventions and initiatives.

The element of end users seems to always be found in conjunction with another element, primarily focus. Examples of this are the following: "(social innovation) *varies social action, and is socially accepted and diffused throughout society*" [1], "*a new set of creative solutions to unmet social needs – from environmental degradation to homelessness and global poverty*" [4], and

> "*a novel solution to a social problem that is more effective, efficient, and sustainable than existing solutions, and for which the value accrues primarily to society as a whole rather than private individuals. A social innovation can be a product, production, process, or technology (much like innovation in general), but it can also be a principle, an idea, a piece of legislation, a social movement, an intervention, or some combination of them*" [4]

In the first definition, the focus is a variation in social action, and the end users is everyone in the society where the social action is changed. The second definition has a focus on providing solutions for unmet social needs, and end users include the homeless and poverty communities, as well as Mother Nature. The third definition has a focus on novel solutions that accrue value, and the end users are society as a whole.

The combination of the elements of context and focus can also be found, as shown in the following: "*a process through which social value is extracted from knowledge… to produce new or significantly improved social, economic, or environmental processes*" [4] and "*an initiative, product, process, or program that profoundly changes basic routines, resource and authority flows, or beliefs of any social system*" [4].

The above definitions have focuses on extracting social value and profoundly changing basic routines and authority flows respectively. The contexts are social, economic, and environmental processes and social systems respectively.

There was one definition found that included all three elements: "(social innovation is a) *process by which value is created for individuals and communities through public and private sectors*" [4]. The context is public and private sectors, the end users are individuals and communities, and the focus is creating values.

2.2 Conflicting Meanings

As can be observed above, the definitions all vary in their contexts, end users, and focuses. There is no found significant difference in acceptance for one meaning over another, which leads to confusion and misunderstandings. The primary area of variation is in the focus, and a debate over whether the social innovation should be completely new, or a modification of what already exists (found prevalent above). It was noted by Edwards-Schachter and Wallace that a differentiation in focus can also be related to geography: North America puts a focus on social innovation as the process of social change, whereas Europe has a focus on social innovation as it relates to sustainable development [3]. Additionally, another debate of focus is the focus on social challenges versus social needs (demonstrated above).

The contexts are consistently related to a 'social' component of society, however there is a fair amount of diversity within that category. This again relates back to the use of the term 'social' and its versatility and complexity. The specificity of the contexts defined above varies as well, which can again lead to confusion. With too broad a context, the its determination is rendered void, and therefore has no relevance to the designers and actors participating in the design process. With too narrow a context, a group of innovations that are commonly considered social ones are excluded. This conflict does not improve understanding.

The end users seem to commonly be defined as everyone, however there were direct conflicts. In one definition, individuals as end users were specifically excluded, and in another, individuals were specifically included as an end user. Once again, the conflict does not improve understanding.

Thus, the concept of social innovation is ill-defined, with significant variation across all definitions in multiple dimensions. An understanding of what SI is cannot be had.

3 The Design Process

There are two main elements of a design project that are common to many types of design: a design brief, and an end user(s). Within the field of design for social innovation, there are three elements that designers should consider throughout their process: an aim, a context, and an output. The aim and context of the project can be outlined in a design brief, often in sections describing the situation being addressed and the context of use. The output is determined somewhat by what the end user needs or wants, and by the brief.

A design brief is an outline of the conditions of the project that must be met by the design output. With respect to design for social innovation, the conditions can be broken down into several categories: the problem or situation being addressed, the context of use of the output, and the functional requirements of the output.

3.1 The Situation

When describing the situation being addressed, the key factors are the current functional issue(s) or aim, the reasons behind them, and the aspects of the situation that specifically need to be addressed (which are often related to the functional issues but may also include other aspects).

The functional issue or aim of the project can have been studied formally and documented, observed informally, or generated through imagination alone. It is typically a broad statement, with specifics later addressed in the brief.

The reasons behind the functional issue or ideal function are of high importance to a designer, as they can either justify the aim of the project, or in the case of a functional issue, indicate the underlying causes of an issue. With an understanding of the underlying cause, the designer can address it directly rather than just addressing the consequences of the cause (which is often the functional issue).

The specific aspects of the aim or functional issue that need to be addressed by the design are also critical as these typically become the bases for the criteria against which the outcome is evaluated to determine its success. These are typically determined by breaking down the broad issue or aim into smaller functional units. For example, a broad functional issue could be not being able to contact a person, and the reason for which is that the internet does not work. The specific aspects of this issue could be the access point to internet and its function, the internet router, and programs on the access point to internet and their functions. All these smaller aspects can then be used to evaluate the success of the outcome in addressing each of these smaller functional units.

3.2 The Context of Use

Sometimes argued as the most important element of a design outcome and consideration throughout the design process, the context of use is critical to the full understanding of the end user, their preferences, and the situation being addressed by the design. It determines the stakeholders involved in the project, the technical/functional and/or spatial requirements of the project, and cultural factors and values that could influence the project. The context of use itself is determined by the end user, by qualities or quantities such as where they are geographically, their culture, their values, their social status and their physical and mental capabilities. The end user can also be defined by the context of use, depending on the situation being assessed.

3.3 Functional Requirements

The functional requirements are essentially a framework within which the designer can work to address the situation and specific aspects. They can be technical requirements such as sizes or electrical outputs, action requirements such as performing a specific action or accessing a certain piece of information, etc. These functional requirements are often described out of context and are non-biased criteria against which to formally evaluate the project output. They are evaluative descriptions of the specific aspects of the situation that need to be addressed (outlined in the situation section of a brief).

3.4 The End User

The end user is the primary stakeholder, the person or population who will use or benefit directly from the design outcome.

4 The Mismatch

Designers need the situation, end user, and context of use defined for them to fully comprehend what they are designing for, and how best to go about it. These elements need to be defined in order to compose the design brief. Each element designers require can be matched to an element of the definitions as outlined above: the context of use matches with the context of a definition, the focus equates to the situation or aim, and the end users match up in both.

At present, designers' needs are not being met by the information provided in the definitions given to them. There was only one definition found that defines all three elements – context, focus, and end user – and each element was defined poorly and vaguely, which as mentioned earlier renders the definition pointless, as it can't categorize things very well. The variation within each element leads to more confusion.

The lack of understanding can truly cause issues due to the complexity of the topic, where vastly diverse situations all fall under the same term of 'social innovation'. These situations require different design strategies and design processes to accommodate the contexts. Misunderstandings can lead to designs that are inappropriate addressing its situation. Therefore, social innovation should be defined in a better way, to accommodate the diversity of contexts where there are considered to be social innovations. Social innovation can be a powerful concept to explore, when defined in an appropriate fashion.

5 Discussion

Stemming from the perspective of encouraging diversity [2], a recommendation in response to the variation across the definitions is to view SI as a field with enough diversity to create categories of social innovation with their own defining elements. This would enable a universal understanding of social innovation and the many different scenarios in which it can manifest. Each category would have a set of subdivisions that each have their own criteria to follow for designing for social innovation. These categories come together to form a framework from which social innovation can be better understood. With this view, scholars and designers should rewrite the definitions for this framework to better help innovators understand the meaning of 'social innovation'.

The categories would be scope, social area of application, and outcome. These terms have been chosen to respond to the need for information to define the context, focus, and end user. To define context, one must first define the general scope of the subject, from which a narrower context can be determined. This narrower context can be called the area of application, to avoid confusion of using the terms 'scope' and 'context' – they are often used interchangeably. To keep the context relevant to social innovation, it must be a social area of application, hence the term used above. The third category, outcome, is not so much a category to have definitions associated with it, but a consideration for designers to keep throughout their process – it is the directly related manifestation of the aim of the project. It is the final step in determining the design or social innovation to occur.

These categories have an order to them, indicated above – they go from general scenario to specific details about the scenario. The social innovation design process starts mostly by determining the end user, from which the general scope can be understood. Once the general scope is defined, the social area of application should be decided to better inform the designer about the other stakeholders. Finally, the outcome can be appropriated to the detailed area of application and its stakeholders, to best support their situations. This order does not indicate a level of priority to any of the categories, but simply a direction for innovators to follow the framework.

5.1 Subdivisions

The subdivisions of scope for social innovation can be determined by geographic scales: global, national, provincial, and municipal. These four scales follow government levels, allowing for the government to be a stakeholder in the social innovation if needed or wanted. In addition, designing for social innovation at a chosen scale informs the designer of obstacles to be considered in the design outcome, such as how to ensure full acceptance of the outcome at their chosen scale, and which social challenges the design is addressing.

The main contexts found in the existing definitions were businesses and their infrastructure, communities, and legislation. These are suitably specific areas to apply the social innovations to, and thus can be used as subdivisions for the social areas of application. These subdivisions, in combination with those of the category of scope, help furthering of comprehension or recognition of the scenario for which a social innovation is being designed, as well as other stakeholders.

The subdivisions of the category of outcomes are the many different ways an idea can manifest itself. Each outcome subdivision can be considered during the design process and one decided for the final solution. These subdivisions can include products, services, legislation, principle, production, social movement, etc. Thus, the outcome another factor of separation for SI that can be used to better understand how it comes about.

6 Conclusion

The concept of social innovation, if defined in a proper way, can help populations enormously, and generate better ways and qualities of life. Enabling universal understanding of SI as a concept can help the world in many ways. This framework is proposed in order to do so.

References

1. Howaldt, J.: The unanswered question: social innovation and social change. The social innovation landscape, pp. 88–91 (2017)
2. Bitencourt, C.C., Marconatto, D.A.B., Cruz, L.B., Raufflet, E.: Introduction to special edition social innovation: researching, defining and theorizing social innovation. Revista de Administracao Mackenzie **17**(6), 14–20 (2016)
3. Edwards-Schachter, M., Wallace, M.L.: 'Shaken, but not stirred': sixty years of defining social innovation. Technol. Forecast. Soc. Change **119**, 64–79 (2017)
4. Goldenberg, M., Kamoji, W., Orton, L., Williamson, M.: Social innovation in Canada: an update. Canadian policy research networks (2009)
5. Ionescu, C.: About the conceptualization of social innovation. Theoret. Appl. Econ. **3**, 53–62 (2015)

The Cork Thread, A Sustainable Material: Branding and Marketing Implications from A Cultural Perspective

Theresa Lobo[1,4(✉)], William Afonso Cantú[2,3,4], and Nelson Pinheiro Gomes[2]

[1] CIAUD, Centro de Investigação em Arquitetura, Urbanismo e Design da, Faculdade de Arquitetura, Rua Sá Nogueira, Pólo Universitário, Universidade de Lisboa, Lisbon, Portugal
theresa.lobo@universidadeeuropeia.pt

[2] CEAUL, Centro de Estudos Anglísticos da Universidade de Lisboa, Faculdade de Letras, Alameda da Universidade, Lisbon, Portugal

[3] IPAM, Insituto Português de Administração e Marketing, Quinta do Bom Nome, Estr. da Correia 54, Lisbon, Portugal

[4] UNIDCOM, Unidade de Investigação em Design e Comunicação, Av. D. Carlos I 4, Lisbon, Portugal

Abstract. The present study addresses the textile technological innovation developed by Cork-a-Tex in terms of the cork thread, which presents itself as a sustainable object that answers to sociocultural changes and current mindsets, underlining a positive perception from the consumer's point of view. The development of a thread using a natural and noble material, cork, allows for the creation of solutions that meet the growing need for sustainable production. Something that is clearly visible in today's market and demands greater attention. This encourages and stimulates the development of new products based on this material, which plays an important role in the Portuguese context. Our paper will articulate the context and potential of this innovation with the main public access socio-cultural trend listings/descriptions that are the result of both empirical/business and academic research. Content and textual analysis of these texts and results will guide the critical approach and process to reach a justification and framework for the development of proposals and insights. This in terms of a cultural strategy that articulates the branding and marketing actions of this Portuguese company. Also, the result of the bibliographic review and systematization of this project aims to stimulate technological research and development for a new approach to this material in the context of fashion design.

Keywords: Cork · Trends · Sustainability

1 Introduction: Cork as a Material

In the 21st century, APCOR [2] refers to cork as one of the most prominent materials for having diverse applications such as in NASA and ESA shuttles, in highly competitive kayaks, tennis balls and cricket. Also, it is present in world-renowned design

E. Markopoulos et al. (Eds.): AHFE 2020, AISC 1218, pp. 94–101, 2020.
https://doi.org/10.1007/978-3-030-51626-0_11

pieces. Cork is a natural material (cork oak bark) and "has a range of applications and attributes that no technology has, until today, been able to imitate, match or surpass" [2, 8].

Cork, a product of the cork oak, is a vegetable and natural material considered part of the Portuguese heritage and a product with unique characteristics and extraordinary potential for the development of new products and the Portuguese economy [2, 8, 17, p. 53, 18] indicates several alternative uses for cork other than the manufacture of stoppers. Many of the developed materials are patented, although they are not on the market, such as rigid panels; granules and residues from the industry resulting from the crushing of cork; condensates of steam from cooking the manufacture of expanded cork agglomerate; cork powder; parts for the automotive industry and coating for various types of transport. Still, it is possible to observe that cork as a membrane can have a different number of applications, which comes close to the use of textiles. Some projects take shape mainly in the field of product design, as is the case, for example, Corque Design; Pelcor, Gencork, Coruche Fashion Cork, which innovatively explores the material using even the texture or the smell that is pleasant and conveys a natural feeling of comfort [1].

It is important to note that the use of cork fabric has, in many cases, been represented through the cork skin which is composed by a thin layer of cork glued in a thin layer of fabric, as we can see in Cunha [5]; Corkpel [3] or Corksribas [4]. However, the differences of this material for a new fabric that was developed in the market, are notorious: we refer to the Cork-a-Tex Yarn. The need to work with a cork membrane that takes the form of fabric was explored by the Cork-a-Tex brand, created from two other founding companies: Têxteis Penedo and Sedacor. Regarding this new material, Pinto [19] tells us that this "high-performance textile […] incorporates the properties of textile substrates in terms of comfort, touch and appearance" [19]. It also takes advantage of the cork properties as "lightness, thermal insulation, anti-mites, dirt and water repellency and impermeability, which are traditionally only obtained by the addition of functional chemical agents." [19].

We see that this new material, unlike cork skin, is a new solution for the creation of cork fabrics without requiring the application of additives, such as glue or other fabrics that are glued to a layer of cork, as this becomes soaked in the wire through a paste. It is also relevant to mention the importance of this new development for the cultural and creative industries, where fashion design is inserted. This textile project with cork and natural-based additives contributes to the development of new products and goods with a strong design component, intertwined with the cultural and creative industries, specifically the fashion industry, not least because the use of the cork yarn still is not known in terms of the industry [19]. From a formal point of view, this material uses waste from the cork industry and contributes to the sustainability and reuse of waste. In terms of culture, creation and fashion, the Cork-a-Tex Yarn brings advantages to the sector, with a strong design component (where sustainability and ecology are combined). This new material brings new functionalities that can be used in clothing design, home textiles, technical textiles, decor, leather goods, among others [19].

2 Sociocultural Trends

Dealing with socio-cultural trends is to work on a malleable concept that operates through several different discourses and that can sometimes be contradictory. However, it is certain that the term is closely related to the ideas of "change" or "shifts", as we can see in several authors, such as Powers [20, p. 10]; Gomes et al. [9, p. 56]; Kongsholm and Frederiksen [16, pp. 25–37]; Dragt [6, pp. 3–39]; Raymond [22, pp. 14–15]; Higham [10, pp. 13–16]; Vejlgaard [29, pp. 7–9]. All of these authors seem to agree that trends are related to the dynamics of society (constantly changing) and the impulses that stimulate cultural development. Powers [20, p. 14] suggests that the term 'trend' circumscribes and synthesizes the present, illustrating future discourses. With this, the author discusses the importance of researching trends to understand events that happen in a socio-cultural environment. In turn, Kjaer [14, p. 3] argues that the observation of the past and the present allows for the stimulation of possible outcomes that are to come. We see that Powers [20] and Gomes et al. [9] propose an articulation between trend studies and culture management, associating the need to detect trends through a practice of cultural analysis [9, pp. 54–56], to understand the present. Not to predict the future, but to propose actions.

Socio-cultural trends convey meaning and studying them allows us to work in a perspective that meets the changes that occur socially and continuously [20]. They have a variable impact on the different social layers. Some spread more intensely than others in communities. This way, we can structure them by groups that reflect the intensity of dissemination (confront with the definitions of fads, micro-trends, macro-trends, megatrends, giga-trends, paradigms, which several authors expose, such as Konghsholm and Frederiksen [16], Vejlgaard [29], Higham [10], among others.

Powers [21, p. 3, 20, p. 21] correlates trends as ways of representing cultural patterns. This draws from Raymond Williams and brings the idea that culture can be seen as a living organism, allowing to create and represent different ways of living from different groups, what the author calls the "way of life" [30]. This agglutinating concept contributes to the understanding of different ways of living in society, paving the way for an exploration of trends through the artefacts that are produced and the mindsets behind them.

This way, studying trends implies an analysis that intends to find mindset patterns that become visible in the social environment [9, p. 53], that is, it requires the decoding of messages implicit within the scope of collective mindsets. To analyze trends, it is assumed that the prior identification of these involves the recognition of signals that are exposed to the scrutiny of specialists (see Gomes et al. [9]). These signs are reflections of behaviour and represent society to the extent that they are mirrors of its culture. The reading of these objects (cool signs) follows a proposal based on the contributions of Carl Rohde [23], among others, regarding the definition of this concept and what Gomes et al. [9] propose as a guide for the observation and description of the respective manifestations, to facilitate cultural readings and trendspotting. Within the scope of this work, the socio-cultural trends present in the analysis are already identified by international research and consulting platforms/networks. Using the data that is already available we can review it in terms of the object of study (Cork-a-Tex Yarn) by means of a content analysis, as we see next.

3 Methodology and Case Study

The literature review contextualized the cork thread and the concept of sociocultural trends. To answer our question and to understand the importance of the cork thread as a sustainable option, we reviewed the description of socio-cultural trends developed by major networks that approach coolhunting and the study of trends. This in order to see if they underline a sustainable mindset as a sociocultural trend. Content analysis presents itself as a starting point that will inform our methodological approach. As Gillian Rose underlines, "content analysis was concerned to analyse cultural texts in accordance with 'the ideals of quantification and natural science methodology' [...] every stage of content analysis, from formulating the research question, to developing coding categories, to interpreting the results, entails decisions about meaning and significance" [24, p. 54, 66]. In more specific terms, and in a visual context, the author underlines that the method "is based on counting the frequency of certain visual elements in a clearly defined sample of images, and then analysing those frequencies. Each aspect of this process has certain requirements in order to achieve replicable and valid results" [24, p. 56]. The topic of systematization and replication is very important. Following Gillian Rose's [24] review of the method process, we can summarize it as: the identification of the problem; the definition of a representative and significant sample [24, p. 57]; the creation of a code and categories on the elements to be analysed; the analysis and counting of references and the interpretation of results that can be crossed based on different codes that could be related.

In turn, Wimmer and Dominick listed the steps related to data-collection in a content analysis: "1. Formulate the research question or hypothesis. 2. Define the universe in question. 3. Select an appropriate sample from the population. 4. Select and define a unit of analysis. 5. Construct the categories of content to be analyzed. 6. Establish a quantification system. 7. Train coders and conduct a pilot study. 8. Code the content according to established definitions. 9. Analyze the collected data. 10. Draw conclusions and search for indications" [31, p. 160]. As we can see, the process is similar and it revolves around the question and the population, as well as the sample, that allows to study the phenomenon; followed by the definition of the unit of analysis (the specific object, like a photo, or in this case a report) and the categories of analysis; which allow for the gathering of data and counting and analysis of results.

Inspired by this method and process, we build our approach in both a quantitative and qualitative perspective, trying to find a recurrence of categories related to sustainability and how trend platforms/network understands them. Regarding the research question, it relates to the main objective of this study and is summarized in the following sentence: "Is the Cork-a-Tex Yarn a manifestation of the sustainability trend?" Regarding the decision of the analysis units, these are directly related to the research theme in terms of textual description of socio-cultural trends. The texts that describe trends will be the (textual) objects used in this analysis, being the target of critical exploratory research of their contents. The decision regarding the sample of this research is related to the choice of trend agencies that had open access to socio-cultural descriptive data and that, in a certain way, are articulated with the presented concept of Trend Studies (Cf. [9]). The texts under analysis are original from trend research platforms and networks (the information contained in their websites in January 2020). Our sample is related to the reference of macro socio-cultural trends from the major platforms/networks: Trends Observer (Portugal) [25], Trend One (Germany) [26], Trend Watching (UK) [28], Faith Popcorn/Brain Reserve (USA) [7], The Science of the Time (Netherlands) [27] and Kjaer Global (Denmark/UK) [14, 15].

The construction of the categories that will guide the classification of the sample consists of the following set of keywords, which represent sustainability and are in line with the scope of the proposed question: 'sustainability', 'ecology', 'environmentalism', 'circular economy', 'green'. The table with the research results was synthetized and organized in the following terms: Name of the Platform/Network; Title of the Related Trend (title of the trend identified by the network that has in its description one of the related keywords); Referenced Keywords (the specific keywords that are present in the description); Trend Summary (our summary and reading of the trends' description provided by the entity):

Table 1. Trend Contents Analysis. Developed by the authors.

Name of the platform/network	Referenced keywords	Title of the related trend	Trend summary
Trends Observer	Sustainable	"The Full Gaze of Sustainability"	More than recycling, it is about the consciousness that resources are finite and we must have sustainable lifestyles, underlining simplicity and natural goods
Trend One	Sustainability, Sustainable	"Sustainability"	Sustainability is now very present in global debate and has an impact in the global economy. Topics of ethical consumption and circular economy are underlined
Trendwatching	Green, Sustainable	"Green Pressure"	"From eco-status to eco-shame". Sustainable consumption reaching a critical moment and eco-consumption as a must
Brain Reserve	Environmental	"Atmosfear"	"Consumers need to sure about "the purity and cleanliness" of produced, and consumed goods

(continued)

Table 1. (*continued*)

Name of the platform/network	Referenced keywords	Title of the related trend	Trend summary
Science of the Time	Circular, Environment	"Sustainable and Responsible"	Underlines three topics: going circular, specially in terms of cities; the impact of climate change in terms of droughts and floods; clean tech and ethical consumption, as well as climate engineering technologies to support an environmental turn
Kjaer Global	Circular	"Circular Economy"	People arc aware of the finitude of the resources and the growth of circular economy, "[they) are increasingly favouring access over ownership and 3 in 5 think that 'sharing is better' for the environment"

4 Data Articulation

Trends, being representations of socio-cultural changes, represent the collective mindset. As proposed in Table 1, the analysis of data across different trend networks leads us to the understanding that sustainability issues are relevant in the current context, as all of them, directly or indirectly, try to describe different related issues such as recycling, natural resources, simplicity (lifestyle) or conscious consumption. The trend titles themselves are attempts to directly reference sustainability. We see keywords such as 'sustainability' and 'sustainable' in three cases, and other words addressing the same issues but with parallel terminologies ('environment', 'green' and 'circular'). The results show that the analyzed trend reports underline that 'sustainability' is present in the collective mindset and plays an important role in the development of new products regarding this concept. The networks also point to the need of transparency in the processes of production and promotion. This is pertinent to provide consumers with information regarding what they purchase. It is also noteworthy the notion that resources are finite and consumers play an expressive role in ethical consumption issues, underlining the importance of honest purchases (consumer empowerment). These guidelines seem to reinforce the issues and strengthen the idea for the cyclicality of products. In this way, consumers become the owners of their choices, generating a change in habits with an impact around them. Regarding ethical consumption, this seems to be part of a reflection that is marked by 'responsibility' and 'conscience', inevitably through the idea of belonging and sharing spaces between collectivities. With this, the content that the networks present express the need to create alert calls on sustainability issues in the context of environmental care and the promotion of spaces that show careful management, both in terms of ecology and the inherent processes. There is also a need for new developments (for example technological ones) linked to the search for generalized well-being, as suggested in the reports. In this context, the search for materials and innovation surrounding sustainability has to be encouraged. Regarding marketing issues, it is important to refer to cork as an important sustainable material. In this regard, the acquisition of this product seems represents added value since, underlining recycling and circularity process. It seems to be important to obtain a material that meets the consumer's mindset and the Cork-a-Tex Yarn fits broadly into the concept of 'sustainability', reflecting the emerging trend identified. Trends, from this point of view, are structures that facilitate intermediation between brands and audiences, as they translate mindsets allowing the creation of more succinct applied strategies [14]. The branding and the strategies to be addressed in the brand should reinforce even more this material that reflects change and is endowed with national significance. This path is relevant in the context, for example, of the use of culture and its representations as an element of brand communication [11, 13]. In this way, creating a narrative around the material permits to associate characteristics to the brand such as 'innovation', 'responsibility' and 'heritage', as well as the issues that transpose the medium where they are inserted[1]. Thus, the relevance of trend analysis for strategic creation, allows brands to identify opportunities that are essential for their affirmation.

5 Conclusions

Trend analysis is advantageous for the development of strategies for the Cork-a-Tex brand, as it allows to reinforce the brand's positioning with regard to audiences and consumers. Globally, trend reports point to the emergence of sustainability as a predominant trend and the rising of a mindset that is widely propagated in society, which guides brand strategies in this context.

The Cork-a-Tex Yarn product, made from cork, has properties that fall within what sustainability stands for (both in its physical properties and production process). Thus, it seems appropriate to communicate the Cork-a-Tex Yarn as a product within the 'green' context to reach an audience that participates and values the cyclicality processes, as observed in the studied trend descriptions. Regarding cork, as a peculiar material, its application in new contexts is important in order to disseminate this material with specific and relevant characteristics in the current scenario. The creation of new products (in this case, fabrics) is stimulated through the cork and Cork-a-Tex Yarn as well as the research for new materials, which can bring new readings and applications.

[1] This practice is reflected in Holt and Cameron [12] when addressing the theory of cultural innovation, bringing an approach to mythological construction of brands from cultural and representative elements of them.

References

1. Amorim. http://www.amorim.com/
2. Apcor: o que é a cortiça (n.d-a). https://www.apcor.pt/cortica/o-que-e/
3. Corkpel. https://www.corkpel.com/
4. Corkribas. http://www.corkribas.pt/
5. Cunha, R.: O tecido de cortiça é um dos materiais sustentáveis para a moda e decoração (2016). https://www.stylourbano.com.br/o-tecido-de-cortica-e-um-dos-materiais-mais-sustentaveis-para-moda-e-decoracao/
6. Dragt, E.: How to Research Trends. BIS Publishers, Amsterdam (2017)
7. Faith Popcorn/Brain Reserve. https://faithpopcorn.com/trendbank/AtmosFear/
8. Gil, L.: Usos alternativos da Cortiça. Naturlink (2009). www.naturlink.sapo.pt
9. Gomes, N., Cohen, S., Flores, A.: Estudos de Tendências: Contributo para uma abordagem de análise e gestão da cultura. Moda Palavra, vol. 11, no. 22 (2018)
10. Higham, W.: The Next Big Thing. Kogan Page, London (2009)
11. Holt, D.: How Brands Become Icons: The Principles of Cultural Branding. Harvard Business School Press, Boston (2004)
12. Holt, D., Cameron, D.: Cultural Strategy - Using Innovative Ideologies to Build Breakthrough Brands. Oxford Press, Oxford (2010)
13. Kapferer, J.: New Strategic Brand Management. Kogan Page, London (2008)
14. Kjaer, L.: The Trend Management Toolkit. Palgrave MacMillan, London (2014)
15. Kjaer Global. https://kjaer-global.com/
16. Kongsholm, L., Frederiksen, M.: Trend sociology 2.0. Denmark, Pej Gruppen (2018)
17. Mestre, A., Gil, L.: Cork for sustainable product design. Ciência e Tecnologia dos Materiais **23**, 52–63 (2011)
18. Niemeyer, L., Cantú, W.: A cortiça como desafios para novos usos no design. Blucher Des. Proc. **2**(9), 4272–4283 (2016)
19. Pinto, C.: Cork-a-Tex Yarn: Fio com elevada incorporação de cortiça, 18 April 2017. https://www.compete2020.gov.pt/noticias/detalhe/Proj17714-CorkaTex-Yarn
20. Powers, D.: On Trend: The Business of Forecasting the Future. University of Illinois Press, Champaign (2019)
21. Powers, D.: Thinking in trends: the rise of trend forecasting in the United States. J. Hist. Res. Mark. **1**(1), 2–20 (2018)
22. Raymond, M.: The Trend Forecaster Handbook. Laurence King Publishing, London (2010)
23. Rohde, C.: Serious trendwatching: 25 of the best coolhunts. Fontys Applied University of Applied Science and the Science of the Time, The Netherlands (2011)
24. Rose, G.: Visual Methodologies. Sage, London (2001)
25. Trends Observer. http://www.trendsobserver.com/
26. Trend One. https://www.trendexplorer.com/en/trends/sustainability/
27. The Science of the Time. https://scienceofthetime.com/portfolio/trend-5/
28. Trend Watching. https://trendwatching.com/quarterly/2019-11/5-trends-2020/
29. Vejlgaard, H.: Anatomy of a Trend. McGraw-Hill, New York (2008)
30. Williams, R.: The Long Revolution. Penguin Books, Harmondsworth (1961)
31. Wimer, R., Dominick, J.: Mass Media Research. Wadsworth, Boston (2011)

Creativity and Entrepreneurship in Education

Playful Approaches to Entrepreneurial Competencies in University Teaching: Introducing the 4Cs Model

Riikka Franzén[1(✉)], Katriina Heljakka[2], and Lenita Nieminen[1]

[1] Turku School of Economics, Pori Unit, University of Turku, P.O. Box 170, FIN-28101 Pori, Finland
{riikka.franzen, lenita.nieminen}@utu.fi
[2] Degree Programme in Cultural Production and Landscape Studies, P.O. Box 124, FIN-28101 Pori, Finland
katriina.heljakka@utu.fi

Abstract. A playful approach used as part of a teaching method in any university subject enhances university students' entrepreneurial mindset and working-life skills. Playfulness has become more accentuated, and playful approaches in the working-life context have been studied, for example, within management and organizations. In university teaching, playful learning contributes to students' problem-solving skills, creativity, foresight, as well as their ability to see things that do not yet exist. These skills are linked to entrepreneurial competences, such as creativity, vision, tolerance of uncertainty and ambiguity. We collected our data through participatory observation in a workshop organized for teachers and researchers at a business school. The workshop utilized playful artifacts as tools for idea generation and to support embodied and spatial co-creation and presentation. The findings indicate that playfulness is useful in teaching an entrepreneurial mindset and attitudes, offering first-hand experiences, invoking emotions, and sharing them with others.

Keywords: Playful approach · University teaching · Entrepreneurship education

1 Introduction

This paper reports the results of our study of playfulness as an approach in teaching that enhances students' entrepreneurial mindset and working-life skills. Students are no longer satisfied with a form of higher education restricted to transferring knowledge and skills and now expect to benefit from a broader form of personal development [1]. Consequently, universities now pay more attention to producing employable graduates with the skills appropriate to working life. The skills required to be competitive in the 21st century employment market include critical thinking, creativity, collaboration and communication skills, literacy skills related to information, media, and technology, flexibility, leadership, productivity, and ability to take the initiative. To answer the call for a more entrepreneurial mindset that is useful whether working in the private, public, or the third sector [2, 3], the list could be extended to include competences such as

E. Markopoulos et al. (Eds.): AHFE 2020, AISC 1218, pp. 105–112, 2020.
https://doi.org/10.1007/978-3-030-51626-0_12

coping with uncertainty and complexity, having a positive attitude toward change and a willingness to take action. Accordingly, entrepreneurship education should reach students on courses that are not specifically entrepreneurship oriented, meaning embedding entrepreneurship in any subject [4].

Play, creativity, and art have acquired an elevated status in the twenty-first century and work and play can no longer be considered polar opposites. Play is now recognized as being an important element in value creation and the management and organization of the workforce, which are central to the knowledge economy [5]. Consequently, it is no surprise that the world's leading enterprises, such as Google, are incorporating playful elements into their working environments (e.g., playful co-working facilities and entertainment spaces), artifacts (toys, games and other equipment, but also play-oriented furniture) and increasingly, encourage the cultivation of playful organizational cultures. In these new working environments, workers are expected to act openly and creatively, by expressing their playfulness in various ways. This development should inspire playful approaches even in the context of higher education.

In entrepreneurship education, serious games have proved a valuable means of simulating entrepreneurial learning by encouraging entrepreneurial preparedness among students in a risk-free environment. However, creating uncertainty and generating interaction between the gameplayers can be problematic [6] and because entrepreneurial learning is social in nature [7, 8], it is vital to build environments enhancing interaction. Applying art practices can be a powerful way to expose students to uncertainty, emotions, creativity, extreme time pressures, and making choices, and can thus prepare them for the subversive side of entrepreneurship [8, 9].

Inspired by the elements enhancing learning in play and art practices, we investigated how playful approaches could support entrepreneurial competences in university teaching. The playful approach implemented was based on interaction with shared physical materials and co-players interacting in the same space. Our study highlights the role of creativity, collectiveness, spatial embodiment, and working with physical tools and materials in developing an entrepreneurial mindset.

2 Theoretical Background

2.1 Practicalizing Entrepreneurship in University Teaching

According to its wider definition, entrepreneurship is often seen as a mindset and behavior that is useful not only in running and creating businesses, but also when working as an employee within business organizations and in the public or third sector [2, 3]. Definitions reflecting that interpretation of entrepreneurship include, "a way of thinking for approaching problems, implementing innovations, finding solutions, sharing ideas, and making change happen." [10], or value creation for others, acknowledging that the value can be financial, cultural, or social [11]. In fact, entrepreneurial competences are now considered important for all citizens [12].

This contemporary view of entrepreneurship has also shifted the focus in entrepreneurial pedagogics from new venture establishment to developing an entrepreneurial mindset and attitudes among students [13]. For example, Gibb [14] has elegantly defined the purpose of entrepreneurship education as developing "behaviours, skills and attributes applied individually and/or collectively to help individuals and organisations of all kinds, to create, cope with and enjoy change and innovation involving higher levels of uncertainty and complexity as a means to achieving personal fulfilment and organisational effectiveness."

Recognizing the call to encourage an entrepreneurial mindset as well as earlier findings indicating that playful learning contributes to university students' problem-solving skills, creativity, foresight, and ability to see things that do not yet exist [15], we became curious about how university teachers could be motivated and trained to use pedagogical approaches that enhance students' entrepreneurial competences and skills. Learning by doing and experiential learning are widely thought to offer a useful pedagogy for entrepreneurship education, as they encourage action and not just entrepreneurial intent [16], and acknowledge the important role of communities and emotions [17]. As all learning is social in nature, entrepreneurial learning also occurs through interaction with others [7]. As entrepreneurship is considered a "journey of the heart" [17], exposing students to emotions involved in the entrepreneurial process is important [4, 8].

Recognizing the linkages and opportunities with playful approaches inspired us to experiment with a playful workshop with business school teachers and researchers with the purpose of providing them with an experience of an emotional learning context: learning by doing, co-creation with others, and using playful approaches to working under pressure in an informal environment [18].

2.2 Theorizing Playful Approaches

In this paper, we follow established theoretical definitions of play, such as those by Caillois [19] and Burghardt [20]. According to Caillois [19], play is always free, but to some degree rule based. Playing is connected to time and environment and based on imagination. Moreover, Caillois sees play as a source of joy and humor. Again, following Burghardt [20], playing is associated with spontaneity, voluntariness, goalorientedness, and enjoyment. Van Oers [21] combines playing with rules, and the degrees of freedom and involvement. Whereas playfulness manifests as a state of mind, playing refers to the activity of play. Nevertheless, the two often intertwine and occur simultaneously; *playfulness* functions as a predisposition to play; and *playing* as an activity always requires a playful state of mind.

The role of playfulness in adult life has been accentuated in recent decades [22]. Consequently, playful approaches employed in working life have been studied for example within management and organizations. For many adults it is easy to recognize a certain playfulness in themselves, yet it seems difficult for adults to admit that they are playing; for example, even in the context of activities with toys, adults more often refer to their activities using euphemisms, such as collecting or pursuing hobbies [23]. For this reason, we chose to use the term *playful approaches* in this paper. To continue, playfulness is in a way flirting with the possibility of play, a mindset necessary to *make use of play*, for example, in a serious context, such as university pedagogics.

However, the role of bodily engagement and use of physical materials and space have seldom been examined in scholarly work on university teaching. This is surprising, as playful approaches to teaching seem to benefit from co-creation and imaginative use of artifacts, tools, and environments, in which bodily engagement and spatiality play significant roles. In order to bridge this gap in research into the uses of play in adult learning, and particularly in university pedagogics, we have chosen to examine the potential of playful approaches in this context. When we refer to playful approaches then, we mean any artifacts, tools, technologies, or environments used in combination with a playful mindset—that is, a predisposition to a task that involves open-mindedness, an entrepreneurial mindset, and the related competences.

3 Methodology

In order to answer the research question: *How can playful approaches support the teaching of entrepreneurial skills in university education?* we collected data from a three-hour-long workshop offered to teachers/researchers at a business school as part of a two-day scientific research seminar. The work, group presentations, and immediate feedback were video recorded.

Open-ended play (less rule-bound play) enables activities in which players devise their own rules, goals, and meanings [24]. In our workshop, we used the *Comicubes* prototyping tool that promotes open-ended, creative play by breaking away from established patterns and combining actions or thoughts in new ways [25]. The giant version of the physical Comicubes consists of foldable cardboard cubes (40 × 40 40 cm) with six information layers to fill with either images or text. As a multisensory play concept, the Comicubes concept also invites players to manipulate/rotate/ organize stack/build sequences of the cubes in either random order or according to the player's wishes.

For this first-round analysis, we used our field notes and the video recordings of the group/participant reflections on their experiences about the workshop. We did not utilize any preset theoretical framework, but tried to listen to and understand the experiences of the participants and the meanings they assigned to their work [26]. All three researchers participated in the analysis and discussed their interpretations of the information gathered.

4 Results

The immediate feedback collected from the workshop participants about the playful approach adopted spurred ideas and comments on creativity, courage, collaboration or co-creation, and conversation, as summarized in Fig. 1, and briefly elaborated upon in the following.

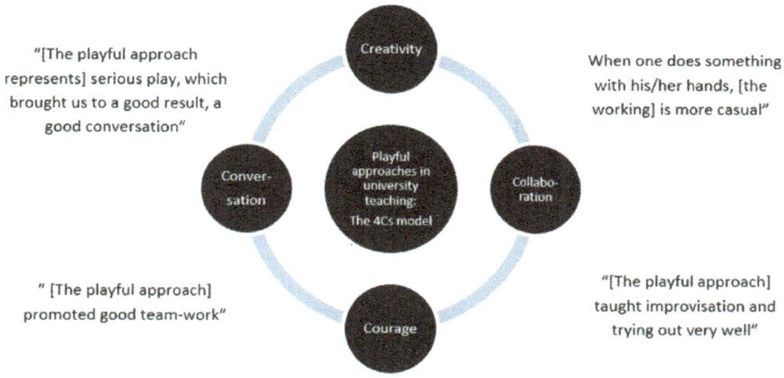

Fig. 1. The 4Cs of playful approaches: a summary of comments of workshop participants.

The first theme to emerge from the data was *creativity*. Whereas critical thinking is regarded as a central approach in university teaching, sometimes being too critical can hinder creative thinking. The development of creativity has been shown to connect with entrepreneurial intentions [17]. The comments on creativity addressed the hands-on and casual qualities of the playful approach used in the workshop. As Branscombe [27] states, "The role of the body is neglected in education, yet it is a growing area of interest and study". The playful approach employed in our study also accentuated the bodily and sensory aspects of the workshop: when introduced to the Comicubes tool, the participants were required to handle the tools and move about when playing with them.

As the workshop pushed the participants to step away from their comfort zone and improvise without clear instructions or an idea of the outcome, it required *courage*. The participants had to trust on the group's ability to create ideas and solutions, even though they would not believe to be successful in the task by themselves. Thus, the workshop provided an arena for learning how to cope with uncertainty and complexity. The workshop also offered an opportunity to fail in a safe environment. Dealing with the feelings caused by failure and recovering from failure are considered as an important part of entrepreneurship education [28]. Failure is often undervalued as a learning experience, as attention is focused more on the outcome. One participant highlighted the role of a process that was even more important than the outcome: "Sometimes the journey is more important than where it ends."

The participants assessed the workshop to have facilitated *conversation* and *collaboration* because it utilized group work. A common task required participants to communicate, speak and listen, and to build a dialogue with trust that encourages participation and sharing. The participants already knew each other as colleagues, but this workshop helped them to learn about each other, which will enhance future collaboration.

Building on our findings and the required twenty-first century skills [29] we tabulated the twenty-first century entrepreneurial skills framework in Fig. 2 below.

Playful mindset	Playful approach	Entrepreneurial mindset
creativity	tools & techniques	enjoying uncertainty and complexity
collaboration	learning environments	creating change
communication		creating value

Fig. 2. Twenty-first century entrepreneurial skills

Our conceptualization of the entrepreneurial skills needed in the twenty-first century incorporates a playful mindset (creativity, collaboration, and communication), a playful approach (to tools, techniques, and environments), and an entrepreneurial mindset (enjoying uncertainty and complexity and creating change and value).

5 Conclusions and Impacts

The purpose of this paper was to report how playful approaches could be used in a higher education context to enhance entrepreneurial pedagogics. We particularly wanted to focus on entrepreneurial competencies and to evaluate the suitability of an embodied, playful approach combined with employing a three-dimensional tool, Comicubes, in experiential learning.

Our analysis indicates that playfulness is useful in teaching the entrepreneurial mindset and attitudes, in that it offers first-hand experiences of coping with uncertainty and complexity, being creative, collaborating, and co-creating. In addition, participating in a playful workshop evokes emotions, and thus provides an opportunity to learn how to cope with them. Playfulness as an approach may also be used when reflecting on experiences and emotions and when sharing them with others—starting a conversation. Reflection is considered an integral part of entrepreneurial learning [30].

To sum up, the participants demonstrated how a playful approach including use of large-scale physical materials can facilitate improvisation, collaboration, creativity, physical movement and collective problem-solving—activities that are considered beneficial to the mental flexibility considered crucial to successful entrepreneurial individuals of the twenty-first century.

References

1. Crayford, J., Fearon, C., McLaughlin, H.: Affirming entrepreneurial education: learning, employability and personal development. Ind. Comm. Train. **44**, 187–193 (2012)
2. Rae, D.: Universities and enterprise education: responding to the challenges of the new era. J. Small Bus. Enterp. Dev. **17**, 591–606 (2010)
3. Neck, H., Greene, P.: Entrepreneurship education: known worlds and new frontiers. J. Small Bus. Manag. **49**, 55–70 (2011)
4. Pittaway, L., Cope, J.: Simulating entrepreneurial learning: integrating experiential and collaborative approaches to learning. Manag. Learn. **38**, 211–233 (2007)

5. Kavanagh, D.: Work and play in management studies: a Kleinian analysis. Ephmera. Theory Polit. Organizat. **11**, 336–356 (2011)
6. Fox, J., Pittaway, L., Uzuegbunam, I.: Simulations in entrepreneurship education: serious games and learning through play. Entrepr. Educ. Pedagogy **1**, 61–89 (2018)
7. McKeever, E., Jack, S., Anderson, A.: Embedded entrepreneurship in the creative reconstruction of place. J. Bus. Ventur. **30**, 50–65 (2015)
8. Bureau, S.P., Komporozos-Athanasiou, A.: Learning subversion in the business school: an 'improbable'encounter. Manag. Learn. **48**, 39–56 (2017)
9. Tunstall, R., Nieminen, L., Jing, L. Hjorth, R.: Out of the blue: teaching creativity and entrepreneurship through Elashmobs. In: Papadopoulos, P.M., Burger, R., Faria, A. (eds.) Innovation and Entrepreneurship in Education (Advances in Digital Education and Lifelong Learning 2), Emerald Group Publishing Limited, pp. 31–56 (2016)
10. Nadelson, L.S., Nageswaran, A.D., Benton, T., Basnet, R., Bissonnette, M., Cantwell, L., Jouflas, G., Elliot, E., Fromm, M., Lanci, S.: Developing next generation of innovators: teaching entrepreneurial mindset elements across disciplines. Int. J. Higher Educ. **7**, 114–126 (2018)
11. Lackéus, M.: Value creation as educational practice - towards a new educational philosophy grounded in entrepreneurship? Doctoral thesis, Chalmers University of Technology (2016)
12. Bacigalupo, M., Kampylis, P., Punie, Y., Van den Brande, G.: EntreComp: the entrepreneurship competence framework. Publication Office of the European Union (2016)
13. Arpiainen, R.-L.: Tiimityötä, toimintaa ja tunteita – näkökulmia yrittäjyyskasvatuksesta ja yrittäjämäisestä oppimisesta kokemuksellisen toimintaoppimisen viitekehyksessä. Aalto University Doctoral Dissertations 46/2019 (2019)
14. Gibb, A.A.: Entrepreneurship and enterprise education in schools and colleges: insights from UK practice. Int. J. Entrepr. Educ. **6**, 48 (2008)
15. Kangas, M., Siklander, P., Randolph, J., Ruokamo, H.: Teachers' engagement and students' satisfaction with a playful learning environment. Teach. Teach. Educ. **63**, 274–284 (2017)
16. Smith, A.M.J., Paton, R.A.: Embedding enterprise education: a service based transferable skills framework. Int. J. Manag. Educ. **12**, 550–560 (2014)
17. Nabi, G., Liñán, F., Fayolle, A., Krueger, N., Walmsley, A.: The impact of entrepreneurship education in higher education: a systematic review and research agenda. Acad. Manag. Learn. Educ. **16**, 277–299 (2017)
18. Gibb, A.A.: Small firms' training and competitiveness, Building Upon the Small business as a Learning Organisation. Int. Small Bus. J. **15**, 13–29 (1997)
19. Caillois, R., Barash, M.: Man, Play and Games. Urbana (Ill.). University of Illinois Press, Chicago (2001)
20. Burghardt, G.M.: Defining and recognizing play. In: Nathan, P., Pellegrini, A.D. (eds.) The Oxford Handbook of the Development of Play, pp. 9–18. Oxford University Press, NewYork (2011)
21. Van Oers, B.: Is it play? Towards a reconceptualisation of role play from an activity theory perspective. Eur. Early Childhood Educ. Res. J. **21**, 185–198 (2013)
22. Raessens, J.: Homo Ludens 2.0, The Ludic Turns in Media Theory. Universiteit Utrecht, Utrecht, Netherlands (2010). http://bit.ly/2WjGxNK
23. Heljakka, K.I.: More than collectors: exploring theorists', hobbyists' and everyday players' rhetoric in adult play with character toys. Games Culture **13**, 240–259 (2018)
24. De Valk, L., Bekker, T., Eggen, B.: Leaving room for improvisation: towards a design approach for open-ended play. In: 12th International Conference on Interaction Design and Children, pp. 92–101 (2013)

25. Heljakka, K., Ihamäki, P.: Comicubes–a playful tool to stimulate (design) creativity, Celebration Contemplation. In: 10th International Conference on Design & Emotion, pp. 27–30 (2016)
26. Leitch, C.M., Hill, F.M., Harrison, R.T.: The philosophy and practice of interpretivist research in entrepreneurship: quality, validation, and trust. Organizat. Res. Method. **13**, 67–84 (2010)
27. Branscombe, M.: Showing, not telling. Read. Teach. **69**, 321–329 (2015)
28. Cope, J.: Entrepreneurial learning from failure: an interpretative phenomenological analysis. J. Bus. Ventur. **26**, 604–623 (2011)
29. 21St century skills, Applied education systems. https://www.aeseducation.com/career-readiness/what-are-21st-century-skills
30. Cope, J.: Entrepreneurial learning and critical reflection: discontinuous events as triggers for "higher-level" learning. Manag. Learn. **34**, 429–450 (2003)

Developing a Method for Understanding How to Empower Creativity Through Digital Technologies: The Case of AI

Carmen Bruno[(✉)] and Marita Canina

IDEActivity Center, Design Department, Politecnico di Milano, Via Durando 38/a, 20158 Milan, Italy
{carmen.bruno,marita.canina}@polimi.it

Abstract. The paper presents a method aimed at empowering human creativity to generate innovation by exploiting the opportunities and potentialities of the emerging digital technologies. The method, through the application of a Creativity 4.0 Model and Framework, aims at observing and understanding the impacts of a specific digital technology on the main dimensions of the human being, understanding how those impacts could positively or negatively influences the creative design process. The human centred method, currently in its early stage of theoretical development, aims to benefit design research, education and practice supporting them in understanding the ongoing human transformation and in making a wiser and consciously use of digital opportunities addressed to human creative enhancement for innovation.

Keywords: Digital maturity · Creative process · Artificial intelligence · Design method · Human centred

1 Introduction

This specific phase of technological progress seems to be very different from all the preceding ones: it involves a combination of transformative and cognitive digital technologies, tools and processes and most importantly people, in terms of culture, skills and mindset [1]. Therefore, the interconnection between technology, human cognition and human life in general will become much more intrinsic, invisible, and impactful. The human being is co-evolving with digital technology [2] as it modifies our "relationships to ourselves (who we are), the interaction with others (how we socialize), our conception and interaction to the real world" [3]. A digitally enhanced generation is growing up accepting digital enhancement an integral part of everyday lives [4].

In this digital era, a fundamental and pressing concern address the role of humans in this abundance of new, smart and very efficient technologies, which progress and evolution cannot be stopped and or even slowed down.

Humans are transforming the society through machines and technologies, and they must learn how to live well with them, and how to manage a radical change that will occur in less than a generation without suffering them. This means, reaching a **digital**

E. Markopoulos et al. (Eds.): AHFE 2020, AISC 1218, pp. 113–120, 2020.
https://doi.org/10.1007/978-3-030-51626-0_13

maturity defined as a "continuous and ongoing process of adaptation to a changing digital landscape" [5].

Reaching a digital maturity allow to continuously understand and possibly anticipate the foreseeable opportunities as well as the threats that will be offered by the technological evolution, developing a strategic approach to the adoption and application of such technology. It is therefore necessary to put human creativity at the center of the digital transition which has been recognized as one of the most distinctive human skills to nurture and develop in order to manage at best the powerful collaboration between human and machine. Indeed, creativity helps people conceive novel and useful ideas [6], representing the intangible substrate for innovation [7] and is, therefore, a key to economic growth and social transformation [8].

Human existence will be related one-to-one with our ability to generate ideas to successfully exploit the opportunities that technology is offering us. Hence the responsibility of studying, understanding and developing creativity skills, and defining how digital technology and human evolution influence those abilities and how to exploit the new opportunities to digitally enhance them, becomes a sort of multidisciplinary mission.

In this scenario facilitate and guide the creative process of every human being become, therefore, a mission and a new professional role for designers that will play a leading role in the process of people transformation toward digital maturity.

Therefore, as design researchers, our main contribution in this scenario of transition is the definition of a Human-Centred Method aimed at empowering creativity to generate innovation by exploiting the opportunities and potentialities of the emerging digital technologies.

We think that the design discipline should therefore prepare the next digital generation of designers and providing new creativity-driven design tools that will form the future designer toolbox for digital maturity.

Therefore, the method aims to benefit design research, education and practice supporting them in understanding the ongoing human transformation and in making a wiser and consciously use of digital opportunities addressed to human creative enhancement for innovation.

2 Building the Method's Tools

The Creativity 4.0 method has been structure throughout a three years long research projects whose main objective was to explore the main positive and negative influences brought by the digital transition oh human creativity within the design process.

The research project main results were the definition of what have been called The Creativity 4.0 Model and Framework which supported the exploration of the influences on human creativity and their relationship with the design process, becoming therefore the two conceptual tools of the method.

The two tools are briefly described in the consequent sub-sections and have been developed and validated through several phases, the first of which was to identify and select, among the broad psychology literature in creativity, the approach through which deal with the creativity domain in this research.

Given the complexity and the multitude of approaches that can be adopted to study and explore creativity [9] it was essential to frame a clear definition and above all to choose which aspects to consider.

Drawing on the 7 C's creativity framework [10], the research aimed at investigating creativity by studying the integration and interconnection of three of its different facets which are:

- **creator**, referring to the digitally enhanced generation, the research adopts an individual perspective, considering the cognitive, personality, motivational and emotional factors of the actors that engage in creative activities. Those factors have been called factors of creativity;
- **creating**, referring to Stage and Componential Process theories [7] that study the creative process from the macro perspective as a form of activity or action;
- **context**, referring to the actual scenario of digital transition that influences several aspects of the human being.

Therefore, I adopted a more extended and elaborated definition of creativity [11] which claim: "Creativity is the interaction among aptitude, process, and environment by which an individual or group produces a perceptible product that is both novel and useful as defined within a social context" (p. 90).

The identification of the three aspect of creativity, allowed to frame the creativity boundaries and build the frame of reference for the analyses of the impacts brough by the digital age which is represented by the Creativity 4.0 model.

2.1 The Creativity 4.0 Model

The Creativity 4.0 Model (Fig. 1) [12] is structured through three main elements: the *digital transition*, that is the ground on which the model is growing and feeding; the *human being* analysed according to the three levels on which the digital transition has an impact (cognitive, individual, social) and, finally, the *creative process*, namely human ability, which is transversal to all levels of the human-being.

These three elements also correspond to the three aspects of the 7 C's creativity framework on which the research is settled – context, creators and creating.

The Creativity 4.0 model built enables to include the many dimensions and factors of creativity identified that intervene within the human-being on the three different levels, which include **creativity as a mental process**, **creativity as an individual practice**, and **creativity as a social process** that unfolds with others.

The intersection of those three levels provides a comprehensive understanding of the complexity and multifarious aspects of the creative process comprising several steps, activities as well as the motivational, cognitive, attitudinal, technical constituents and the social and environmental components influencing these constituents during the process. The three levels also influence each other as they are part of the same person.

The Creativity 4.0 model is not static, but it represents a tool that allows to observe the impacts of the digital transition on the three levels of the human-being understanding how they influence the creativity factors. The model has been validated and refined through three main exploration which allowed on one side to keep only the factors of creativity that are considered more relevant in this digital age, and on the other side, to collect a set of influences brought by the digital transition on the human creative abilities. For design research and practice, this analysis becomes strategic if it is structured within a theoretical framework that associates the impact identified and the influenced factors with the design process. Therefore, a Creativity 4.0 Framework has been built within the research, representing the second tool of the method.

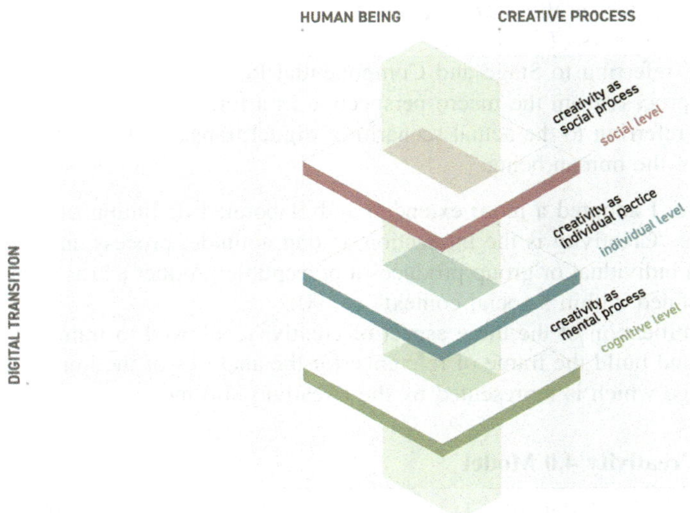

Fig. 1. Creativity 4.0 model

2.2 The Creativity 4.0 Framework

The Creativity 4.0 Framework [13] has been developed to properly understand how the digital transition is influencing the cognitive, emotional and social factors of the digitally enhanced generation that intervene in the design process that lead to the production of new, original and useful ideas.

By **deconstructing the** design **process in stages, steps, activities** and **thinking style,** the framework allowed to map the less rationale creativity factors emphasizing their interconnections and showing how the impacts of the digital transition, identified through the model, are influencing the activities, the creative factors and the overall design process.

The Creativity 4.0 Framework represent a tool to analyze the influence of the digital transition on the multiple aspects of the design process as well as to define design actions and tools to empower it.

3 The Creativity 4.0 Method

The Human-Centred Design method has the main aim of exploring the changes and opportunities of the digital evolution for augmenting human creativity, empowering the design processes for innovation in different context of application, putting people at the centre of the change.

The method aims to **facilitate and support the emotional, motivational, cognitive and social factors of the human being that intervene in the creative design process as well as the process itself.** These factors are the levers to be activated and enhanced to allow the human being to express his maximum creative potential.

The Creativity 4.0 method is structured according to four main steps where the model and the framework become operational tool that guide each step as follow:

Step1_Observe: the Model, in its static version, guide in the observation of the impacts that could affect the human being as well as in the observation of how technologies can support the creative process.

Step 2_Understand: the Model in its dynamic version, allow to understand how the impacts identified as well as the technologies observed, could influence the creativity factors and therefore the human creative abilities.

Step 3_Map: in this step, the Framework become the guiding tool. It allows to see the effects on the design process and to define action to empower creativity through the design process.

Step 4_Design: this represents the active step in which the identified and defined action is designed and implemented. It may refer to new tools, processes, activities to enhance single steps or the whole creative design process.

The Model and the Framework represent the fundamental knowledge needed to design actions to empower creativity abilities in in the digital age, exploiting the opportunities provided by the digital transition and the digital technologies.

In the next sub-section, we envisage the application of the method to explore the influences of a specific digital technology, Artificial Intelligence (AI), on creativity within the design process. This approach has been called **Technology-driven** since is based on the application and experimentation of the potentialities of a certain technology as a powerful source of competitive advantage.

3.1 Application of the Method on Artificial Inteligence

Step 1_Observe: in this step is fundamental to study the potentialities of AI and observe how it can support the creative process. This understanding can be reached by collecting researches projects, cases best practice that uses AI to enhance the creative process in its positive and negative aspects. For example, from a deep investigation of the literature, emerges that AI can be used as a co-creator within the creative process, supporting humans in fastening and amplifying activities and tasks in specific moment of the process. AI can also be used as a feeding partner, able of reproducing human cognitive processed to feed human thinking with inspiring divergent information.

Step 2_Understand: in this step is fundamental to analyze the cases with the lens of the factors of creativity trying to understand in which part of the process AI can perform its best as partner. One of the main recognized potentialities, as co-creator, is that it helps to accelerates research and discovery by freeing people from repetitive tasks, empowering teams to focus on more creative, higher-value work. Since its declared function in accelerating discovery and research, one of the opportunities identified is that AI could support in *Augmenting and speeding up human in exploring information and knowledge.* Therefore, the Explorative stage of the process could be its chance.

Step 3_Map: in this step is fundamental to relate the opportunity identified with the framework to identify which steps or activities within the process can be influenced by the adoption of AI as a co-creator partner to augment and speeding up in the exploration of information and knowledge.

It emerges that AI can have a fundamental role in managing a huge amount of data and information, by analyzing and interpreting them. The computational technology can support human to expand their initial information, speed up the analysis of large amounts of information, identify and make associations through information from different context of exploration and quickly identify both expected and unexpected findings. In this case the human partner will have the important role of analyze the result provided by the machine, generate hypothesis and identify novel directions. An important reflection emerges regarding the changing role of the designer that become the facilitator of the process and the designer of the process that allows the machine to contribute and co-create.

Step 4_Design: from this analysis, an hypothesis of action that goes in the direction of empowering creativity emerge, addressing the design practice.

From the analysis emerge that the system could supported the designers in the earlier steps of exploration and interpretation of the information. Indeed, during the process, the cognitive system become knowledgeable, as an expert, on the specific topic. By reproducing the learning and synthesis processes of a human being, it amplifies and extend the human cognitive potential providing meaning and insight out of a vast mass of unstructured data and information (i.e. specific information about materials, forms and design elements to be transferred in the final product) which would have taken a great deal of time to be done by human. Moreover, AI can discover unusual details that helped shape the designer creative vision.

Therefore, for the design practice, it could be interesting to develop a design tool based on AI, Natural Language and Visual Recognition tools, that acting as co-creator could analyze and clusters all the information that a designer collects on a specific domain and the information on the field such as research notes, interview transcripts, inspirational images and so forth, to speed up the explorative phases of the design process and to support the designers in extracting meaning and insights out of that vast mass of information facilitating the generation of alternative hypothesis and its strategic thinking.

4 Discussion and Further Development

The paper presented a method developed to deeply understand the main positive and negative influences that the current scenario of digital transition is bringing to multiple levels of human creativity to inform and design actions to empower the creative design process for innovation.

The method can be used for:

- Enhancing the activities and steps of the design process by experiment, implement and adopt the new emerging digital technologies in a conscious way, in order to strengthen human factors within the creative process of individual and/or collaborative design aimed at innovation. This could result in the development of creativity support tools to inspire cognitive processes, to speed up design activities or that can augment human physical and cognitive capabilities.
- Defining and strategically apply new processes for designing with specific digital technologies. This allow to consciously drive the technological evolution putting technology at the service of human needs.
- Empowering and strengthening the comprehension of the overall design process and the mechanisms behind the creative expression by developing training module for creativity that can fit and adapt to the new habits of the digitally enhanced generation.

The method has been theoretically validated by applying it to explore in a structured way the knowledge produced in the emerging domain named "Digital Creativity", which is a wide and rapidly evolving realm, where multiple disciplines – psychology, sociology, computer science, HCI, etc. – already investigate the influence of and relationship between creativity and digital technology from several and fragmented perspectives. A total of 17 influences and their positive and negative influences on some factors of creativity have been identified and mapped on the creative design process, highlighting opportunities and threats brought by the digital transition.

At this point of the research the method needs an implementation and evaluation based on practical researches and the development of guidelines that allows everyone to put the method in practice.

References

1. World Economic Forum. The future of jobs: Employment, skills and workforce strategy for the fourth industrial revolution, Geneva, Switzerland (2016)
2. Corazza, G.E.: Organic creativity for well-being in the post-information society. Eur. J. Psychol. **13**(4), 599–605 (2017)
3. Floridi, L.: The Onlife Manifesto. Springer, Cham (2015)
4. Prensky, M.H.: sapiens digital: From digital immigrants and digital natives to digital wisdom. Innovate J. Online Educ. **5**(3), 1–9 (2009)
5. Kane, C.: Digital Maturity, Not Digital Transformation, MIT Sloan Management Review (2017)

6. Amabile, T.M.: A model of creativity and innovation in organizations. Res. Organizat. Behav. **10**, 123–167 (1988)
7. Kozbelt, A., Beghetto, R.A., Runco, M.A.: Theories of creativity. In: Kaufman, J.C., Sternberg, R.J. (eds.) The Cambridge Handbook of Creativity. Cambridge University Press, Cambridge (2010)
8. Florida, R.: The Rise of the Creative Class, Revisited. Basic Books, New York (2014)
9. Kaufman, J., Sternberg, R.: The Cambridge Handbook of Creativity. Cambridge University Press, Cambridge (2010)
10. Lubart, T.: The 7 C's of creativity. J. Creat. Behav. **51**(4), 293–296 (2017)
11. Plucker, J.A., Beghetto, R.A., Dow, G.T.: Why isn't creativity more important to educational psychologists? Potentials, pitfalls, and future directions in creativity research. Educ. Psychol. **39**(2), 83–96 (2004)
12. Bruno, C., Canina, M.: Creativity 4.0. Empowering creative process for digitally enhanced people. Des. J. **22**(1), 2119–2131 (2019)
13. Bruno, C., Canina, M.: Designing a framework to investigate creativity enablers and inhibitors in the digital era. In: Meanings of Design in the Next Era. 4D Conference Proceedings, pp. 162–173 (2019)

Brain Dominance and Learning Style Preference of Quantity Surveying Students in South Africa and Malaysia

Andries (Hennie) van Heerden[1(✉)], Michelle Burger[2], and Elzane van Eck[2]

[1] School of Built Environment, College of Sciences, Massey University, Auckland, New Zealand
a.vanheerden@massey.ac.nz
[2] Department of Construction Economics, EBIT Faculty, University of Pretoria, Pretoria, South Africa
{michelle.burger, elzane.vaneck}@up.ac.za

Abstract. The purpose of this study was to investigate the brain dominance of quantity surveying students in South Africa and Malaysia to determine the learning style preference to which they best relate. It further aims to establish if there is any correlation in brain dominance of quantity surveying students in different countries. A quantitative research methodology was applied, making use of a questionnaire consisting of short questions. The research respondents were third-year undergraduate quantity surveying students at the University of Pretoria and Universiti Teknologi Malaysia. Data collection and analysis was undertaken at the University of Pretoria. The most dominant brain quadrant for South African quantity surveying students was the lower left quadrant B (organised, sequential, controlled, planned structured, detailed, and scheduled). For the Malaysian students, it was the upper left quadrant A (analytical, technical, logical, rational, and precise). A comparative analysis between the two student groups indicated that there was a difference in brain dominance, although not significant. Adapting teaching techniques according to these findings may help students to better process and engage with module content at a higher educational level.

Keywords: Quantity surveying · Brain dominance · Learning styles · Higher education

1 Introduction

The construction industry-focused traditionally on hard skills and less on the soft skills. Although industry still acknowledges the importance of the much needed technical requirements, employers are seriously seeking students with employability skills or soft skills [1]. The construction industry moved to a service industry that is competing in the international arena [2]. There is a soft skill gap among construction school graduates [3]. They further elaborated on the fact that the construction industry requires

E. Markopoulos et al. (Eds.): AHFE 2020, AISC 1218, pp. 121–127, 2020.
https://doi.org/10.1007/978-3-030-51626-0_14

people who are well versed with a deep level of soft skills, as this increases the efficiency of work, processes, and technology.

The students enrolled both at the University of Pretoria (UP), and the Universiti Teknologi Malaysia (UTM) has a multicultural background. A previous research study demonstrated significant ethnic group differences in the learning style preferences of Southeast Asian and White students [4]. This study further substantiated statistically significant differences in the learning style preferences within diverse Southeast Asian groups.

The traditional Quantity Surveying (QS) programme was mostly taught methodically and logically around left-brain activities. It is necessary to have basic knowledge about specific topics to have a comprehensive understanding of the preferred learning methods. These topics include quantity surveying as a profession, the University of Pretoria's quantity surveying curriculum, and brain dominance. First, it is necessary to understand what quantity surveyors must be able to do in their profession. Secondly, the need to investigate if there is a correlation in brain dominance of quantity surveying students from different countries and lastly, it is necessary to establish if there is a need to increase the right brain activities within the various courses presented in the quantity surveying curriculum at the University of Pretoria.

2 Literature Review

2.1 South African Quantity Surveying Profession

The modern quantity surveyor can be seen as the "construction cost advisor or construction cost consultant with a broad knowledge of construction economics" [5]. The Royal Institution of Chartered Surveyors (RICS) elaborated further and stated that a quantity surveyor is "an expert in the art of costing a building at all its stages".

The professional quantity surveyor should be able to provide services within the six-stages of construction [6]. The first stage refers to the "Inception," and this is where the initial brief is prepared and needs approval from the client before cascading down to the next stage, "Concept and Viability". During the second stage, the concept design is completed, and the client receives an initial cost estimate and feasibility for approval. After approval, the third stage, "Design Development," is reached. During this stage, the architect will develop more detailed designs that will enforce more accurate cost estimations. Again the client needs to sign approval to move into the next stage, "Documentation and Procurement". Bills of quantities and comprehensive tender documentation are prepared during this stage. The tender evaluation process follows and contract documentation between the successful tenderer and the client is signed. The "Construction" stage follows where the quantity surveyor will become responsible for the cost management of the project. During the last stage, "Close-out," the quantity surveyor will be responsible for the financial close-out of the project [5].

2.2 University of Pretoria's Quantity Surveying Programmes

Students at the University of Pretoria can enroll for both a degree and honours degree in quantity surveying. Several subjects are taught at the university in preparation for the quantity surveying profession. At the undergraduate programme, these subjects can be divided into building essential modules, such as building sciences, building services, building drawings, and site surveying. A robust numerical basis is instilled through subjects such as mathematics, various structure modules, and statistics. Further to these, the quantity surveying students complete subjects such as economics, financial management, and their most essential year modules, quantities. During their honours degree, they focus more on contract law, construction management, and cost-and feasibility studies [7].

As indicated above, these subjects taught at the University of Pretoria aims to provide quantity surveying students with a comprehensive knowledge of building methods, procedures, and materials. Further to this, the student is equipped with economics, finance, contractual matters, and detailed measuring of quantities. The subjects aim to sufficiently prepare students to face the realities of the quantity surveyor profession. Further to the honours degree programme, the University of Pretoria alsorequires quantity surveying students to accumulate 240 h of work experience before obtaining their honours degree [7].

2.3 Hermann's Whole Brain Model

The Whole-Brain Model divides the human brain into four major quadrants, which are exemplified by the upper left A (analytical, technical, logical, rational and precise), B (organised, sequential, controlled, planned structured, detailed and scheduled. C forms part of the lower right and D the upper right quadrant. Quadrant C is sensory, emotional, people-oriented, teamwork, personal relationships and communications. Quadrant D refers more to visual, holistic, creative, integrative, conceptual, intuitive, entrepreneurial and future-oriented thinking [8, 9].

A table with the different associations of skills, actions, and perceptions with the four Whole Brain Model quadrants were created. The table highlighted words in quadrant A such as, mathematical, accounting, technical, working alone, running objects, processing numbers, logical operations, problem-solving, analysis, financial management, and rational thinking. Planning, scheduling, implementation, organising, mechanical, legal, administrative, controlling, disciplined and management in quadrant B. Quadrant C, producing ideas, synthesis, strategy building, innovation, risky, communication and ability to grasp the whole. Interpersonal communication, understanding people, supporting, writing, participant, speaking and helping people just to mention a few from quadrant D [9] (Fig. 1).

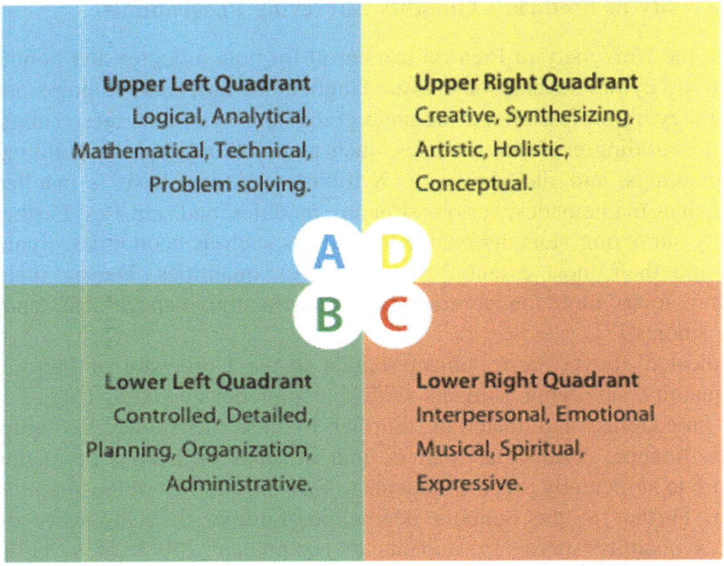

Fig. 1. Hermann Whole-brain model [13]

It was confirmed that there is "compatibility between the processes associated with each of the four modalities that constitute whole-brain learning and the processes associated with reaching outcomes-based education's (OBE) critical outcomes". They indicated that outcomes such as critical thinking, problem-solving, appreciation, evaluation of information, and communication are all associated with all four quadrants. It further indicated quadrant B (application) quadrant A (analyzing) and quadrant D (synthesizing) [10].

A research study further revealed that there is a noteworthy difference in the learning styles of White and Southeast Asian students. Quantity surveying students are industry-specific and, even though from different ethnic backgrounds, should fulfill the same job descriptions, thus indicated similar brain dominant profiles [4].

"Deep structured learning takes predominately place when all four quadrants are engaged [10]." Although each quadrant is different, it is regarded as equally important, and to be more industry ready, the quantity surveyor students need to be well balanced in all four quadrants [11, 12].

3 Research Method

Data collection and analysis was undertaken at the University of Pretoria. A quantitative research method was applied, making use of a questionnaire consisting of 24 short questions. Each question was associated with a different brain quadrant, and questions were presented in random order. Students answered each question by using a 1 to 5 Likert scale. The numerical representation of the Likert scale ranged from strongly disagree (1 on Likert scale) to strongly agree (5 on Likert scale). Empirical research was conducted in South Africa and Malaysia. The research respondents were third-year undergraduate quantity surveying students at the University of Pretoria and Universiti Teknologi Malaysia. 18 UTM students completed the questionnaire, but only 17 responses were valid. 45 UP students completed the questionnaire, and all responses were valid.

4 Data Analysis and Findings

Data were analyzed by firstly calculating an average percentage per question. Once this was done, each question was related to the relevant brain quadrant and an average percentage per quadrant was calculated. These percentages are indicative of the preference students have towards utilizing a particular brain quadrant.

Figure 2 indicates that the most dominant brain quadrant for South African QS students was quadrant B (86%), followed by A (80%), D (71%), and C (69%), respectively. The average of each quadrant was calculated and ranged between 69% and 86% for South African students. Figure 2 further indicates that the most dominant brain quadrant for Malaysia QS students was quadrant A (81%), followed by D (80%), B (79%), and C (77%). The average of each quadrant was calculated and ranged between 77% and 80% for Malaysian students. A comparative analysis between the two student groups indicated that there was a difference in brain dominance but not significant. The South African QS students are inclined towards the left side of the brain (quadrant A & B) more than the right side (quadrant D & C), were it seems that the Malaysian students are more balanced between the left and right side of the brain. Both QS student groups seemed to least prefer quadrant C.

In both groups, the average of each brain quadrant scored a high percentage (above69%), indicating that there is a strong preference for utilizing all four

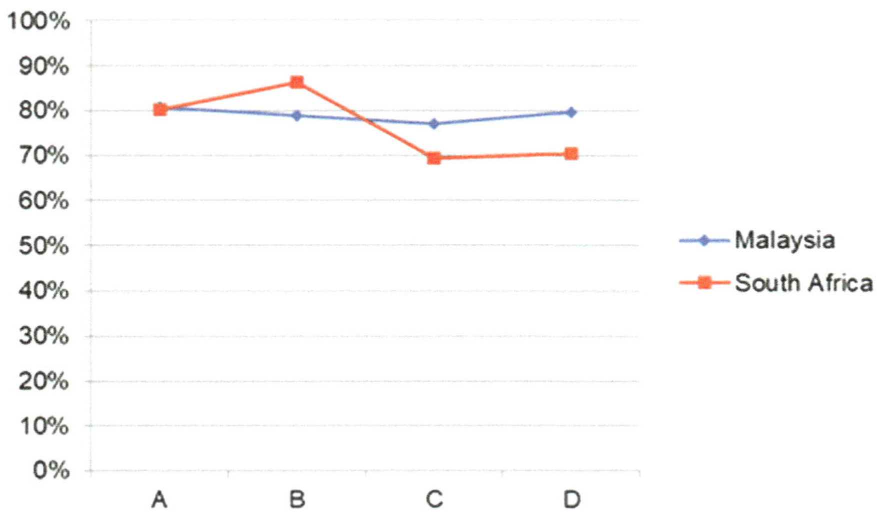

Fig. 2. Brain dominance comparison – UP and UTM QS students

brainquadrants. Quadruple dominant profiles constitute to approximately 3% of the population and often makes excellent CEOs [14].

QS students should thus relate best to teaching and learning strategies that are balanced between all four brain quadrants. When it comes to facilitating learning, there needs to be a shift from the predominant left-brain approach to a balanced whole-brain approach.

5 Conclusion

The traditional quantity surveying programme was mostly taught in a methodical and logical manner structured around left-brain learning preferences. The findings of the study indicates that QS students have a strong preference to utilize all four brain quadrants. Quadrant C is the least preferred and thus can effect brain processes that includes other brain quadrants where all four is needed, such as, critical thinking, problem solving, appreciation & evaluation of information. Some of the other important activities laking from quadrant C that is very important in the construction industry is; teamwork, passion, explain ideas, supporting, the ability of persuasion, interpersonal communication, teaching, socializing and speaking fluently.

Brain dominance is thus balanced between all four brain quadrants. It can be concluded that there is a need to increase right brain activities in order to achieve a balanced approach to whole brain learning. Quantity surveying lecturers should present module content in a way that is representative of all four brain quadrants.

References

1. Sweeney, E.: Beyond the technical: the soft skills needed to succeed in construction (2018). https://jobsite.procore.com/beyond-the-technical-the-soft-skills-needed-to-succeed-in-construction/
2. Shakir, R.: 'Soft skills at the Malaysian institutes of higher learning', Asia Pacific. Educ. Rev. **10**, 309–315 (2009). https://doi.org/10.1007/s12564-009-9038-8
3. Mahasneh, J.K. Thabet, W.: Rethinking construction curriculum: a descriptive cause analysis for the soft skills gap among construction graduates. In: 51st ASC Annual International Conference Proceedings, viewed 11 March 2016 (2015). from https://ascpro0.ascweb.org/archives/cd/2015/paper/CEUE391002015.Pdf
4. Park, C.C.: Learning style preferences of southeast asian students. In: URBAN EDUCATION, vol. 35 No. 3, September 2000, 245–268 © 2000. Corwin Press, Inc. (2000)
5. Maritz, M.J. Sigle, H.M.: Quantity Surveying Practice in South Africa. Construction Economics Associates (Pty) Ltd., Pretoria (2010)
6. van Heerden, H., Eck, E., Burger, M.K.: Learning style preferences of undergraduate quantity surveying students. Quant. Surv. 118–127 (2016). https://doi.org/10.18638/hassacc.2016.4.1.191
7. Faculty of Engineering, Built Environment and Information Technology (EBIT). Rules and regulations part 2 (Excerpt). University of Pretoria (2019)

8. le Roux, I.: New large class pedagogy: developing students' whole brain thinking skills 1877–0428 © 2011 Published by Elsevier Ltd. (2011). https://doi.org/10.1016/j.sbspro.2011.03.116. Procedia Social and Behavioral Sciences 15 (2011) 426–435 WCES-2011
9. Sarikaya1, B., Söylemez, Y.: Evaluation of acquisition of comprehension skills according to whole brain model. Univ. J. Educ. Res. **6**(11), 2444–2452 (2018). https://doi.org/10.13189/ujer.2018.061107, http://www.hrpub.org
10. Munro, M., Coetzee, M.-H.: Mind the gap: beyond wholebrain learning. South Afr. Theatre J. **21**(1), 92–108 (2007). https://doi.org/10.1080/10137548.2007.9687856
11. Coetzee, M.-H., Munro, M., de Boer, A.: Deeper sites through various lines: LMS and whole-brain learning in body/voice training for performers in the HET band. South Afr. Theatre J. **18**(1), 135–158 (2004). https://doi.org/10.1080/10137548.2004.9687784
12. Hughes, M., Hughes, P., Hodgkinson, I.R.: In pursuit of a 'whole-brain' approach to undergraduate teaching: implications of the Herrmann brain dominance model. Stud. Higher Educ. **42**(12), 2389–2405 (2017). https://doi.org/10.1080/03075079.2016.1152463
13. HBDI (2019). <http://www.hbdi.com/online-reference-and-activity-guide/a/index.html>. Accessed 7 Dec 2019
14. Herrmann International: How does your profile compare to others. Lake Lure, North Carolina: The Ned Herrmann Group (2000)

Hotspots and Trends in Creativity and Design Research: A Knowledge Graph Analysis

Yao Song and Yan Luximon(✉)

School of Design, The Hong Kong Polytechnic University, Hung Hom, Kowloon, Hong Kong
yao.song@connect.polyu.hk, yan.luximon@polyu.edu.hk

Abstract. Design creativity is the process or result of extending, presenting and interpreting creative ideas and concepts in a design way, the role of design creativity in corporate creativity activities has drawn increasing attention from the academic community. However, limited research has tried to holistically summarize the research trends and hotspots in the context of creativity and design. Through knowledge graph analysis, this paper analyzed the academic literature from Web of Science and visualized the knowledge domain in creativity and design research from 2010 to 2019. It has been found that the number of creativity-related design papers is increasing rapidly with a total of 498 papers. The top 15 co-occurring keywords include "design creativity", "innovation", "design making", "engineering design", "design education", "divergent thinking", "representation", "student", "knowledge", "aesthetics", "design process", "perception", "product design", "fixation" and "quality". The research that focuses on this topic could be summarized into eleven clusters namely industrial design, disability, design activity, explanatory design, design education, assessment, users, design fixation, design theoretic, TRIZ, and art education.

Keywords: Creativity · Design research · Itespace · Graph analysis

1 Introduction

Design creativity is the process or result of extending, presenting and interpreting creative ideas and concepts in a design way. It includes industrial design, architecture design, packaging design, service design, graphic design, fashion design and so on [1]. Take the role of creativity in product design as an example, either product functions or forms have undergone a dramatic change [2]. Regarding smart products that have ushered in new development opportunities, it is necessary for traditional products to consider the current rapidly changing market needs [2]. Exploring new development directions of traditional products under the background of rising manufacturing costs and innovation, we might reconsider the significant role of creativity in enterprise development and design process [3]. In the process of transformation and upgrading, enterprises get rid of the factor-scale development mode, and creativity in the design process plays an imperative role in transmitting business into a more sustainable model [3]. In the long run, the continuous improvement of production factors will reduce the

E. Markopoulos et al. (Eds.): AHFE 2020, AISC 1218, pp. 128–135, 2020.
https://doi.org/10.1007/978-3-030-51626-0_15

cost difference between enterprises, and creativity and innovation might become the key impetus for enterprises to shape their new core competitiveness [4].

Another example is promoting creativity in design process and education. Kids, students, or junior designers tend to have a higher intention to pursue the difference in their works' performance, form, connotation and style in their design process. In other words, not following the established thinking patterns, they try to break the routine, expand thinking outward, give their works unique connotations, and show their insight into art [5]. Thus, the vitality of design is the unconventional expression of personality, where the value of design creativity lies [6]. When applying creativity into enterprise development, it could help enterprises to include unconventional expression of personality as well as integrating roles of technology and market factors in design activities. Dosi [7] proposed that market pull and technology promotion are the main driving forces for creativity and innovation. Based on the perspective of product development, Vergant et al. [8] considered design creativity as the third driving force behind the market and technology and explored the relationship between three factors. To be more specific, creativity has a directional guiding role in the choice of corporate development strategy. While the level of technological innovation reflects the driving force of the company's development, the design creativity reflects the decision of the company's product development direction.

Although creativity in the design research is an interdisciplinary field which contained design framework, computer science, management & marketing, and innovation system [9, 10], limited research has tried to holistically summarize the research trend and hotspots. Indeed, it is theoretically significant to visual the knowledge graph of the role of creativity in the design research, which facilitates relevant researchers to understand this research hotspot and the development trends. Thus, this paper tries to fill in this research gap by analyzing and visualizing the network of co-occurring keywords and the research clusters via Citespace. Based on the data from Web of Science Core Collection and a visualization tool, Citespace, we tried to get a whole picture of the patterns and trends of the role of creativity in design research from 2010 to 2019 [11].

2 Paper Preparation

Regarding the data collection process, the articles were collected according to the Boolean logic of Web of Science search using the subject index (TS) = "design" and "creativity"; the document type is "article"; the language is English, and the timespan is 2010-2019. All the documents are collected from the Web of Science (WoS) Core Collection of Thomson Reuters including SCI-Expanded, SSCI, A&HCl, CPCI-S, CPCI-SSH, ESCI, CCR-Expanded, and IC. A total number of 498 articles were retrieved and exported to Citespace for further analysis [12].

As a Java application for scientometric processing and visualizing scientific works [11], Citespace could analyze academic literature on their co-citation references, co-authors, and co-occurring keywords [12]. In order to explore the nature of a knowledge domain, co-occuring keywords, betweenness centrality, and time-related cluster networks, are used to describe the research trends and the research focus [12]. More specifically, there are mainly six steps to process the data: time span allocation, thresholding, pruning, clustering, and visualization. It is normal to not set pruning in the program, because the nature of the scientometric data set is usually very sparse [12].

Therefore, we can check the knowledge graph of a specific topic by visualizing the relationship between different articles. Retrieve article records of titles, keywords, abstracts, and identifiers and use them to analyze current research frontiers. Therefore, by identifying nodes with higher intermediateness, it is easier to access the connections and relationships between different articles [12]. The relationship between keywords is highlighted with solid line in Fig. 1. Lighter lines indicate earlier times, while darker lines indicate later times.

The betweenness centrality is calculated by the equation below:

$$\text{Betweenness Centrality}\ (Node_m) = \sum\nolimits_{m \neq n \neq p} \frac{\rho_{np}(m)}{\rho_{np}}$$

To specify, ρ_{np} refers to the quantity of shortest paths between node n and p, and $\rho_{np}(m)$ refers to the quantity of those paths that go through the $Node_m$.

In this way, knowledge graph analysis can help illustrate the current research priorities and emerging areas of design research related to creativity. Unlike scholars' traditional summary of creative-related design research, knowledge map analysis can more objectively test the research frontier on the role of creativity in design research and application.

3 Results and Discussion

Keywords in the field of creativity in design research can reveal the focus of this field. As shown in Fig. 1, all papers with a time span from 2010 to 2019 were analyzed, and the co-occurrence network of keywords was checked by the minimum spanning tree (MST) algorithm. Each node displays a keyword whose size indicates the frequency of common key-words of all the papers. Colored links between nodes show their chronological order: dark (purple) is related to later trends, while light (red) is related to previous trends. The network shows the relationship between each keyword and the top 15 cited concurrent keywords. More specifically, keywords surrounded by dark (purple) colors show higher intermediateness and importance. In general, nodes with high frequency and intermediate centers are considered as key nodes in the network. [11]. As suggested in Table 1, "creativity" enjoyed the highest frequency of 203, followed by "innovation" (52), "decision making" (39), "engineering design" (23), "design education" (21), "divergent thinking" (19), "representation" (17), and "student" (13). Thus, "creativity", "innovation", "decision making", "design education", "divergent thinking", and "representation" are considered as the key factors in the network.

Table 1. Co-occurring keywords centrality and count for creativity in design

Key words	Betweenness Centrality	Starting Year	Counting
Design creativity	0.07	2010	203
Innovation	0.11	2010	52
Decision making	0.14	2010	39
Engineering design	0.07	2011	23
Design education	0.24	2010	21
Divergent thinking	0.09	2012	19
Representation	0.02	2015	17
Student	0.30	2011	13
Knowledge	0.09	2013	13
Aesthetics	0.03	2013	13
Design process	0.03	2011	13
Perception	0.03	2015	12
Product design	0.02	2014	9
Fixation	0.14	2014	8
Quality	0.01	2013	7

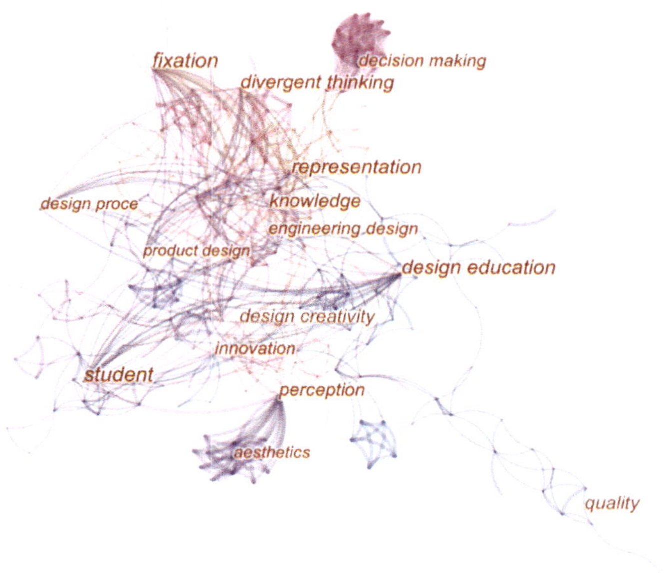

Fig. 1. Co-occurring keywords for creativity-related design research

Silhouette is a measure of how similar a node is to its genus cluster compared to other clusters. Its value ranges from -1 to 1. A larger value indicates that this node more closely matches its genus cluster rather than adjacent clusters. If most nodes have a high silhouette value (silhouette score > 0.5), then the number of clustering is appropriate [12]. Thus, all papers can be summarized into eleven clustering labels according to time series, as shown in Fig. 2. Detailed descriptions and typical paper summaries of each cluster are as follows.

Based on the latent semantic index (LSI) clustering algorithm, eleven clusters were showed via processing sematic abstract of all the papers in Fig. 2: industrial design, disability, design activity, exploratory design theory, design education, assessment, users, design fixation, design theoretics, TRIZ, and art education. To be more specific, as for industrial design, the typical paper is "Managing Creativity: A Gap Analysis Approach to Identifying Challenges for Industrial Design Consultancy Services" [13], which emphasizes the creativity application in an abstract system-level modelling and simulation framework (ARTS) for industrial design consultancy. As for disability, the research mainly discussed the creative application in ergonomic design for disability. For example, Wignjosoebroto et al. [14] designed a wheelchair for aging disable people. For design activity, the research mainly analyzed the personal creativity modes from design perspective and even introduced the way to exploiting creativity modes for team members [15]. Different from design activity, creativity in user-driven creativity treated user as the center for successive design strategy and further discussed the process how user-driven creativity helped design application [16]. As for exploratory design theory and design theoretics, the former one concentrated on examining the innovative exploration in different design structure [17] while the latter one generally discussed the framework for design theory and creativity practice and introduced a set of knowledge transformer between them [18]. Regarding design education and art education, they are all paying attention to education and related activities; however, the emphasized fields are different: while the former one examined the process to foster creativity of people through design education, the latter one analyzed how design interacted with art, music, drama, and design to support students' creativity [19, 20]. Both assessment and design fixation focused on design practice and process. Assessment focused on the feedback strategies in promoting creativity but design fixation mainly discussed its unintentionally limit and mechanism in design research and practice [21, 22]. Last, as for TRIZ, a typical paper, "Research and Practice on the Theory of Inventive Problem Solving (TRIZ): Linking Creativity, Engineering and Innovation", mainly discussed a framework which could balance different factors in the design process, such as creativity, engineering, innovation, cost, and trade-off [23].

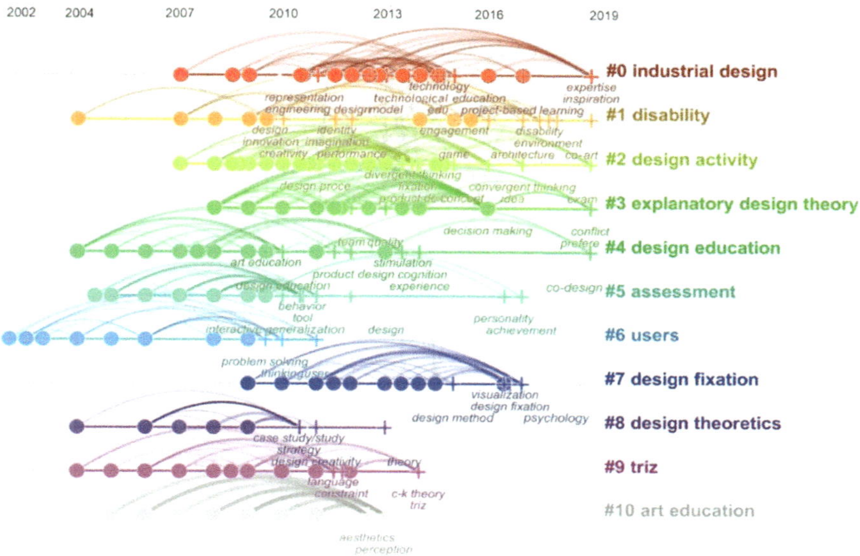

Fig. 2. Research clusters for creativity related design research in time series

4 Conclusion

In conclusion, the current paper tries to discuss the role of design creativity in creativity related researches which have drawn increasing attention from the academic community. However, limited research has tried to holistically summarize the research trend and hotspots. Indeed, it is theoretically significant to visual the knowledge graph of the role of creativity in the design research, which could facilitate relevant researchers to understand this research hotspots and the development trends. Thus, this paper tries to fill in this research gap by analyzing and visualizing the network of co-occurring keywords and the research clusters via Citespace. Through knowledge graph analysis, this paper analyzed the academic literature from Web of Science and visualized the knowledge domain in creativity and design research from 2010 to 2019. It has been found that the number of creativity-related design papers is increasing rapidly, with a total of 498 papers.

Results from the top 15 co-occurring key-words showed the research hotspots on the role of creativity in design research mainly lay in concept exploration ("creativity" and "innovation"), creativity representation ("perception", "representation" and "aesthetics"), thinking pattern ("design making" and "divergent thinking"), product design ("engineering design", "product design" and "quality"), and education ("design education", "design process", "student", "knowledge" and "fixation"). As for the research trends, this topic could be summarized into eleven clusters namely industrial design, disability, design activity, explanatory design, design education, assessment, users, design fixation, design theoretic, TRIZ, and art education. Time series data showed two trends: before 2016, the research trends might be more focused on assessment, users, design fixation, design theoretic, TRIZ, and art education, while the research focus begin to shift to industrial design, disability, design activity, explanatory design and design education after 2016. It suggested research trends might evolve from theoretical exploration and assessment determination to theory application and explanation, which is consistent with the theory of cognitive evolution on creativity [24].

Acknowledgments. The authors declare no conflict of interest. This work was funded by The Hong Kong Polytechnic University (99JV).

References

1. Sarkar, P., Chakrabarti, A.: Assessing design creativity. Des. Stud. **32**, 348–383 (2011)
2. Gans, J.S., Stern, S.: The product market and the market for "ideas": commercialization strategies for technology entrepreneurs. Res. Policy **32**, 333–350 (2003)
3. Chen, C.H.V., Li, H.H., Tang, Y.Y.: Transformational leadership and creativity: exploring the mediating effects of creative thinking and intrinsic motivation. Int. J. Manag. Enterpr. Dev. **6**, 198–211 (2009)
4. Alves, J., Marques, M.J., Saur, I., Marques, P.: Creativity and innovation through multidisciplinary and multisectoral cooperation. Creativ. Innovation Manag. **16**, 27–34 (2007)
5. Druin, A., Solomon, C.: Designing Multimedia Environments for Children: Computers, Creativity, and Kids. Wiley Computer Publishing, John Wiley and Sons, Inc., One Wiley Drive, Somerset (1996)
6. Dorst, K., Cross, N.: Creativity in the design process: co-evolution of problem–solution. Des. Stud. **22**(5), 425–437 (2001)
7. Dosi, G.: Innovation, evolution, and economics: where we are and where we should go. Fagerberg et al., pp. 111–133 (2013)
8. Verganti, R.: Design driven innovation: changing the rules of competition by radically innovating what things mean. Harvard Business Press, Boston (2009)
9. Bannon, L.J.: A human-centred perspective on interaction design. In: Pirhonen, A., Saariluoma, P., Isomäki, H., Roast, C. (eds.) Future Interaction Design, pp. 31–51. Springer, London (2005)
10. Ueki, A., Kamata, M. and Inakage, M.: 'Tabby: Designing of coexisting entertainment content in everyday life by expanding the design of furniture'. In: ACM International Conference Proceeding Series, pp. 72–78 (2007)
11. Xiao, F., Li, C., Sun, J., Zhang, L.: Knowledge domain and emerging trends in organic photovoltaic technology: a scientometric review based on Citespace analysis. Front. Chem. **5**, 67 (2017)
12. Chen, C., Hu, Z., Liu, S., Tseng, H.: Emerging trends in regenerative medicine: a scientometric analysis in Citespace. Expert Opin. Biol. Therapy **12**, 593–608 (2012)
13. Hakatie, A., Ryynänen, T.: Managing creativity: a gap analysis approach to identifying challenges for industrial design consultancy services. Des. Issues **23**, 28–46 (2007)
14. Wignjosoebroto, S., Sudiarno, A., Rakhmaniar, A.: Ergonomics Wheelchair Design for the Aging People. World Congress on Ergonomics (2009)
15. Kim, Y.S., Kim, M.S., Wilde, D.J.: Toward the management of desig n creativity: personal creativity modes, design activity, and team interaction. Des. Manag. J. **3**, 45–52 (2010)
16. Luh, D.B., Chang, C.L.: Incorporating users' creativity in new product development via a user successive design strategy. Int. J. Comput. Appl. Technol. **32**, 312–321 (2008)

17. Chamakiotis, P., Dekoninck, E., Panteli, N.: The creativity of virtual design teams: an exploratory case study. In: Mediterranean Conference on Information Systems 2011 Proceedings, p. 44 (2011)
18. Kotzé, P., Purgathofer, P.: Designing design exercises—from theory to creativity and real-world use. In: Kotzé, P., Wong, W., Jorge, J., Dix, A., Silva, P.A. (eds.) Creativity and HCI: From Experience to Design in Education, pp. 42–59. Springer, Boston (2009)
19. Wong, Y.L., Siu, K.W.M.: A model of creative design process for fostering creativity of students in design education. Int. J. Technol. Des. Educ. **22**, 437–450 (2012)
20. Griffiths, F.: Supporting Children's Creativity through Music, Dance, Drama and Art: Creative Conversations in the Early Years. Routledge, Abingdon (2010)
21. Visser, I., Chandler, L., Grainger, P.: Engaging creativity: employing assessment feedback strategies to support confidence and creativity in graphic design practice. Art Des. Commun. Higher Educ. **16**, 53–67 (2017)
22. Crilly, N.: Fixation and creativity in concept development: the attitudes and practices of expert designers. Des. Stud. **38**, 54–91 (2015)
23. Chechurin, L.: Research and Practice on the Theory of Inventive Problem Solving (TRIZ): Linking Creativity, Engineering and Innovation. Research and Practice on the Theory of Inventive Problem Solving (TRIZ). Springer, Cham (2016)
24. Finke, R.A., Ward, T.B., Smith, S.M.: Creative Cognition: Theory, Research, and Applications. MIT Press, Cambridge (1992)

Research and Construction of the Innovation and Entrepreneurship Collaborative Education Pattern of Design Discipline

Wei Ding, Wenxin Hu, and Junnan Ye(✉)

East China University of Science and Technology, No. 130, Meilong Road, Xuhui District, Shanghai 200237, China
dw.7689@163.com, hwx960723@163.com, yejunnan971108@qq.com

Abstract. Today's economic development of China is facing changes in the industrial structure adjustment and upgrading. Design innovation has become an important engine to promote the development of the industrial revolution. However, compared with other countries, the Education about design discipline in China set up relatively late. With more than 40 years of development, the evolution speed of Chinese design education is not synchronized with the needs of industrial development. As a significant place of design education, there are many problems in Universities, such as unitary teaching pattern and inadequate students' innovation and entrepreneurship training. Based on the analysis of several typical cases of innovation and entrepreneurship education in universities located in Shanghai, this paper explores to establish a "Five-in-one cooperative education pattern", which means university, society, industry, scientific research institutions, and enterprises are supposed to cooperate closely. Moreover, by the construction of the "LCS training model" for innovation and Entrepreneurship talents, this research finds a way to optimize the teaching mode and training mechanism for innovation and entrepreneurship in the design discipline. Ultimately the design innovation will be integrated into the industrial chain which can make the design industry play a more critical role in the national innovation system.

Keywords: Design discipline · Innovation and entrepreneurship · Collaborative education mode · LCS education mode

1 Introduction

The pattern of international industrial division is being reshaped. A new round of world technological revolution, industrial transformation and economic development of China are forming a new convergence. From a global perspective, design innovation has become a necessary engine to promote the development of a new round of industrial revolution. Moreover, it has become an important part of the national innovation strategy. In 2011, the Ministry of Education of China launched the "Collaborative Innovation Project" program, which takes the collaborative innovation model as a cooperative link. These program aims to promote the reform of the

E. Markopoulos et al. (Eds.): AHFE 2020, AISC 1218, pp. 136–142, 2020.
https://doi.org/10.1007/978-3-030-51626-0_16

personnel training mechanism, through interdisciplinary integration. In 2015, The International Council of Societies of Industrial Design (ICSID) released the latest definition of industrial design, which clarified that the content of design services is facing the fundamental change [1]. As an important place for cultivating industrial designers, there are many problems among colleges and universities, such as the disconnection between design education and social development needs, and the lack of collaborative innovation and cooperation among students. How to update the design education concept and cultivate interdisciplinary composite professionals that meet the needs of contemporary society is an important issue [2]. East China University of Science and Technology seized the opportunity to make full use of the accumulation of school running to tap into its resources. The School of Art Design and Media, under the guidance of the five-in-one collaborative education training concept, has launched the LCS education model education practice.

2 The Main Problems of Modern Education

2.1 Single Education Model

The rapid development of the Chinese economy has promoted the transformation of the manufacturing industry, among the society needs a large number of innovative and entrepreneurial designers. The United States is one of the earliest countries in the world to start entrepreneurship education. As early as 1998, American universities collaborated with federal laboratories in the industry to conduct curriculum research [3]. The "dual-track system" teaching model, which originated in the German Bauhaus, is rooted in the institutional soil of the German market economy. However, the Chinese education model is monotonous. Chinese curriculum system is mainly composed of public basic courses and professional courses.

2.2 Weak Awareness of Entrepreneurship

Influenced by the current Chinese education model, college students lack of entrepreneurial experience, capital funds and in-depth understanding of the market. They are easy to be limited by less venture capital. These reasons have affected college students' choice and attitude towards innovation and entrepreneurship. At present, the division of disciplines in Colleges and universities is becoming more and more sophisticated. After entering the University, college students are exposed to more professional knowledge, which is easy to form a solid mode of thinking. It is easy to form barriers between disciplines because of the lack of linkage.

2.3 Separated Innovation Practice Projects

The design project is the specific implementation object of innovation and entrepreneurship, which is an important way for students to connect theoretical knowledge with practice. Generally, practical projects in schools mainly include academic competitions, teaching researches, school-enterprise cooperation projects, etc.

Governments, enterprises and universities collect a lot of good ideas and plans through competitions, scheme solicitation, and other ways. However, these plans cannot form deeper cooperation and implementation with enterprises. Schools should constantly refine innovation and entrepreneurship projects from competition and scientific research systems to improve the practicability and feasibility of the projects [4].

2.4 Backward Curriculum System

The evolution speed of society may be changing in a cycle of three months or six months, while the evolution of the education system in Chinese universities is a slow process, which will not be changing greatly in five or even ten years. This phenomenon leads to a mismatch between the demand for society and the supply of universities [5]. For instance, around 2010, with the rapid development of interaction design in China. Many industrial designers change their careers to become interactive designers. However, because most schools do not offer interactive majors, graduates have to relearn professional skills from the work.

3 Exploration of Design Education

3.1 Synchronization with Industrial Structure Reform

China is traditionally a big manufacturing country. The evolution of design patterns is given in Table 1. In the development stage of the manufacturing industry, designers must have strong knowledge of modeling design, model making, material technology to cooperate with the manufacturing production. After turning to the information age, a large number of consumer products focus on information products. The development of the mobile Internet brings more possibilities and opportunities for design. The third transformation has changed to a service-oriented society. The service industry puts more emphasis on the service operation. According to the experience of developed countries, the proportion of service economy should be more than 70%, and the proportion of the service economy in Shanghai has reached more than 60% [6].

Table 1. Evolution of design patterns

Time	Period	Leading pattern	Concerns
1800s	Industrial age	Industrial design	Physical logic
1900s	Experience economy	Interaction design	System logic
2010-	Service economy	Service design	Design thinking
2020-	Integration development	Design driven society	Social innovation

The development of design education is closely related to the changes of society, economy and technology. There are four stages of change in design education to cope with industrial changes, while each stage required various knowledge structures. The design discipline of East China University of science and technology is a comprehensive specialty. Throughout its history, the concept of design discipline is deeply rooted in the rich soil of industrial technology, art design, humanities, and social sciences. Students have a solid theoretical foundation, international vision, and innovation ability. In addition to focusing on the education of basic professional theories and technical knowledge, students are also guided to mine the users' essential needs, understand the business operation and pay attention to social development.

3.2 Interdisciplinary Collaborative Education Mode

With the development of online and offline integration, the boundary of the designer's responsibility is fuzzier, and the design link is longer and longer. The dimension of design is constantly enlarged, which inevitably leads to new cross fusion (Fig. 1 shows an example). An excellent designer should not only understand visual design, interactive design, but also understand product operation, user research, etc. The increasing transition of design from node value to system value, from focusing on individual design to the whole industrial chain. The innovation and entrepreneurship projects emphasize the realization of project closed-loop, which will inevitably lead to deep integration and intersection with various industrial chain disciplines.

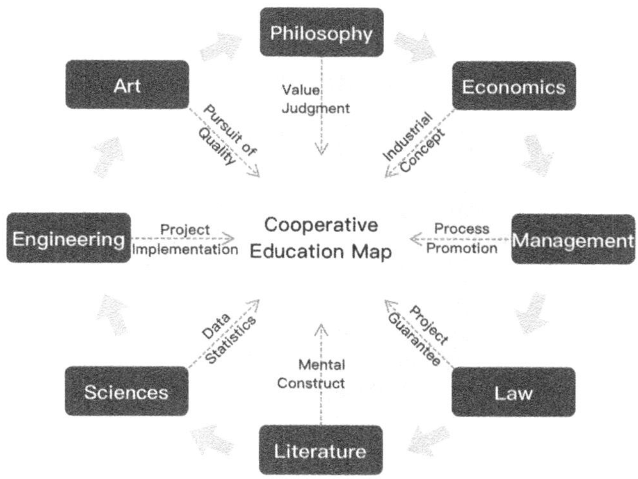

Fig. 1. Interdisciplinary collaborative education model map

Philosophy knowledge guides the value judgment of innovators. The thinking logic of economics links all segments of the project. The knowledge method of management is helpful for the smooth progress of the project. Legal content provides guarantee for enterprise operation. User psychology research of liberal arts is helpful to the success of the project. The application of mathematical logic in science contributes to the team to make decisions. Engineering guarantees the implementation of the project from the technical level. Art science uses aesthetic advantages to improve product quality.

With the help of interdisciplinary knowledge content, enriching each node of the project practice process.

3.3 Construction of New Knowledge System

In the new round of innovation environment, our society has higher requirements for people who need to focus on both the range and depth of their knowledge. The complete design process includes discovery, definition, development, design, transformation, integration and communication [7]. Take innovative seven-step as the overall path of innovative design development. In the first stage, art collides with philosophy and economics to perceive the next outlet of the industry. The collision between art and science engineering enables designers to predict industry trends more accurately. In the definition stage, the theory and method of management are used to develop the design strategy and business plan. In the process of actual project development, it requires deep cooperation between developers and designers to understand the current technology boundary and various innovation possibilities. In the integration stage, the integration of science and technology is applied to the nodes of the supply chain, so as to open up the commercialization path of innovative products.

Based on the vertical dimension of innovation seven-steps, combined with eight disciplines including philosophy, economics, management, law, literature, science, engineering and art as the horizontal dimension, a knowledge system of innovative personnel training under the new industry mode is established, hoping to redefine the criteria for the evaluation of new personnel [8]. In recent three years, East China University of science and technology has actively explored the teaching of innovation and entrepreneurship education for students, which has completed many undergraduate innovation and entrepreneurship projects, such as we have published <Integrated Innovation Design Based on Red Dot Competition> and <Exploration of "1 + 1 + 1" International Design Innovation and Entrepreneurship Personnel Training Mode>.

3.4 Redefining Teaching Scene

In the classroom, students need to discuss together and practice in person, besides speaking and listening. For design education, both collaboration and sharing are key points to adapt to the new transformation. Renovating training mode and establishing innovation and entrepreneurship training mechanisms require universities to redefine the teaching scene [6].

(1) Collaborative Group: the organizational form of curriculum changes from division to integration. The new form of curriculum organization should have a complete curriculum goal. For example, to complete a subject of intelligent travel, a cross-border group should be formed at this time. The group members should include students with different professional backgrounds, such as brand, machinery, computer, interaction, design and so on. We can also call it "collaborative group" which can accomplish the target tasks in collaboration [6]. (2) Sharing Classroom: create a new teaching scene. The boundary of the classroom will be demolished to create a more common and integrated flow space and shared space, and to break the barriers of different disciplines. (3) Application of new technology: Combine with the

development of the technology of the times and apply some new technical means. For example, through AR, VR and MR, stimulate students' interest in learning and innovation. (4) Learning ecology: schools should not stand alone. Schools should form a dynamic relationship with enterprises, R & D institutions and social organizations, thus let social resources enter schools. Schools participate in the whole process of social evolution, also can turn the curriculum into a real topic.

4 LCS Education Mode

Based on the "construction of multi-level innovation and entrepreneurship training platform for design majors", East China University of science and technology researched LCS education model framework which accord with the policy background and technological development needs of "innovation and entrepreneurship" development in the new era (Fig. 2).

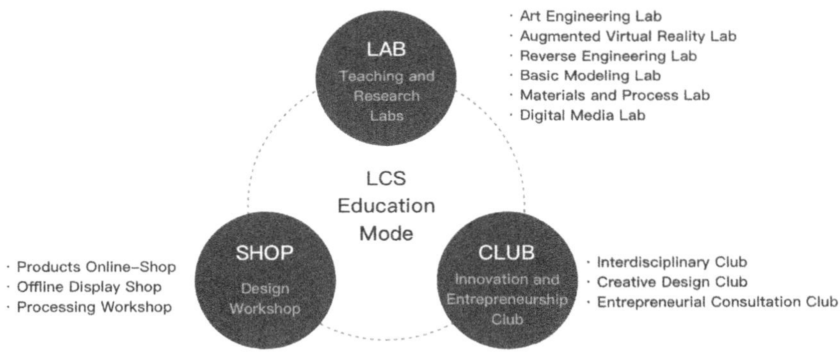

Fig. 2. LCS education mode

(1) The art design and media teaching and research laboratory are based on the establishment of the Art Engineering Experiment Center, Augmented Virtual Reality Lab (Manheng co-construction), Reverse Engineering Lab and the others to provide pilot service for outstanding projects. These labs can accelerate the transformation of achievements and reduce the risk of innovation and entrepreneurship. (2) Through the construction of effective interdisciplinary collaborative education club, strengthen the independent entrepreneurship and training of design and innovation talents, and form a good teaching supplement. (3) Through cooperation with more enterprises, the project platform will become a practice base for professional students before graduation.

The school of art design and media of East China University of science and technology cooperates with the network platforms, such as 51 design and Puxiang industrial design station, and the Industry organizations, such as Shanghai Industrial Design Association and Shanghai Creative Workers Association. Through the interlinkage of online to offline, this mode promotes enterprise docking and achievement transformation of innovation and entrepreneurship projects. At present, these labs that more than 90% of the comprehensive and designed experimental courses are offered are all open to the students of our university.

5 Summary and Prospect

This paper presents an Interdisciplinary collaborative education model that breaks the traditional boundary, including discipline boundary, industry boundary and region boundary. This mode plays a significant role in promoting the design industry in the national innovation system [9]. East China University of science and technology, based on the multi-disciplinary integrated knowledge system, carried out the project of “professional cognition and innovation ability training system construction guided by design principles” and won the second prize of the 2017 Shanghai teaching achievement award. As of 2019, the school has arranged students to participate in more than 60 practice activities and organized students to contribute to more than 100 innovation competitions. The whole process and multi-dimensional construction of LCS education mode for innovation and Entrepreneurship of design discipline will contribute to regional planning, industrial development, discipline construction, research progress, and technological development, etc. [10].

References

1. Deng, R.: The exploration of reconstructing the curriculum system of industrial design. J. Creativity Des. **04**, 74–77 (2016)
2. Ye, J., Zhang, Q.: Exploration and practice of innovative industrial design personnel training mechanism. J. Chem. Higher Educ. **34**(04), 42–47 (2017)
3. Gu, Z.: Collaborative innovation and entrepreneurship education mode of foreign universities and its enlightenment. J. Lianyungang Normal Univ. **35**(01), 67–70 (2018)
4. Xie, Z.: Innovation and entrepreneurship education mode of universities in the New Era. J. Cult. Educ. Mater. (30), 120–121 (2018)
5. News Network of East China University of Science and Technology. https://news.security.edu.cn/news/48712?Category?Id=21
6. Ding, W.: Transformation of Contemporary Design Education Paradigm. http://www.sohu.com/a/144117823_
7. Xu, B., Ding, W., Zhang, Y.: Research and thinking on future design system based on industry 4.0 information physics system. J. Des. (13), 56–57(2017)
8. Yang, B.: Research on the Cultivation Mode of Innovative Undergraduate Personnel. Southwest Jiaotong University, Beijing (2011)
9. Shaonan, S.: Design and theoretical research on education system of university-wide innovation and entrepreneurship. In: Institute of Management Science and Industrial Engineering. Proceedings of 2019 3rd International Conference on Education Technology and Economic Management (ICETEM 2019). Institute of Management Science and Industrial Engineering: Computer Science and Electronic Technology International Society, pp. 1155–1159 (2019)
10. Yu, W.: On the reform of practical teaching mode of the industrial design course. J. Popular Lit. Art (08), 222 (2015)

Research on the Kernel and Model of Shanghai Design Science Development

Wei Ding, Bingyao Xu, and Zhang Zhang(✉)

East China University of Science and Technology,
Shanghai 200237, People's Republic of China
dw.6789@163.com, 462300838@qq.com,
zhangzhang@ecust.edu.cn

Abstract. In Shanghai, 5 local universities jointed to constructing design science, which are Tongji University, Donghua University, East China University of Science and Technology, Shanghai University of Engineering Science, and Shanghai Institute of Technology. Then the Shanghai International Institute of Design and Innovation (SIIDI) was established as the actual operating carrier for this project. East China University of Science and Technology takes mainly responsible for the research direction on Intelligent Sustainable Design for Industries Transformation and Urban Life. Under this background, the kernel and model of Shanghai design science development are explored. Four teams have been united, their research respectively targets on "Urban Intelligent Public Facilities and Equipment Research", "Intelligent Products, Human-computer Interaction and Service Design Research", "Regional Characteristics and Urban Brand Research", "Innovation Environment and Urban Renewal Research". Five master studios have been constructed. From 2017 till 2019, 31 teams were funded to do relative research. In this exploration process, the research model of Shanghai design Science development is built from four dimensions, they are exploring the frontiers of science and technology, meeting consumption upgrades, serving the regional economy and country-oriented strategy.

Keywords: Design science · Shanghai · Intelligent Sustainable Design

1 Challenges and Opportunities for the Development of Design Science in the New Era

The focus of design science development needs to be oriented towards industry issues and new trends in the industry, basing research on real problems.

1.1 The Development Orientation of Design Science Which Faces Country-Oriented Strategy

State Council (2016) "13th Five-Year" National Strategic Emerging Industries Development Plan points out that Strategic emerging industries represent the direction of a new round of sci-tech revolution and industrial transformation, which are key areas for cultivating new momentum and gaining new competitive advantages in the future.

E. Markopoulos et al. (Eds.): AHFE 2020, AISC 1218, pp. 143–151, 2020.
https://doi.org/10.1007/978-3-030-51626-0_17

In the 13th five-year plan, China proposes to put strategic emerging industries in a more prominent position in economic and social development, vigorously build a new system of modern industries, and promote sustained and healthy economic and social development. The national strategy always has a profound influence on the positioning and development of design science.

1.2 The Development Extension of Design Science Which Meets Consumption Upgrades

Market demand is the key factor to drive the development and growth of strategic emerging industries [1]. The state is strengthening policy guidance on the demand side, and enterprises are accelerating the application demonstration of new products and services—to transform potential demand into real supply, and drive industrial upgrading with consumption upgrading. The middle class refers to families with a monthly disposable income between RMB 12,500 and 24,000. According to a study by Boston Consulting Group and the Ali Institute, middle-class consumption will grow at a rate of 17%, [2] and the population of the middle class is expected to reach 630 million by 2022. This shows that most Chinese families will move towards a well-off and prosperous life. In the next few years, the improvement of consumption quality and demand in the middle class will contribute to the majority of consumption in the overall consumption. Profound changes are taking place in the lifestyle, behavior habits and consumption culture of users at all consumption stages. How to make use of the characteristics of consumption upgrade to upgrade design empowerment experience has become a common research topic of academia and business. Tentacles of design extend continuously while nurturing and stimulating the kernel and development of the design science.

1.3 The Development Integration of Design Science Which Explores the Frontiers of Science and Technology

The next 5 to 10 years will be a critical period for a new round of global sci-tech revolution and industrial transformation from gathering momentum to a mass outbreak. With the continuous and rapid evolution of information revolution, technologies such as the Internet of things, cloud computing, big data, and artificial intelligence have widely penetrated all sectors of the economy and society; the prosperity of information economy has become an important symbol of national strength. Breakthroughs have been made in additive manufacturing (3D printing), robotics and intelligent manufacturing, metamaterials, nanomaterials and other fields [3], which will promote the differentiation and transformation of the traditional industrial system and reshape the international division of labor in the manufacturing industry. With the rapid development of genomics and related technologies, new models such as precision medicine, biosynthesis, industrial breeding are accelerating their evolution and popularization; the new biological economy is expected to lead human production and life into a new world. Coping with global climate change boosts the tide of green and low-carbon development; the scale of cleaner production technology applications continues to expand, and the new energy revolution is changing the existing landscape of international resources and energy. With the deep integration of digital technology with cultural creativity and design services, the digital creative industry has gradually become an intelligence-intensive industry that promotes the effective supply of high-quality products and services; the creative economy is emerging as a new development model. Innovation-driven emerging industries have gradually become the main driving force for global economic recovery and growth, triggering the reconstitution of the international division of labor and international trade pattern, and ushering a new era of global innovation economy development.

1.4 The Development of Design Science Which Serves the Regional Economy

On May 13, 2019, the Political Bureau of the CPC Central Committee held a meeting to review the Outline of the Yangtze River Delta Regional Integration Development Plan. This undoubtedly pushed the "fast forward" button for the integration process of the Yangtze River Delta, and also defined a clearer roadmap for its development. The Yangtze river delta is one of the regions with the most active economic development, the highest degree of openness and the strongest innovation capacity in China. It plays an important role in the national economy. The state puts forward two keywords of "integration" and "high quality" for the regional economy of the Yangtze River Delta. The design science development of the college is always positioned in the scope of serving the regional economy. The high-quality design of integrated industries is the mission of the discipline serving the regional economy.

2 A New Look of Chinese Design Science

In 2011, the State Council Academic Degrees Committee and Ministry of Education (The people's republic of China) published the newly revised file "Catalogue of Degrees and Talents Cultivation (2011)". Art became the new 13th subject category, moreover, under the category of art, for the first time, design science has become one of the five first-level disciplines. It consists of seven basic majors (art design, visual communication design, environmental design, product design, clothing, and apparel design, public art, arts and crafts, digital media art) and one ad hoc major (art and technology). As shown in Fig. 1-design science category in the new era.

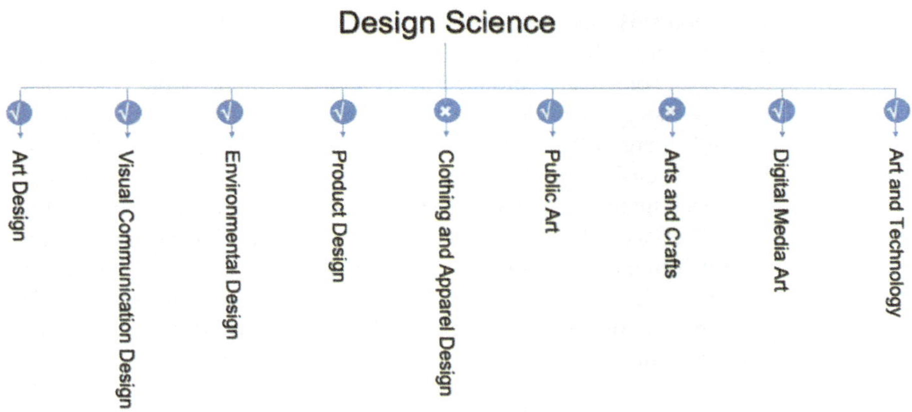

Fig. 1. Design science category in new era.

3 New Extension of Design Science of East China University of Science and Technology

3.1 Background of the Class IV Summit Discipline Construction of Shanghai Design Science

In 2014, Shanghai Municipal Education Commission issued the file "Planning for Discipline Development and Optimized Layout of Shanghai Higher Education Institutions (2014–2020)" (no. 44 of Higher Education Department of Shanghai Municipal Education Commission (2014)). It plans to carry out the construction of summit and plateau disciplines in Shanghai universities: focusing on supporting a group of advantageous disciplines to take the lead in establishing summit disciplines, consolidating and enhancing a group of potential disciplines to form plateau disciplines, and making them landmark disciplines in Shanghai universities with important influence at home and abroad. Among them, the construction goal of Class IV summit discipline is to become a core discipline to support national strategy and meet the regional economic and social development, at the same time, to make the overall level of this category of disciplines in Shanghai rank the top two in China through dislocation competition and collaboration within the city. First-level disciplines meeting one of the following principles can be selected: supporting strategic emerging industries, urgently needed for Shanghai's economic and social development but not yet laid out; or good discipline foundation, able to improve the overall discipline strength of Shanghai. Through research analysis and scientific demonstration, 32 first-level disciplines were selected as Class IV summit disciplines for construction, of which design science (discipline code: 1305) was selected.

Therefore, the construction of the Class IV summit discipline of Shanghai design science is led by Tongji University, and jointly built with Donghua University, East China University of Science and Technology (hereinafter referred to as "ECUST"), Shanghai University of Engineering Science, and Shanghai Institute of Technology. Then the Shanghai International Institute of Design and Innovation (SIIDI) was established as the actual operating carrier for this project. See Fig. 2 below. Construction goal of SIIDI are: adhering to problems and needs as the guidance, closely meeting the requirements of Shanghai Science and Technology Innovation Center, establishing a synergy mechanism within the city, further giving full play to the advantages of resource gathering of attended universities, gathering and cultivating a number of international first-class research teams, striving to become a research institution with international influence and competitiveness by 2020, and promoting Class IV summit discipline of Shanghai design science to become an important component and construction carrier of Shanghai Science and Technology Innovation Center.

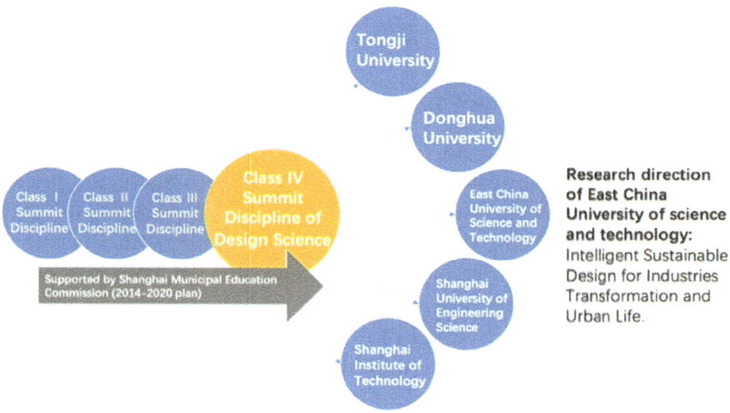

Fig. 2. Structure chart of class IV summit discipline construction of Shanghai design science.

3.2 Construction Model and Characteristics of Design Science in East China University of Science and Technology

School of Art Design and Media, East China University of Science and Technology (hereinafter referred to as "ECUST Arts") takes mainly responsible for the research direction on Intelligent Sustainable Design for Industries Transformation and Urban Life. ECUST Art has set up 4 research teams, namely: a team of Urban Intelligent Public Facilities and Equipment Research; a team of Intelligent Products, Human-computer Interaction and Service Design Research; a team of Regional Characteristics and Urban Brand Research; and a team of Innovation Environment and Urban Renewal Research. As shown in Fig. 3 below.

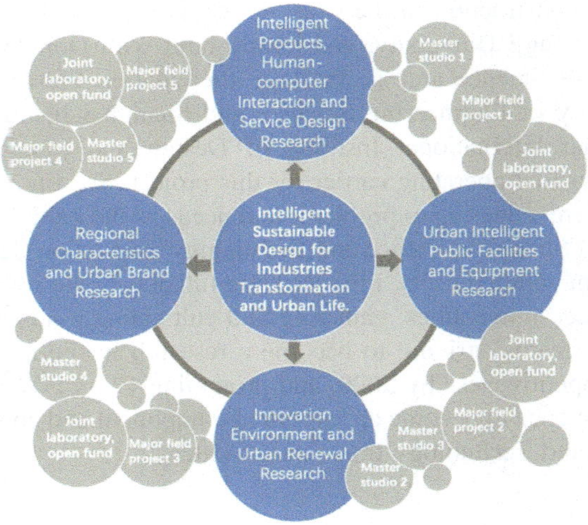

Fig. 3. Structure of research teams.

3.3 Research Features of the Four Teams

Team 1: Team of Urban Intelligent Public Facilities and Equipment Research. As an important part of smart city construction, the construction of intelligent public facilities and equipment systems is an interdisciplinary and multi-professional event, which needs the joint efforts of many fields to promote its development. The team is committed to using virtual reality technology, collaborative innovation design and other theoretical methods to carry out research. The research content can be divided into five parts: the preliminary research stage of the background, technology, development direction, and other investigations; the design research stage of the establishment of the design model; the design practice stage of industrializing the design product; the stage of establishing and maintaining product database; the stage of establishing evaluation system and feeding back the preliminary design. The research will better integrate smart city planning, design and construction technologies with the Internet and communication technologies; focus on the comprehensive research and application of the full life-cycle integration, design, construction, and operation and maintenance of intelligent public facilities and equipment; and provide comprehensive research from top-level design to practice management and the integration of industrial chain resources.

Team 2: Team of Intelligent Products, Human-computer Interaction, and Service Design Research. By analyzing the modern design methods and related theories, the team redesigned and restudied the modern mechanical products based on the intelligent interactive design method and the system service design concept. Specific studies are as follows: the performance of rolling bearing and the kinematics and dynamics of the rotor it supports and the vibration performance of the medium-sized cylinder in the process of centrifugal casting were researched through methods of mechanical modeling and computer numerical simulation to provide theoretical references and basis;

the research of mechanical intelligent products design in automobile collision test combined artificial intelligence technology to develop and apply intelligent products of automobile collision test; the human-computer interaction experiment in automobile collision test was conducted based on G-Magic Virtual Reality Lab; the research combined the theoretical knowledge of virtual reality and human-computer interaction technology, and was applied in vehicle crash test; research on the application of service design concept in modern intelligent mechanical design took the whole process of service system factors into account in the selection of design concept. The team will focus on modern mechanical design and modern intelligent products, human-computer interaction, theory integration of service design, system design and integrated application.

Team 3: Team of Regional Characteristics and Urban Brand Research. At present, China's economic development has entered a new normal. The CPC Central Committee and the State Council attach great importance to brand work and have made a series of major deployments for brand building. During the "13th Five-Year Plan" period, Shanghai faces some obvious shortcomings, mainly including that innovation ability needing to be improved, core technologies and components needing to be broken through, and lack of independent brands with international influence. It is urgent to comprehensively strengthen the brand building, accelerate the building of some well-known brands with strong internationally competitive and high value-added, and promote sustainable and stable economic development. The team adopted a variety of research methods, focusing on the development needs of the construction of Shanghai Science and Technology Innovation Center and industrial transformation and upgrading, to layout research directions and recent key research fields from two dimensions of regional characteristics and urban brands. Vigorously cultivate the brand's temperament, promote the Shanghai brand from culture to spirit and simplicity to diversification. It is expected that these new understandings can provide the think tank support for Shanghai to choose the top-level design of brand strategy, the international layout of industrial development, and enterprise transformation and upgrading strategy.

Team 4: Team of Innovation Environment and Urban Renewal Research. With the rapid advancement of urbanization, urban renewal and urban spatial environment construction have become one of the important ways to promote urban development and shape urban characteristics. However, the extensive renewal mode and the environment construction for quick success and instant benefit have led to a series of vicious events of urban renewal and a shoddy spatial environment. Taking Shanghai as an example, the team hopes to re-examine the connotation of urban renewal and environmental design from the depth and breadth by exploring the "dual innovation" mode of urban renewal system innovation and environmental design method innovation based on "meeting the needs of citizens" and "building urban characteristic environment". Under the background of the negative growth of construction land in Shanghai, it is related to the future connotative development of Shanghai that creating incremental value through urban renewal and characteristic environment construction. Therefore, this project will be of practical significance. Secondly, urban renewal and environment construction are related to the life, work, and study of citizens. Therefore, the research results of our team will have practical guiding value for improving the living environment and living quality of citizens.

3.4 New Orientation of Design Science in East China University of Science and Technology

Draw the project positioning diagram of Fig. 4 by constructing the above team, studio and project and finding the positioning served by each work.

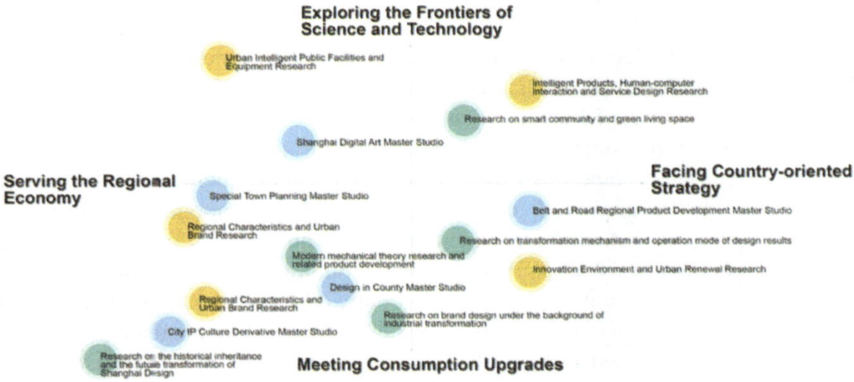

Fig. 4. Project positioning map.

4 The Future of Chinese Design Science

The focus of art-design science development needs to be oriented towards industry issues and new trends in the industry, basing research on real problems. Today, information dissemination is characterized by fragmentation, vertical segmentation, value extension and so on. Industrial concentration begins to improve and gradually enter the era of all-media. Fragmentation is a typical feature of the mobile Internet and the Internet of Things era. Information is bound to people and things in real-time; the efficiency of information generation and transmission is further improved. It is critical to dig out the effective value from the seemingly disordered big data. Vertical segmentation is another important trend. The masses will form different ethnic groups according to their habits, regions, behaviors and other characteristics, while value extension based on ethnic groups and specific scenes is another trend of the creative industry and art design. New products and new media spread the value proposition of advocating new lifestyles, and "Life Aesthetics" social media e-commerce is the commercial value extension of this value proposition. The boundary between new and old information and value transmission will eventually dissolve, and information will be effectively spread in multiple alternations of the user journey, forming system influence. Disciplinary development is based on new changes and trends in the art design and creative industries, forming a new summit discipline system. For example, digital transmission for intangible heritage is one of our key research areas, which displays the cultural value of intangible cultural heritage through digital means. Let traditional culture re-enter contemporary life through cultural archeology, cultural interpretation and cultural innovation. Service system design breaks down traditional barriers, which is guided by the concept of service design instead limited to product design itself; it effectively integrates brands, spaces, information, and products to create a new business model and user experience. Urban IP focuses on the reproduction of urban culture, improving urban experience and reputation of the urban brand through the excavation and recreation of urban cultural connotation.

Acknowledgments. Project "Research on innovative design of teaching and research space" (DA19306) supported by Shanghai Summit Discipline in Design.

References

1. Zhou, K.Z., Li, C.B.: How strategic orientations influence the building of dynamic capability in emerging economies. J. Bus. Res. **63**(3), 224–231 (2010)
2. Consumers M.: Consumer Behavior in Emerging Markets. The Oxford Handbook of Management in Emerging Markets, 2019: 219 (2019)
3. Ngo, T.D., Kashani, A., Imbalzano, G., et al.: Additive manufacturing (3D printing): a review of materials, methods, applications and challenges. Compos. Part B: Eng. **143**, 172–196 (2018)

Interest in Design Studios: A Theoretical Framework of Teaching Creativity for Millennials and Generation Z

Hoa Vo(✉)

Interior Design Program, College of Design, University of Minnesota,
1985 Buford Avenue, St. Paul, MN 55108, USA
voxxx248@umn.edu

Abstract. Millennials and Generation Z make up the new wave of students in design studios. These generations, while being tech-savvy, are undermined in creativity due to risk-aversion and collaboration-avoidance. Design educators need to consider these characteristics and create pedagogical approaches that can engage the students in learning creativity. The author proposes a theoretical framework to tackle the challenges Millennials and Generation Z bring to design studios by bridging the literature of design education and educational psychology. Scholars from both fields have highlighted the positive link between students' interest and their performance in academic and creative tasks. Hence, it is plausible to hypothesize that task interest plays an important role in helping Millennials and Generation Z improve in learning creativity. This paper will discuss prominent studies that support the proposed theoretical framework and suggest pedagogical approaches for design studios with new-wave students.

Keywords: Creativity · Design education · Design studio · Design pedagogy · Interest · Millennials · Generation Z

1 Introduction

Millennials (born between 1981 and 1996), followed by Generation Z (born from 1997 onward), are two generations that dominate the future educational, economical, and political sceneries [7, 37]. These "digital natives" differ from previous generations in how they learn, work, and interact either in the physical or virtual environment [26]. Being tech-savvy and multi-tasking but lack of creativity are their prominent characteristics [11, 20].

The Learning Challenge. Digital prosperity allows these two generations to accumulate ample information via the internet. Millennials and Generation Z, hence, regulate their own learnings using online resources such as Google and YouTube. However, this approach becomes counterintuitive with information overload, inaccuracy, and distraction [25, 36]. Scattering attention to multiple digital gadgets and online sources of information, the speed and accuracy in performing tasks of the "digital natives" decline [20]. In academic settings, distraction also impedes focus and thus, interferes with the learning of students from these generations [36].

E. Markopoulos et al. (Eds.): AHFE 2020, AISC 1218, pp. 152–158, 2020.
https://doi.org/10.1007/978-3-030-51626-0_18

The Creative Challenge. Risk-aversion and collaboration-avoidance disrupt the creativity of Millennials and Generation Z. For instance, a study of 106 Millennial college students using the Scale of Attributes and Behaviors (SCAB) showed that risk-taking behaviors scored the lowest [19]. Risk-aversion also occurs in Generation Z due to self-doubt and insecurity (i.e., the fear of missing out collective experiences) [24, 41]. Hence, while being creativity requires the willingness to take risks [32], Millennials and Generation Z come short in this aspect.

Another reason for the lack of creativity in these generations is collaboration-avoidance. Regarding interactions with peers in the classrooms, Millennials are self-interested while Generation Z defers the need to collaborate [25]. The reliance on digital gadgets and online sources of information results in the inadequacy of socialize-skills and the loss of social interest in the two generations [1, 40]. Unfortunately, without sufficient collaborations and/or communications, creativity withers as ideas are unable to collide, combine, and evolve [9, 12, 38].

2 Literature Review

Distraction, risk-aversion, and collaboration-avoidance are the challenges that Millennials and Generation Z bring to academic classrooms in general and design studios in particular. As these generations make up the majority of college students, design educators need pedagogical approaches that can counter the challenges and improve the learning of creativity. Current literature in design education highlights technology integrations in studios to adapt to and engage this new wave of students in learning [14, 29]. Design educators can plan and implement technology integrations that reduce distraction. The challenges of risk-aversion and collaboration-avoidance, nevertheless, are intangible and call for a psychological (rather than technical) approach. Design educators, therefore, should look for insights from the profuse research on Millennials and Generation Z in relevant fields such as educational psychology.

2.1 Interest and Learning

Educational psychology researchers have emphasized that interest (in classroom tasks, activities, assignments) leads to extended effort in learning for students across age groups [6, 27]. Studies focusing on Millennials and Generation Z, likewise, indicate the importance of facilitating interest in classrooms as these generations display apathy toward content and activities that they deem unappealing and impractical [2, 4, 35]. A survey of 94 college students (in business, engineering, and entrepreneur) shows that 43% attended 75% of learning activities while 37% attended between 50% and 75%. Given that the activities took place in the online environment and offered real-world meanings. Students expressed high expectations for interesting pedagogical approaches (25.38%), only after the ability to ask questions (25.89%) [3]. These findings call attention to pedagogical approaches that capture the interest of Millennials and Generation Z by responding to their digital proficiencies and practical needs.

This phenomenon is understandable as researchers with fMRI studies found the activations of reward circuitry linked to subjects' interest [30, 31]. As students view interest as a reward for engaging in specific tasks/activities, they become motivated to sustain the tasks/activities, exert effort, and eventually improve in their learning. In a study with 331 middle school students, task interest explained 26% of the variance in mathematic performance [5]. Another study with 218 high school students indicates that classroom interest significantly predicted students' effort, participation, attention, and engagement ($p < 0.001$) [27]. Interest, thus, mitigates the challenge of distraction/scattered attention in students of Millennials and Generation Z.

2.2 Interest and Creativity

Scholars in creativity research, similarly, have highlighted the positive link between task interest and creativity with intrinsic motivation as the mediator. The enjoyment of engaging in personally interesting tasks or intrinsic motivation benefits creativity [15]. According to the model of creative thinking [33], individuals incline to find, reform, and solve problems that are interesting yet novel and foreign to them. This inclination fuels the intrinsic motivation that keeps them engaging with the problems until they achieve relevant and meaningful solutions. Personal interest in the novel problems/creative tasks, in fact, encourages risk-taking behaviors [22]. Individuals spend more time and exert more effort in try-and-error attempts to solve problems without guarantees for solutions. Henceforth, task interest helps counter the tendency of risk-aversion in Millennials and Generation Z.

Empirical evidence from a study with 264 creativity workers presented a positive association between job interest and creativity ($\beta = 0.20, p < 0.05$). The repeated study on 405 college students came up with the same finding ($\beta = 0.12, p < 0.10$) [8]. In a recent study of 238 Millennial students, having interesting and challenging tasks that matched their skills positively correlated with their creativity ($\beta = 0.25, p < 0.01$). A subsequent study of 90 Millennial students found a similar result ($\beta = 0.49, p < 0.05$) [23]. Design educators who wish to promote creativity in their studios should offer tasks that match their students' interest [39].

3 Interest in Design Studios: The Theoretical Framework

Based on the literature discussed above, it is plausible to hypothesize that task interest plays an important role in helping Millennial and Generation Z students improve in learning creativity. Besides integrating technologies in design studios, educators can utilize communications (e.g., feedback) as the means to spark students' interest [13]. However, different phases of interest require specific forms of communication.

Interest evolves through the four phases: (1) triggered situational, (2) maintained situational, (3) emerging individual, and (4) well-developed individual [16]. The first-phase-interest is short-term and relies on external supports. The second-phase-interest sustains if tasks are appealing and meaningful. The third-phase-interest is stable and initiates task engagement. The fourth-phase-interest is personal and long-lasting. Students in the early interest-phases prefer approvals and positive feedback with modest and manageable suggestions of changes. Whereas, those in the later interest-phases prefer challenging conversations and even withstand criticisms [21]. Thus, educators should adapt their communications (e.g., feedback) to students' current interest phases to facilitate learning improvements (Fig. 1 demonstrates this theoretical framework).

Positive communications (including praises, encouragements for task-related progress and behaviors, suggestions for changes) from educators or peers can promote task engagement and attention [28]. Consistent with the General Interest Theory (GIT), these intangible rewards can act as external supports for students with low interest in tasks/activities. As validations for students' task performance, these rewards create an encouraging environment for situational interest to thrive via the feeling of competence (i.e., meaningful outcomes for task engagements) [17, 34].

Challenging communications (including open-ended reactions, honest feedback, constructive critiques) from educators or peers can activate individual interest (i.e., phase (3) and (4) of interest) and improve task performance [18, 21]. Students rely on such communications to evaluate the accuracy of task performance and find the prospect of task improvement. Nevertheless, challenging communications can discourage students with low interest due to the corresponding negative emotions (e.g., shame, boredom) [10].

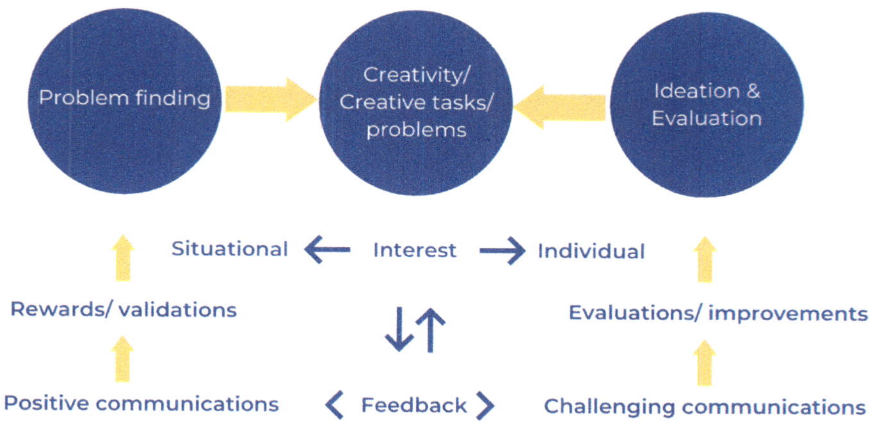

Fig. 1. The theoretical framework of adapting classroom communications (e.g., feedback) to interest-phases.

In the context of creativity teaching, both positive and challenging communications are important. Creative thinking involves the stages of problem finding, ideation, and evaluation [33]. Situational interest (i.e., phase (1) and (2) of interest) encourages students to take risks finding and resolving novel problems without definite outcomes. Individual interest help students to sustain the effort and persistence in try-and-error attempts of ideating and evaluating solutions that are relevant and meaningful to the problems. Educators, therefore, need to regulate classroom communications (of themselves and student peers) to match students' interest phases and, thus, facilitate their creativity.

4 Conclusion

The theoretical framework of adapting classroom communications (e.g., feedback) to students' interest-phases can result in several pedagogical approaches.

Encouraging Environments. This approach responds to the risk-aversion tendency of Millennials and Generation Z. Educators can nurture risk-taking behaviors (that are beneficial to creativity) of students in design studios with encouraging environments. They can define and assess task performance as processes rather than outcomes. For creative tasks, it is vital to withstand ambiguity (i.e., no definite solution) and explore multiple ideas before narrowing down to a final solution. Once students get credits from educators for their time and effort in risk-taking ideations, their chances of being creative will increase.

Peer Interactions. This approach responds to the collaboration-avoidance tendency of Millennials and Generation Z. Educators can enable students' interactions in design studios with peer feedback sessions. They need to remind students of the benefits of presenting ideas and getting feedback from peers for self-improvements (i.e., progress and accuracy checking, etc.). Students should also view grades as self-referential, not normative indications. Design studios that stress on creative developments rather than outcomes are preferable for peer interactions to take place.

The theoretical framework and pedagogical approaches (as discussed above) need the supports of empirical evidence. Experimental and observational studies that either testify hypotheses or provide insights on the subject matter, then, are the next steps.

References

1. Afdal, A., et al.: An analysis of phubbing behaviour: preliminary research from counseling perspective. In: International Conference on Educational Sciences and Teacher Profession (ICETeP 2018), (2019)
2. Arkhipova, M.V., et al.: Reaching generation Z. Attitude toward technology among the newest generation of school students. In: Perspectives on the use of New Information and Communication Technology (ICT) in the Modern Economy, pp. 1026–1032 (2017)
3. Chicioreanu, T.D., Amza, C.G.: Adapting your teaching to accommodate the Net Generation/Z-Generation of learners. In: The International Scientific Conference eLearning and Software for Education, pp. 13–20 (2018)
4. Cilliers, E.J.: The challenge of teaching generation Z. PEOPLE Int. J. Soc. Sci. **3**, 1 (2017)
5. Cleary, T.J., Kitsantas, A.: Motivation and self-regulated learning influences on middle school mathematics achievement. Sch. Psych. Rev. **46**(1), 88–107 (2017)
6. Dewey, J.: Interest and Effort in Education. Houghton Mifflin, Boston (1913)
7. Dimock, M.: Defining generations: where Millennials end and generation Z begins. Pew Res. Cent. **17**, 1–7 (2019)
8. Eisenberger, R., Aselage, J.: Incremental effects of reward on experienced performance pressure: positive outcomes for intrinsic interest and creativity. J. Organ. Behav. Int. J. Ind. Occup. Organ. Psychol. Behav. **30**(1), 95–117 (2009)
9. Florida, R.: The Rise of the Creative Class. Basic Books, New York (2019)

10. Fong, C.J., et al.: "Inside out": appraisals for achievement emotions from constructive, positive, and negative feedback on writing. Motiv. Emot. **42**(2), 236–257 (2018). https://doi.org/10.1007/s11031-017-9658-y
11. Fuentes, G.: Pedagogy With and against the flow: generational shifts, social media, and the gen Z brain. In: 102nd ACSA Annual Meeting Proceedings, Globalizing Architecture/Flows and Disruptions, Miami, FL, pp. 669–677 (2014)
12. Gibson, C., et al.: Creativity without borders? Rethinking remoteness and proximity. Aust. Geogr. **41**(1), 25–38 (2010)
13. Harks, B., et al.: The effects of feedback on achievement, interest and self-evaluation: the role of feedback's perceived usefulness. Educ. Psychol. **34**(3), 269–290 (2014)
14. Harpur, P.: Mobile lecturers, mobile students: an exploratory study in a blended architectural technology context. J. Enterp. Inf, Manag (2017)
15. Hennessey, B.A.: Motivation and creativity. In: Kaufman, J.C., Sternberg, R.J. (eds.) The Cambridge Handbook of Creativity, pp. 374–395. Cambridge University Press (2019). https://doi.org/10.1017/9781316979839
16. Hidi, S., Renninger, K.A.: The four-phase model of interest development. Educ. Psychol. **41** (2), 111–127 (2006)
17. Jovanovic, D., Matejevic, M.: Relationship between rewards and intrinsic motivation for learning–researches review. Procedia-Soc. Behav. Sci. **149**, 456–460 (2014)
18. Lee, M.-K.: Peer feedback in second language writing: Investigating junior secondary students' perspectives on inter-feedback and intra-feedback. System **55**, 1–10 (2015)
19. Lester, M.: A study of the innovation, creativity, and leadership skills associated with the college-level millennial generation. Pepperdine University (2011)
20. Lin, L., et al.: Task speed and accuracy decrease when multitasking. Technol. Knowl. Learn. **21**(3), 307–323 (2016)
21. Lipstein, R.L., Renninger, K.A.: Interest for writing: how teachers can make a difference. English J. **96**, 79–85 (2007)
22. Malik, M.A.R., et al.: Distinct effects of intrinsic motivation and extrinsic rewards on radical and incremental creativity: the moderating role of goal orientations. J. Organ. Behav. **40**(9–10), 1013–1026 (2019)
23. Mihelič, K.K., Aleksić, D.: "Dear Employer, Let Me Introduce Myself"–flow, satisfaction with work-life balance and millennials' creativity. Creat. Res. J. **29**, 4 (2017). https://doi.org/10.1080/10400419.2017.1376503
24. Mohr, E.S., Mohr, K.A.: The ABCs of the XYZs: adding a critical dimension to contemporary teacher education. Question **5**, 2 (2018)
25. Mohr, K.A.J., Mohr, E.S.: Understanding Generation Z students to promote a contemporary learning environment. J. Empower. Teach. Excell. **1**(1), 9 (2017)
26. Palfrey, J.G., Gasser, U.: Born digital: understanding the first generation of digital natives. ReadHowYouWant.com (2011)
27. Patall, E.A., et al.: Daily interest, engagement, and autonomy support in the high school science classroom. Contemp. Educ. Psychol. **46**, 180–194 (2016)
28. Patrick, H., et al.: Early adolescents' perceptions of the classroom social environment, motivational beliefs, and engagement. J. Educ. Psychol. **99**(1), 83 (2007)
29. Pektas, S.T.: The blended design studio: an appraisal of new delivery modes in design education. Procedia-Soc. Behav. Sci. **51**, 692–697 (2012)
30. Renninger, K.A., Hidi, S.: Revisiting the conceptualization, measurement, and generation of interest. Educ. Psychol. **46**(3), 168–184 (2011)
31. Richards, J.M., et al.: A systematic review of fMRI reward paradigms used in studies of adolescents vs. adults: the impact of task design and implications for understanding neurodevelopment. Neurosci. Biobehav. Rev. **37**(5), 976–991 (2013)

32. Runco, M.A.: Meta-creativity: being creative about creativity. Creat. Res. J. **27**(3), 295–298 (2015). https://doi.org/10.1080/10400419.2015.1065134
33. Runco, M.A., Chand, I.: Cognition and creativity. Educ. Psychol. Rev. **7**(3), 243–267 (1995)
34. Sansone, C.: A question of competence: the effects of competence and task feedback on intrinsic interest. J. Pers. Soc. Psychol. **51**(5), 918 (1986)
35. Schwalbe, C.B.: Leveraging the digital media habits of the millennials: strategies for teaching journalism courses. Southwest. Mass Commun. J. **25**, 1 (2009)
36. Sebire, K.B., et al.: Learning in the age of distraction. In: EdMedia + Innovate Learning, pp. 467–472 (2018)
37. Shatto, B., Erwin, K.: Teaching millennials and generation Z: bridging the generational divide. Creat. Nurs. **23**(1), 24–28 (2017)
38. Siraj-Blatchford, I., et al.: Creativity, communication and collaboration: the identification of pedagogic progression in sustained shared thinking. Pacific Early Childhood Education Research Association (2007)
39. Steele, L.M. et al.: Intrinsic motivation and creativity: opening up a black box. In: Handbook of Research on Leadership and Creativity. Edward Elgar Publishing (2017)
40. Turner, A.: Generation Z: technology and social interest. J. Individ. Psychol. **71**(2), 103–113 (2015)
41. Yetzke, A.: Fueling the creative agency of generation Z. Danc. Educ. Pract. **5**(3), 29–36 (2019)

Engineering Design Thinking and Making: Online Transdisciplinary Teaching and Learning in a Covid-19 Context

Jinge Huang, Wenjie Pan, Yao Liu, Xiaohan Wang, and Wei Liu(✉)

Faculty of Psychology, Beijing Normal University,
Beijing 100875, People's Republic of China
wei.liu@bnu.edu.cn

Abstract. This paper introduces an engineering design thinking and making course that has been taught at Beijing Normal University since 2019. In its 2-year journey and iterations, both teachers and students learn to dance with ambiguity, collaborate in teams, build to think, and make ideas real. They embrace engineering design thinking and making and experience the maker culture of the China-US young maker competition in this 16-week semester-long course. This year because of the Covid-19, the innovative course changed to online teaching. The course focus on people's basic needs during the Covid-19, including study, fitness, shopping, entertainments and long-distance relationships and communication with family members. Student teams collaborate online to solve the special challenges of Covid-19 in innovative ways and deliver functional proof-of-concept prototypes along with in-depth documenta-tion that not only captures the essence of designs but the learnings that led to the ideas.

Keywords: Transdisciplinary education · Online teaching · Covid-19 · Chinese context

1 Introduction

Entrepreneurship education is gaining momentum in today's world. In recent years, entrepreneurship education has become increasingly important in engineering education [4]. Interdisciplinary, as a transdisciplinary interaction method, has a positive influence on the knowledge, ability, and quality structure of teachers and students [1]. Carrying out interdisciplinary education in universities helps to optimize students' knowledge structure, cultivate innovation ability. Moreover, improve overall quality, which is an inevitable choice for training innovative talents [2]. The cultivation of college students should emancipate the mind and update ideas [3]. With the help of interdisciplinary education. It can effectively improve the quality of teaching, integrate superior majors, broaden the knowledge of students. Moreover, strengthen the improvement training of students' comprehensive ability. As an effective way to cultivate innovative thinking, design thinking is more and more carried out in the form of courses in colleges and universities. As a methodology, the characteristics of integrating creative thinking into action have made it more and more popular among

E. Markopoulos et al. (Eds.): AHFE 2020, AISC 1218, pp. 159–166, 2020.
https://doi.org/10.1007/978-3-030-51626-0_19

educators. Moreover, it has also exerted a significant influence on education in the 21st century [9]. In China, more and more colleges and universities offer courses related to Design Thinking. Applying this way of thinking to the cultivation of undergraduates [10]. Undergraduate students are in the stage of learning socialization. The learning of undergraduate higher education and the thinking mode they cultivate all affect their socialization process and life path. At present, the cultivation of undergraduates mainly focuses on professional training, and there are shortcomings in innovation and practice [11]. The introduction of the concept and method of Design Thinking in undergraduate education is conducive to the cultivation of students' innovative consciousness. And the promotion of students' all-round development.

Since the early spring of 2020, Chinese universities have been experiencing an unprecedented massive "migration" from traditional in-class, face-to-face education to online education [8]. The outbreak of novel coronavirus pneumonia has brought significant challenges to higher education in China. How to complete the teaching in the epidemic is a real problem in front of every teacher. In the field of education, there is also thinking about "Internet + Education". The large-scale distance online teaching activities have put forward new requirements for the platform operators—the education management of colleges and universities, also teachers, and students. The "Design Thinking" course focuses on the user, context, emotion, interaction, technology, and human factors, through practicing engineering design thinking and making. In this year, because of the Covid-19, the innovative course changed to online teaching. The course takes the most basic life needs of people during Covid-19 as project topics and encourages students to use design thinking as a method to solve practical problems around them to optimizes the life experience during the epidemic through innovative solutions.

2 Related Work

Highly innovative ideas are largely the result of the flexible use of interdisciplinary thinking models [7]. A study results show that the characteristics of the interdisciplinary study is the multidisciplinary focus on the integration of knowledge [18]. Critical thinking ability and metacognitive skills are also learning to improve the other subjects in the process of development [12]. At the same time, the team cooperation ability of students in the process of collaborative learning a workout [13]. Contact with other disciplines can guide students through the process of deep learning. Making the knowledge internalization and using critical thinking to promote a higher level of cognitive processing capacity [19]. So, it can extend their focus from a discipline to the development of multiple structures of knowledge, and continue the evolution of the knowledge structure [20]. To help build a strong collective identity within the team is also one of the advantages of developing interdisciplinary teaching [21]. Although such a link would end as the project ended [22]. However, for stimulating and developing interdisciplinary connections between this purpose, it is worth [27]. Over the past two decades, there has been a new understanding of how to teach online. Online teaching is no longer regarded as a way for students to review, but a more personalized learning scene [5]. Therefore, higher requirements are put forward for online teachers. Shift in higher education to online education at the same time, and teachers also need to turntable teaching way [26]. Such a request is not accepted by teachers at the beginning, and they think it is unnecessary [14–16]. According to a survey, the main reason why teachers object to online education is that they have many difficulties in technical operation. The process of preparing online teaching materials will not only consume much time but also lose the opportunity to interact with students. Hinder teaching activities [23–25]. Attitude toward the opposite of people thinks educators need to equip students with the teaching skills to share knowledge [17]. Teachers do not have to answer all students' questions. It is a good way to encourage students to help each other appropriately [28]. Some researchers also give suggestions on the problems that may arise in online teaching. Because of possible cheating, each student can confirm whether it is the student by tracking the user's location when logging into the system with his or her user name and password [29]. Besides, as a teaching form different from traditional education, students' satisfaction with the curriculum will also be one of the criteria to evaluate the quality of the curriculum [6]. A satisfactory attitude can lead to positive learning, attitude, experience, and results [30]. At the same time, teachers with rich teaching experience and high-quality online course resources can stimulate students' moods, enhance students' learning enthusiasm and learning effect [31].

3 Settings

The Co-making the Future China-US Young Maker Competition is an activity supporting the China-US high-level consultation on people-to-people social and cultural exchange. In this spring semester, the research team has mentored and worked with 40 student makers who work in teams of 8 on design briefs concerning study, fitness, shopping, entertainment, and relationships in the Covid-19 context. The undergraduate students participating in the course come from various majors, including 13 different majors such as geography, Chinese language, education, economics, and psychology. The course advocates "learning by doing". This course has three distinctive features: 1) Project-based learning. The course takes the most basic life needs of people during Covid-19 as project topics and encourages students to use design thinking as a method to solve practical problems around them to optimizes the life experience during the epidemic through innovative solutions. 2) Maker-style education. Unlike some traditional courses that separate theory and practice, this course uses the Maker Competition to help students get inspiration in practice and experience the role of design thinking in the creative process. 3) Cooperation in collaborative innovation. Through online communication and collaboration, students are encouraged to expand their ideas and jointly create new solutions. Due to the Covid-19, the course changed from the original offline teaching to online teaching, retaining the form of transdisciplinary teamwork.

4 Online Teaching and Procedure

Empathy is the first stage of the course and also the first stage of the design thinking process. The overall background of the course project was introduced. Unlike the offline observation and interviews of previous courses, in this semester course, students need to use the scope of their own observations or personal experience to conduct scene-based observations to understand and empathize with users. The form of interviews has also become mainly online interviews. Students formulated inter-view outlines through online discussions and understanding target users through online interviews. After the interview, the students completed the production of affinity maps through the online collaboration platform like ZOOM.

After completing the empathy for users, the course enters the Define phase. In the Define stage, the researchers used Persona and User Journey Map to help students understand their users' needs and pain points, thereby defining what problems the team needed to solve. First, students need to complete Persona through group collaboration to form a unified understanding of users. In previous classrooms, researchers handed out paper toolkits to each group. The team members discussed face to face, formed a consensus on their Persona, and filled in the paper toolkit. In the online classroom scenario, the researchers distributed the electronic version to each group. The students marked and modified the toolkit pictures through the "whiteboard" function in the online video conferencing software. In the next section, students still need to complete the User Journey Map in groups. How to complete the User Journey Map in the online teaching situation is a new challenge for researchers. In previous courses, students thought highly of the paper toolkit of the User Journey Map, especially in the part of making the emotional curve. The students analyzed the users' behaviors and emotions by pasting emotion stickers. In this process, they experienced more fun and gained useful inspirations Under this premise, the researchers developed the toolkit. In this toolkit, students can visualize the user's emotions by filling the cell color. The toolkit design received good feedback, so the researchers continued to use a similar format for the Ideate stage of teaching. In the Ideate stage, group cooperation becomes more necessary and closer. Students need to complete brainstorming, C-box, MoSCoW, design brief through cooperation. This requires a lot of discussion and communication among team members. Therefore, the researchers continue to use the form mentioned above for toolkit design. After three rounds of silent brainstorming, each team harvested many solutions that could solve users' pain points. In the scheme screening, the researchers used different color backgrounds to represent the different quadrants of C-box, and used the relative positions of the cells in the spreadsheet to help students compare and screen brainstorming schemes. After that, the students conducted a new round of brainstorming on product functions. General, in the stages of Define and Ideate, researchers used online collaborative spreadsheet formation to design toolkits and implement methods, and finally completed the teaching transfer from offline to online.

The multidisciplinary team has now completed the design stage, further understanding of the users through the creation of a persona, several brainstorming, and refinement of the design, multi-dimensional analysis to obtain the existing design brief.

The next stage of prototyping and testing will focus on building and implementing student programs in a way that combines technology and online education. 1) Interactive logic map. Explore and design more appropriate ways of interacting in conjunction with the needs of the team's target users and other stakeholders, which is followed by a structured disassembly and analysis of the product functionality. In order to solve the problem of unsynchronized information, a logic map is created by implementing a collaborative website online to show product functions, and the visual interactive logic map is a great help in translating programming into specific product displays. 2) Arduino Connection. Arduino and Visual Programming can be used interchangeably. It perfectly addresses the inability of team members from multidisciplinary backgrounds to join a collaborative disengaged group atmosphere, allowing everyone to participate in the hardware design and production process. The online simulation modeling software breaks the limitation of time and space, a simple drag and drop function can add electronic components to the hardware circuit part, can complete the intelligent matching design. Team members can learn from the circuit diagrams and, if necessary, do the soldering and solid construction of the product themselves. 3) Visual programming. Combined with Arduino, visual programming solves the problem of multidisciplinary operators needs to spend much time learning programming languages. After the operator edits the program module, the system will automatically generate the code, write the specific action of the components, which can achieve more convenient hardware construction of the product. Online visual programming solves the problem of not being familiar with the programming language and not being able to find bugs. At the same time, the system can conduct actual programming of specific functions in time, avoiding illegal operations and unnecessarily repeated programming of components. 4) Business Canvas. The sub-sectional business model canvas toolkit, visually demonstrates how the team describes and evaluates the business model and product value of their product. Combined with the detailed presentations made by each team, the team analyze the needs of the target users and other stakeholders, design the corresponding business model, and choose the value proposition and services to be provided to the users. The business canvas has more sections and covers a broader range of business aspects, so in the toolkit, the small sections are aggregated, using a right-to-left research and discussion logic, to fill in and analyze specific modules. 5) Storyboard. After completing the above product design process, the storyboard will be used to create multiple forms of narrative about the product and user interaction, such as a realistic or cartoonish style. There are few online storyboard collaboration software platforms, so the members engage less. The members will use a narrative timeline or write a script to tell a complete and vivid story so that the viewer can empathize with why the product is needed and recognize the output of the team.

5 Conclusion and Future Perspectives

This study is an online multidisciplinary design thinking course in the Covid-19 context, so the students are guided to solve problems innovatively. The research process demonstrated the feasibility of designing thinking online education and also demonstrated the advantages and problems of online education. In the future research, the researchers will better identify and use more appropriate online platforms for online education on design thinking, encourage more multidisciplinary students to join the online course, take advantage of the different professional backgrounds of the team members, and apply design thinking to different fields to help others and society to have a better experience.

Acknowledgments. We would like to thank our teaching team, students, and the Fulbright Research Scholar Grant (ID: PS00284539).

References

1. Piotr, D.A., Michael. B.T.: Supporting multidisciplinary collaboration: requirements from novel HCI education. In: Proceedings of the SIGCHI Conference on Human Factors in Computing Systems, pp. 1073–1076. ACM Press (2007)
2. Ben, S., Philip, H.: A case study of an interaction design approach to pervasive healthcare. In: Proceedings of the 12th EAI International Conference on Pervasive Computing Technologies for Healthcare, pp. 23–244. ACM Press (2018)
3. Wang, J., Yan, Y., Zhang, L.: Research on the design principles for intelligent products. In: Proceedings of the International Conference on Human-Computer Interaction, pp. 351–367. Springer (2019)
4. Roy, N., Schlosser, F., Pasek, Z.: Stimulating entrepreneurial interest in engineers through an experiential and multidisciplinary course collaboration. Entrepr. Educ. Pedagogy **3**(1), 14–40 (2020)
5. Gloria, A.M., Uttal, L.: Conceptual considerations in moving from face-to-face to online teaching. Int. J. E-Learn. **19**(2), 139–159 (2020)
6. Hampton, D., Culp-Roche, A., Hensley, A., Wilson, J., Otts, J.A., Thaxton-Wiggins, A., Fruh, S., Moser, D.K.: Self-efficacy and satisfaction with teaching in online courses. Nurse Educ. (2020)
7. Baran, M.L., Jones, J.E.: Online teaching and learning in higher education settings: focus on team effectiveness. In: Enriching Collaboration and Communication in Online Learning Communities, pp. 137–158. IGI Global (2020)
8. Bao, W.: COVID-19 and online teaching in higher education: a case study of Peking University. Hum. Behav. Emerg. Technol. **2**, 113–115 (2020)
9. Wrigley, C., Straker, K.: Design thinking pedagogy: the educational design ladder. Innovat. Educ. Teach. Int. **54**(4), 374–385 (2017)
10. Bowler, L.: Creativity through "maker" experiences and design thinking in the education of librarians. Knowl. Quest **42**(5), 58 (2014)
11. Brenner, W., Uebernickel, F., Abrell, T.: Design thinking as mindset, process, and toolbox. In: Design Thinking for Innovation, pp. 3–21 (2016)
12. He, J., Freeman, L.A.: Can we trust teaching evaluations when response rates are not high? In: Implications from a Monte Carlo simulation. Studies in Higher Education, pp. 1–15 (2020)
13. Gepp, A., Kumar, K.: How to improve teaching using blended learning. In: Theoretical and Practical Approaches to Innovation in Higher Education, pp. 80–90. IGI Global (2020)

14. Gratz, E., Looney, L.: Faculty resistance to change: an examination of motivators and barriers to teaching online in higher education. Int. J. Online Pedagogy Course Des. (IJOPCD) **10**(1), 1–14 (2020)
15. Tolman, S., Dunbar, M., Slone, K. B., Grimes, A., Trautman, C.A.: The transition from teaching F2F to online. In: Handbook of Research on Creating Meaningful Experiences in Online Courses, pp. 67–84. IGI Global (2020)
16. Moorhouse, B.L.: Adaptations to a face-to-face initial teacher education course 'forced' online due to the COVID-19 pandemic. J. Educ. Teach. 1–3 (2020)
17. Richardson, K.: Online tools with synchronous learning environments. In: Handbook of Research on Online Pedagogical Models for Mathematics Teacher Education, pp. 68–78. IGI Global (2020)
18. Cresswell-Yeager, T.J., Aucoin, J., Budenz, D.: A framework for student engagement: strategies for faculty teaching online. In: Handbook of Research on Creating Meaningful Experiences in Online Courses, pp. 153–171. IGI Global (2020)
19. Zhang, X.: Thoughts on large-scale long-distance web-based teaching in colleges and universities under novel coronavirus pneumonia epidemic: a case of Chengdu University. In: 4th International Conference on Culture, Education and Economic Development of Modern Society (ICCESE 2020), pp. 1222–1225. Atlantis Press (2020)
20. Dimitrov, N., Haque, A.: Intercultural teaching competence in the disciplines: teaching strategies for intercultural learning. In: Multicultural Instructional Design: Concepts, Methodologies, Tools, and Applications, pp. 332–355. IGI Global (2020)
21. Huang, L., Lan, L.: Research on MOOC's teaching method about the category of innovation and entrepreneurship courses. In: 5th International Conference on Economics, Management, Law and Education (EMLE 2019), pp. 1322–1325. Atlantis Press (2020)
22. Beltran, V., Decker, J.C., Matzaganian, M., Walker, N.T., Elzarka, S.: Strategies for meaningful collaboration in online environments. In: Enriching Collaboration and Communication in Online Learning Communities, pp. 1–18. IGI Global (2020)
23. McDougald, J.S.: Quality assurance: breaking through the online learning plateau. In: Handbook of Research on Creating Meaningful Experiences in Online Courses, pp. 101–117. IGI Global (2020)
24. Sofat, R., Sharma, R.: A study on perception of academicians towards online education courses. Online J. Dist. Educ. e-Learn. **8**(1), 1 (2020)
25. Ben-Chayim, A., Reychav, I., McHaney, R., Offir, B.: Mediating teacher for distance teaching and learning model: an exploration. Educ. Inf. Technol. **25**(1), 105–140 (2020)
26. Themelis, C., Sime, J.A.: From video-conferencing to holoportation and haptics: how emerging technologies can enhance presence in online education?. In: Yu, S., Ally, M., Tsinakos, A. (eds.) Emerging Technologies and Pedagogies in the Curriculum, pp. 261–276. Springer, Singapore (2020)
27. Ebner, M., Schön, S., Braun, C.: More than a MOOC—seven learning and teaching scenarios to use MOOCs in Higher Education and Beyond. In: Yu, S., Ally, M., Tsinakos, A. (eds.) Emerging Technologies and Pedagogies in the Curriculum, pp. 75–87. Springer, Singapore (2020)
28. Nwabude, A.A., Ogwueleka, F.N., Irhebhude, M.: The use of virtual learning environment and the development of a customised framework/model for teaching and learning process in developing countries. Education **10**(1), 1–12 (2020)
29. Anand, S.V., Anand, T.S.: Using technology to teach and teaching about technology: synergies for the digital age. In: Handbook of Research on the Efficacy of Training Programs and Systems in Medical Education, pp. 379–403. IGI Global (2020)

30. Anderson, J., Bushey, H., Devlin, M.E., Gould, A.J.: Cultivating student engagement in a personalized online learning environment. In: Handbook of Research on Fostering Student Engagement With Instructional Technology in Higher Education, pp. 267–287. IGI Global (2020)
31. Lopez, A.E., Olan, E.L.: Critical practices for teaching and learning in global contexts: building bridges for action. In: Accessibility and Diversity in Education: Breakthroughs in Research and Practice, pp. 613–629. IGI Global (2020)

Innovative Design

Personal Experience in Concept Generation of New Products

Suresh Sethi(✉) and Simrun Sethi(✉)

University of Illinois at Urbana-Champaign, Champaign, IL, USA
{sethis1, simrun}@illinois.edu

Abstract. The Designers create a new product for attention and differentiation. How do designers start the design process, given that they want to come up with a yet non-existent product? The concept generation phase in design is not a case of problem-solving but as a starting point for a journey of exploration. Designers have started to explore how the use of personal experience can help them accomplish the most important goals of design – that is, communicating their intentions, creating the context for the product form, drawing attention to their product, and constructing something new. The research aimed to explore the distinctive practice of professional designers and to document what inspired them in their creative process. The result of this research provided a basis for developing a design framework using images, stories, and personal experiences to support the concept generation of new products.

Keywords: Personal experiences · Concept generation · New product form

1 The Personality of Product Form

"Souls are mixed with things, things with souls" [2]. The creation of concept and form is seen as an attempt to extend the self in new ways and to customize (personalize) the object. When one embeds personal experiences and narratives in the form of an object, such an object is regarded as part of the self. Sartre suggested that creators retain their identity in their products for as long as the products retain some association with the person who brought them into existence [3]. Designers often create the product form as a vehicle of self- expression (Fig. 1). Designers "seek, express, confirm and ascertain a sense of being through what they create. The idea that we make things a part of ourselves by creating or altering them appears to be a universal belief" [4]. Designers can draw themselves into their own experience of spaces, people, and things. Clues exist in their memory [5]

"Objects can also be seen as embodiment of a person – his world-view, his values." [6]. An old vessel (Fig. 2) at the Shanghai Museum exemplifies how objects can be employed to present ourselves to others – either as an extension of ourselves or as a vehicle of maintaining identities. The value of Zhong Jiang's pan is not only due to its physical form but also to the narrative inscribed into it and the message it conveys: the inscription records that the head of the state made the bronze vessel for his wife, Zi Zhong Jiang, and hoped that his descendants would remember to use this vessel forever.

E. Markopoulos et al. (Eds.): AHFE 2020, AISC 1218, pp. 169–174, 2020.
https://doi.org/10.1007/978-3-030-51626-0_20

Fig. 1. Memory Containers, "una tazza per Marcello, Sethi, S (1984) In claiming that design is "mine," designers also come to believe that the object is "me." Such objects confirm our identities. [1]

Fig. 2. Zhong Jiang's Pan This 7th century BC bronze ceremonial pan was made for princess Zi Zhong Jiang. The intricate design of the vessel, unique to bronze utensils of the time, make this piece of work a real masterpiece. The three- dimensional birds, fish, and frogs in the pan can turn 360°.

As one looks at the pan and reads the accompanying narrative, one connects with the husband's feelings and affection for his wife as well as the personal dimensions of his reason to create the pan. The description and the pan become one – it is impossible to draw a line between the two. Zhong Jiang's Pan materialized as a subjective version of reality, unifying feelings and desires, and communicating the husband's emotions to other members of his family through the form of the pan. [7]

When creating a product form, the designer attempts to make sense of the ambiguous. Both conscious and subconscious ideas about self, play essential roles in designers' conception of products. Product form emerges from the designer's efforts to create order not only to the physical material elements but also to the emotions and subjective factors. Designers "provide or select the formal means that are appropriate for creating the intended" product appearance [8]. Product form is a way of communicating the intention and the meaning of the object; therefore, it is a crucial step in the development of appropriate product solutions.

2 Unearthing Designer's Stories

The principles underlying the way designers visualize product concepts are not well documented. There are two possible reasons why little research has been carried out on how designers embody meaning: one is that designers are reluctant to make their knowledge explicit, and the other is that they perceive form generation as an intuitive activity that is difficult to verbalize [9].

A survey questionnaire was sent by email to design professionals globally, to understand the realities of design practice concerning how designers visualize product form. The survey aimed to establish if designers use images and memories from their own lives to visualize product forms at the earliest stage of the design process. The intention was to understand how they arrived at the product outcome. What means do they use to visualize unique and original product forms? What role do emotions, subjective experiences, and narratives play in this process?

Overall, eighty-one responses were collected. The survey responses were mainly from industrial designers, architects, and a few graphic designers. They worked within corporate design studios or ran their practice. They all had developed products for a minimum of two years (Table 1), and at least one of the products they designed was on the market.

Table 1. Years of experience of the participating designers

Years of experience as a designer	Number of participants
Between 2 and 5	15
Between 5 and 10	24
Between 11 and 15	18
Between 16 and 20	3
21 and above	21

The participants reported their approach and working method preferences (Fig. 3). These results show that the majority of the designers considered the emotional-intuitive approach to be the most effective when tackling design problems.

The 81 designers were also asked about the source of inspiration when visualizing product form. The findings were that drawings and sketches produced at the initial stages of the design process are often drawn based on the designers' experiences (Fig. 4). 95% of designers said they used visual impressions of their everyday experiences or other images to generate ideas, and they claimed that this method improved efficiency.

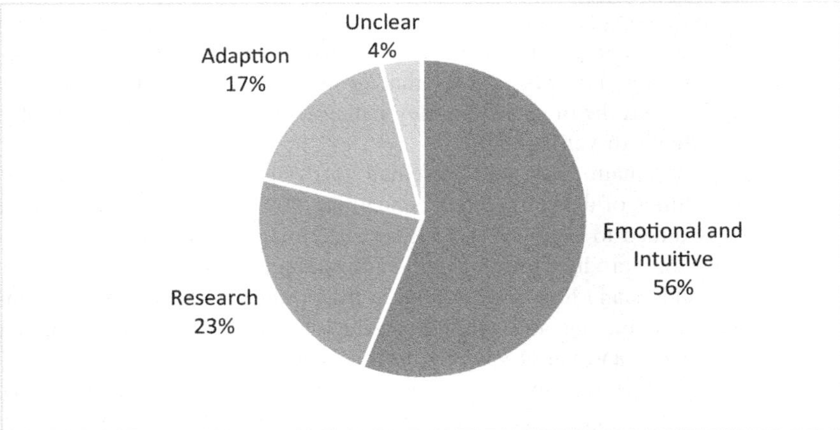

Fig. 3. Designers working method

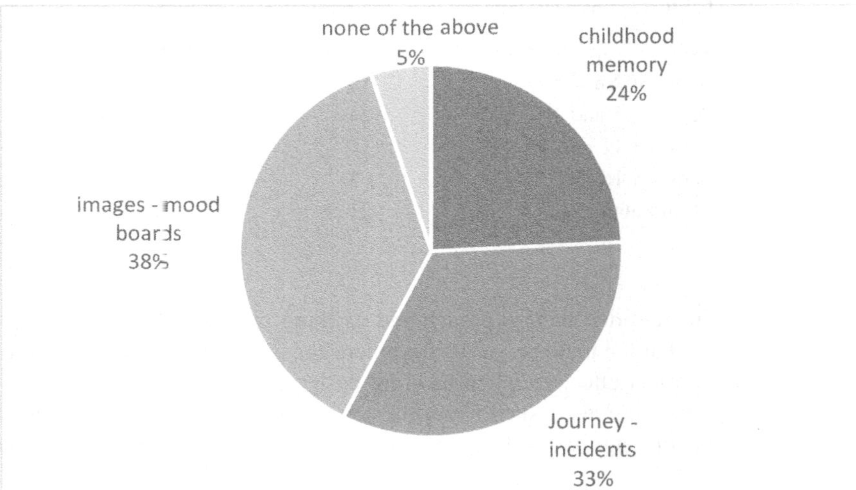

Fig. 4. Survey questionnaire results

Product	Inspiration
	1. The lucky pine cone of Sicily led the moneybox "Soldino" designed by Giovanni Levanti and produced in 2010.
	2. "When I designed the first "Cooler with Wheels, I looked back into my childhood and remembered having a red wagon," Jim Costello.
	3. Mabkhara is the traditional Arab incense burner, an authentic object in daily domestic use in the Arab world, Constantin Boym's time at Doha is visible in this beautiful glass pieces.
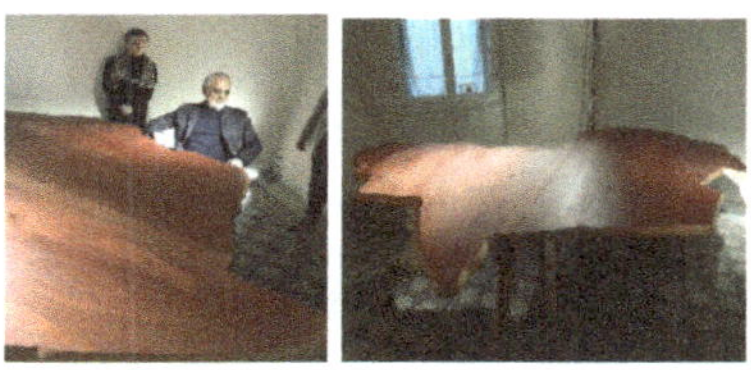	4. Francesco Binfare's love for Africa and his frustration of not being able to talk to someone sitting next to him led him to design this table.
	5. Clare Brass designed these containers for Alessi. The inspiration for her design came from the images and objects she came across during her travels in India.

Fig. 5. Product samples from design practitioners – the designer's capacity to render experience as an instrument for making meaning of the product form.

3 Conclusion

These results of the survey provide a basis for developing a framework using images, stories and lived experiences to support the efficient ideation and realization of product form, some examples from the survey to illustrate the findings can be seen and read in Fig. 5. The study identified the significance of the use of narratives, images, and lived experiences in concept generation during the design process. Being able to see these often invisible or forgotten experiences is a way of creative language use – it allows the designers to play with what otherwise might constrain them. The form is conceptualized as a kind of narrative between the experience and the sketch, conceived as a balancing act of designer' inner tensions. For the designers, impressions of personal travels and life experiences gave them the freedom to pursue their own sense-making goals to formalize the character and elements of the form. The emergence of coherent information from their experience became the guiding principle that created a meaningful and integrated product form. Personal stories gave the designers a pair of wings to transcend limitation and do original work, not set apart from daily life but spring forth from visual impressions of their experiences.

References

1. McCarthy, E.D.: Towards a sociology of the physical world: George Herbert Mead on physical objects. In: Denzin, N.K. (ed.) Studies in Symbolic Interaction, pp. 105–121. JAI, Greenwich CT (1984)
2. Mauss, M.: The Gift. W.W. Norton, New York (1990)
3. Belk, R.: Possessions and the extended self. J. Consumer Res. **15**(32), 139–168 (1998)
4. Belk, R.: Possessions and Extended Self. pp 145–146 (1998)
5. Sethi, S.: Every design tells a story. In: ICERI2015 Proceedings, pp. 6529–6535. IATED Publications, Seville (2015). https://library.iated.org/publications/ICERI2015
6. McCarthy, E.D.: Towards a Sociology of the Physical World: George Herbert Mead on Physical Objects, pp 12 (1984). https://www.researchgate.net/publication/324993740_Toward_a_Sociology_of_the_Physical_World_GH_Mead_on_Physical_Objects_1984_STUDIES_IN_SYMBOLIC_INTERACTION
7. Sethi, S.: A Study on the Role of Narratives in Product Form Visualization (Doctoral Thesis). Indian Institute of Technology Bombay, Mumbai (2019)
8. Schneller, A.: Scratching the surface: "Appearance" as a bridging concept between ontology and design aesthetics. In: Vermass, P.E., Vial, S. (eds.) Advancement in the Philosophy of Design, p. 40. Springer, Cham (2018)
9. Crilly, N., Moultrie, J., Clarkson, P.J.: Shaping things: intended consumer response and other determinants of product form. Des. Stud. **30**, 224–254 (2009)

Designing for Unpredictable Households: Furniture Design Requests for a Flexible Use of Dwellings

Rute Gomes(✉)

CIAUD, Lisbon School of Architecture, Universidade de Lisboa, Rua Sá Nogueira, Polo Universitário do Alto da Ajuda, 1349-063 Lisbon, Portugal
rutelgomes@gmail.com

Abstract. New ways of living and a bigger variation on household's geometry leads to an unpredictability on households' functional needs throughout time, causing sometimes an inadequacy of dwellings' space towards their users' needs. Furniture solutions may enable a more flexible use of dwellings. What are the main requests to design furniture that will increase a flexible use of space? The answer for this question was found with the crossover of primary research, via direct observation/interviews on overcrowded dwellings, and literature review, leading to conclude that for furniture to provide a flexible use of the dwelling along the time, the designer must adopt user centered methodologies that focus on sustainability principles: i) a long usage period, supported by ii) durable materials, iii) services of maintenance and upgrade, iv) formal and/or functional adaptability and v) user inclusion on the design process to better understand the complexity on their present and future needs.

Keywords: Furniture design · Family dynamics · Design requests · User centered design

1 Introduction

Variations on households' geometry and new ways of living contribute for dwellings to become unsuitable to accommodate household needs during its life cycle [1]. This unsuitableness may trigger either a space rearrangement or a quest for a more appropriate dwelling. To avoid this, there have been developed various architectural proposals that enable households to adapt spatial organization according to their changing needs [2–4]. Also, Furniture design through its history had very remarkable examples on flexibility [5, 6].

A research has been carried to understand how furniture design can provide a flexible use of dwellings throughout time, following the variations on household's spatial needs. The study was developed using the «design thinking» methodology [7]. In the first research moment («empathy») the focus is to study and understand the contemporary housing context and to collect examples on design and architecture that provide a flexible use of space. The second moment of the research («definition») has the main objective of identifying the conceptual and functional requests to design

E. Markopoulos et al. (Eds.): AHFE 2020, AISC 1218, pp. 175–184, 2020.
https://doi.org/10.1007/978-3-030-51626-0_21

furniture that allows a flexible use of space. The third moment («ideation») is centered in the generation and filtering of ideas that meet the defined requests. The fourth and fifth moments are dedicated to prototyping and testing of furniture solutions by selected households and design experts to validate the proposal.

This paper presents the results of the second phase («definition»). This phase started with the definition of domestic activities that have more constraints in their development in households that inhabit in overcrowded dwellings. For this purpose, households that were living in overcrowded dwellings were interviewed in their domestic environment. The identified activities were: *being in private (sleeping)*, *living (receive guests)*, and *storing personal clothes and objects* [8].

Besides the identification of the activities with more constraints, the interviews, allowed to collect other valuable information, like the dwellers' routines, and direct observation on their dwellings spatial organization.

Following the achieved results, the research focused in identifying the conceptual and operational design strategies for flexible furniture used in a wider variety of design proposals, for this, a survey on flexible furniture proposals for the identified domestic activities was carried. The results were that the furniture pieces that are more frequently used to ensure flexibility are room dividers for *sleeping privacy*, expandable/collapsible dining tables for *dining*, and container expanding systems for *living*. Multifunctional objects that maximize the use of space were found for the three functions. The strategies that are applied more frequently to promote furniture flexibility are *hinging*, *telescopic* and *assembling*. These strategies are applied in pieces of furniture that have strong structures and are made of longer lasting materials. In contrast, the least used strategies are *inflating*, *rolling* and *compressing*. These strategies are applied in pieces of furniture made of soft materials and associated with objects of short term use. Finally, most used strategies are applied to mass produced objects. In contrast, less frequent strategies are applied on prototypes that serve conceptual or artistic purposes [9].

This paper presents the results of phase two, which led to understand and identify the requests to design furniture design for a flexible use of dwellings.

The following section explains the research methodology and Section three presents the results. Section four describes the conclusions.

2 Methods and Materials

The goal of this part of the research is to identify the requests to design furniture that enables a flexible use of dwellings. This paper will present the crossover of i) the results from the interviews to the selected dwellers, ii) the results from the survey on flexible furniture for the identified constrained activities, iii) premises observed during direct observation, and iv) review on existing literature on durable products.

2.1 Interviews and Direct Observation

The use of the dwelling was classified in ten main domestic functions [10]: *sleeping*, *cooking*, *eating*, *living*, *playing/studying/working*, *clothes care*, *personal hygiene*, *circulation*, *domestic management*, and *being outside in private space*. Each function combines several activities. Who, where and when each activity is takes place depends on household routines. During some periods of the day several activities are going on. Space use conflicts arise when simultaneous activities overlapped or demand different environment conditions.

The research *definition* phase started with the need to understand daily routines, domestic activities constraints in households that live in overcrowded dwellings, and possible strategies adopted to solve those eventual constraints.

There were conducted interviews to nine households of varied structures that lived in overcrowded dwellings (Fig. 1). The interviews were carried in their domestic environments allowing to observe the real usage contexts, and possible problem-solving strategies.

The interview protocol started by asking a dweller to draw to plan of their house, indicating the rooms, their functions, and which activities were developed there.

Fig. 1. Six of the interviewed households: Couple 1, Mono-parental 2, Mono-parental 1, Co-inhabitation 2, Bi-cephalous 2 and Couple with kid1 (from left to right and top to bot-tom).

The protocols were designed to be playful, engaging dwellers on answering about their routines on a drawn plan of their dwelling. Each dweller was represented by a color post-its and they should indicate in their house plan where they developed each activity.

During the interviews, the dwellers would recall on their routines, and sometimes would complain about the others' behavior or activities overlapping (Fig. 2).

The direct observation during the visits to the households' dwellings, was a complement to the interviews, and notes on important were taken during their conversations, and in the observation of their spatial arrangement.

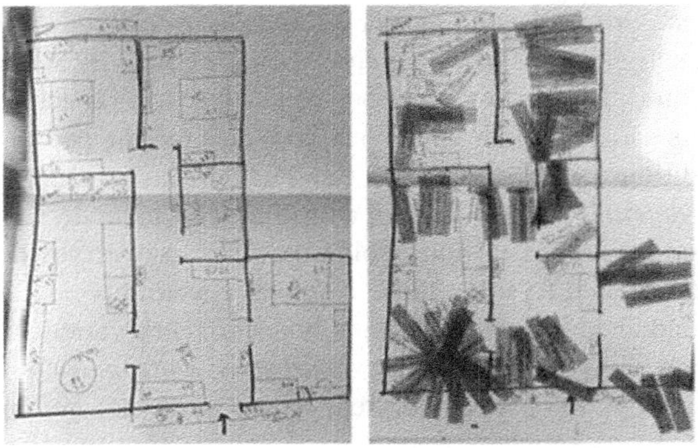

Fig. 2. Drawn plan of household bi-cephalous 2's dwelling. On the left clear drawing, indicating the rooms and the location of the main functions, on the right, the plan with different colored post-its indicating what and where each dweller would develop each activity.

2.2 Conceptual and Operational Strategies to Design Furniture for a Flexible Use

The existing research on flexible artifacts was a reference to categorize operational and conceptual design strategies [6].

Taking as starting point the identified domestic activities with more constraints on overcrowded dwellings. There was the need understand and identify the design operational and conceptual strategies to develop flexible objects used in a wider variety of furniture solutions

A survey was carried on design proposals for the identified activities. The criteria set for the sample selection was: i) *Having variety:* Since one of the objectives of the survey was to understand the variety of existing solutions, furniture with similar solutions was avoided, ii) *Being relevant and inspirational:* The furniture collected had to be relevant and inspirational for the research. Furniture could be mass produced, a limited edition or a prototype, iii) *Keeping an open mind:* No preconceived ideas guided the collection of the sample. A wide range of solutions was pursued.

A product page was designed with three sections: i) product ID, ii) an analysis table indicating with type of object, the activities it would serve, the applied operational strategies, the applied conceptual strategies, and iii) a small contextualization text.

The collected sample was composed by 111 objects: 40 served the function of *sleeping*, 30 for *being in private*, 51 for *storing personal clothes and objects*, 24 for *eating*, 52 for *living*.

2.3 Literature Review Strategies for a Flexible Use of Dwellings

Households spatial and functional needs can change unexpectedly. History has many examples on flexible furniture that came along with the history of human social and technological contexts and achievements. The development of furniture that enables a more flexible use of dwellings throughout time, is linked to strong pillars of design for sustainability. A literature review about strategies for longer use period for products was conducted.

2.4 Data Analysis

The achieved results from each of the three parts are sequential and will be organized in four parts: identified domestic activities with more constraints; the clues from direct observation; the operational and conceptual flexibility design strategies; the develop theories for product adaptability.

The crossed information allowed to determined general and specific requests to design furniture for a flexible use of dwellings.

3 Results

3.1 On Most Constrained Domestic Activities

Nine households were interviewed. The collected data allowed to identify three major domestic activities with more constraints on their development. The constraints are due to the overlapping with other activities in the same space at the same period of day, or only because of lack of space for their normal development (Table 1).

Table 1. Constrained domestic activities in interviewed households.

Household	*Sleeping*	*Cooking*	*Eating*	*Living*	*Playing / Studying / Working*	*Clothes care*	*Personal hygiene*	*Circulation*	*Domestic management*	*Being outside*
Couple 1									■	
Couple withe kid, two adults, one child			■						■	
Mono-parental 1, One adult, one child, two teens	■									
Mono-parental 2, One adult, two children	■			■					■	
Bi-cephalous 1, One couple, two children, one adult		■	■							
Bi-cephalous 2, Two couples one child			■	■					■	
Bi-cephalous 3, One couple, one teen, one child, one adult	■									
Co-ihabitation 1, Five adults									■	
Co-inahitation 2, Two adults		■	■	■					■	

The activities that faced more space use constrains are:

Being in private - Constrains occur when dwellers share their sleeping space with others not romantically involved. None of the households that reported this constraint had a strategy to solve or soften the problem.

Receive guests - *Eating with guests* and *recreation with guests* face constrains due to the lack of space. The dwellings just have enough space to accommodate the regular dwellers. These dwellings are cramped with furniture, appliances, or are too small to receive other people than the households'.

Storing personal clothes and objects - This is the activity with more reported constraints. Dwellers tend to accumulate many personal assets. When households have to move to smaller houses, they realise how many belongings they have stored throughout time.

3.2 On Operational and Conceptual Design Strategies (Survey)

Once all pieces of furniture of the sample were described, the main information was gathered in a synthesis table. Conceptual approaches and operational strategies more frequently used were summed up with this table.

Three conceptual approaches were found in the design of furniture that enable a flexible use of space: *modular, multifunctional* and *reconfigurable*. Most pieces of furniture are reconfigurable (97). The number of pieces that are multifunctional (62) or modular (27) was less significant (Table 2).

Table 2. Conceptual design strategies identified in the sample.

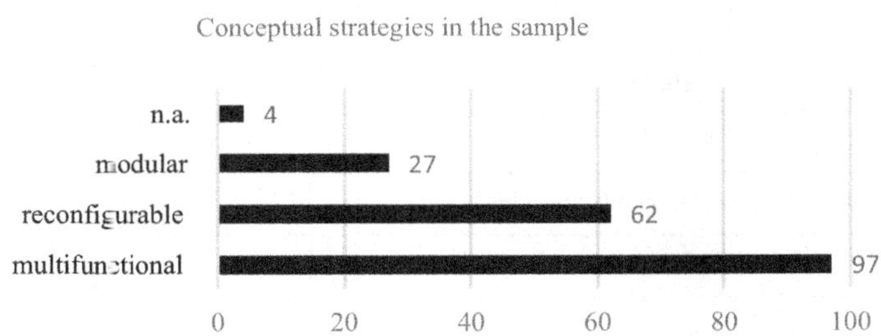

The operational strategies more frequently used are *assembling* (40), *articular* (31), *telescopic* (28) and *pivoting* (18). These strategies are used in pieces of furniture made of hard and long-lasting materials. Therefore, the pieces are designed for long-term use and for being operated frequently without detrition.

Assembling means joining several elements in one piece of furniture, which may be disassembled later for storing or to have a different configuration. *Articular* means joining two elements or more by flexible junctions that can be moved. It is mostly seen with the use of hinges. *Telescopic* means enabling components to slide with minimum detrition in the contact areas. This strategy is often found in drawers. *Pivoting* means that a pivot holds many components overlapped that can be rotated and exposed. The less used strategies in the sample are *compressing, pilling, inflating, bellows, rolling* and *creasing*. These strategies are associated with soft materials. The exempt is *pilling*. Although there are many pieces of furniture that use this strategy, most of them have an identical approach (*e.g.*, pilling chairs) (Table 3).

Table 3. Identified operational design strategies in the sample.

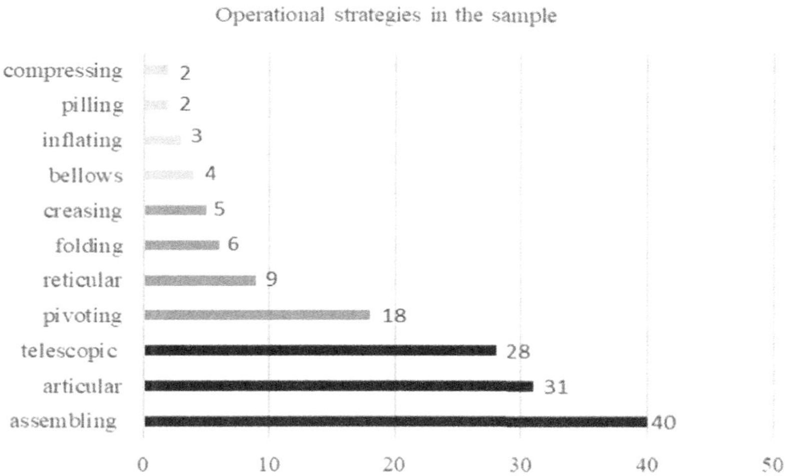

3.3 On Strategies for Products with Long-Time Usage

Throughout history, dwellings and objects have been adapting their societies' needs, aspirations and achievements. The history of flexible spaces and objects is not new, and it recalls the ancient nomads from the deserts. The subject of nomadism and travelling have always been a trigger for new solutions of flexibility [11].

Since this is a very wide theme, and the goal of this part of the research is to identify design requests for furniture that adapts to their households' needs along time, we will focus on the subject of long usage period of objects (life cycle strategy), services design and design for adaptability.

The quest for products that are used in a longer period, than regular ones, lay on principles of *design for sustainability*. The concern for reducing natural resources consumption in the production of goods and services has been gaining weight in various actors of our system.

The *life cycle* of an object is a way of analysing all the phases of a product life cycle from its pre-production to its despoliation. Normally is organized in five phases: i) pre-production; ii) production; iii) distribution; iv) use; v) despoliation. All of them have sub-phases, where there are inputs and outputs to biosphere and geosphere [12]. Keeping in mind our focus on products that have a longer period of utilization, the authors define as some strategies to design using life cycle approach i) the maximization of products and materials consumption; ii) the selection of low impact processes and materials, iii) Optimization of the products' lifetime, iv) the extension of materials lifetime, v) to facilitate the disassembling [13].

For a longer period of usage there can be applied services of maintenance, repairing and upgrade [12, 14–16]. Upgrading a product contributes for its longevity, this allows its adaptation to surrounding contexts, like technologies or domestic environments.

Adaptability comes as another strategy to increase de durability of a product. While upgrading is about substituting a product component to other more actual, adaptability is about adapting the product to new environments and the physical and cultural changes of their user [15, 17].

Chapman [15] points out the need to adopt design strategies that make that users don't throw away product that still are in good functioning conditions, to acquire others that are more recent and have the same function. The strategies are bout encouraging proximity relationships from consumers towards objects, allowing them to adapt their products to their functional and aesthetical needs, increasing this way the products usage period.

Also, Manzini and Vezzoli [12] indicate some strategies to design adaptable and upgradable products: i) Enabling software upgrade; ii) enabling hardware upgrade; iii) the design modular products, that can be configurated in a dynamic way, to facilitate their adaptation to changing environments; iv) To design products that are upgradable insite; v) to design complementary tools and documentation that eases the upgrading and adaptation of products.

Another strategy is about constructing a symbolic value to the product: marketing comes with an important paper in the providing the consumer a conscience and idea of well-being concept. Trying to trigger towards to user the desire of a social categorization from their consumer patterns. This way, focusing on communicating consumer a responsible life-style, designers can influence consumers to question themselves about their options and direct them into more sustainable decisions [18].

Finally, the *user centred design* approach, will allow designers to better understand contexts of usage. The UCD will count with the participation of consumers in the inspirational phase of the project, where there is the need to interpret the contexts and the real neds for products and services. UCD also counts with users' participation for testing the design proposals [19, 20].

4 Conclusions

The crossing of the results led us to identify general and specific requests to design furniture for a flexible use of dwellings.

The general requests to design furniture for a flexible use of dwellings are:

1. Designing objects for a long usage period, this must be supported by:
2. Projecting a long-lasting solution – that besides using durable materials, to develop proposals projecting actual and future needs of their users.
3. To be upgradable and adaptable – to open the possibility for users to adapt their furniture to new tendencies, and to functional, spatial and cultural needs.

4. Using a *User Centred Design* approach – user inclusion in the design process to better understand the complexity on their present and future needs.

The fact that there was used a UCD approach is this research made us identify specific design requests, that are focused in this research in particular:

a. Selected activity: *storing personal clothes and objects* – the choice of this activity is justified by three reasons: i) There are plenty of good and varied solutions in the market that provide privacy in *sleeping*, ii) The constraint on the activity of *receiving guests* for *eating* is due to lack of available space, so, designing more objects for that would possibly not solve the problem; iii) There were not very different solutions for storing clothes and personal objects. Households lack of space is often due to bad storing. If we design an adaptable storage system, it may contribute also to free more space for the activity of *receive guests* and *eating*.
b. The conceptual design strategy to adopt will be *modular*. This will help to better adapt a storing system to different spaces of dwellings, by assuming distinct configurations, according to the needs of the household.

5 Discussion

It is urgent that consumers adopt more responsible and sustainable habits. What specific communication strategies would be necessary to engage households/consumers into using furniture for a «lifetime» or beyond?

To produce a furniture system that is upgradable and adaptable there is the need to build a design management structure that is focused in a transdisciplinary consumer focused service. How could this work?

Acknowledgments. The research was conducted with financial support from *Fundação para a Ciência e Tecnologia* (SFRH/BD/82572/2011). We acknowledge the interviewed households who kindly unveiled their domestic routines. Also, the *"Associação Renovar a Mouraria"*, especially Nuno Franco, for supporting the contacts with households in Mouraria neighborhood. We also acknowledge the centre where the author develops her research: CIAUD (Research Centre of Architecture, Urbanism and Design).

References

1. Guerra, I.: Transformações dos modos de vida e suas implicações no habitat. In: Cabrita, A. R., Coelho, A.B. (eds.) A Futura Habitação Apoiada. LNEC, Lisbon, pp. 179–194 (2000)
2. Paiva, A.: Habitação Flexível: Análise de Conceitos e Soluções. Master thesis. FAUTL, Lisbon (2002)
3. Hertzberger, H.: Lições de Arquitectura. Martins Editora, Lisbon (1999)
4. Schneider, T., Till, J.: Flexible housing: opportunities and limits, https://jeremytill.s3.amazonaws.com/uploads/post/attachment/37/flexible_arq_1.pdf
5. Morellup, P.: Collapsibles, a Design Album of Space-Saving Objects. Thames and Hudson, London (2001)

6. Lemos, S.: Artefactos economizadores de espaço. Master Thesis. Universidade de Aveiro, Departamento de Comunicação e Arte, Aveiro (2006)
7. Brown, T., Wyatt, J.: Design thinking for social innovation. In: Stanford Social Innovation Review, winter 2010, vol. 8, no. 1. Stanford Graduate School of Business, Stanford (2010)
8. Gomes, R., Pedro, J.B., Freitas, M.J., Almendra, R.: Space use conflicts in cramped dwellings a comprehensive analysis of Portuguese households. In: ENHR LISBOA 2015: Housing and cities in a time of change: are we focusing on people? 28th of June to 1st of July (2015)
9. Gomes, R., Pedro, J.B., Almendra, R.: Furniture design for a flexible use of dwellings. In: Sustainability, Design, Urbanism and Architecture. Espaços Vividos Espaços Construídos, vol. 1, no. 4. pp. 110–114. Lisbon School ofArchitecture, Lisbon (2017)
10. Pedro, J.B.: Programa Habitacional: Habitação. LNEC, Lisbon (1999)
11. Rybczynki, W.: Home: A Short History of an Idea, 2nd edn. Penguin Books, London (1986)
12. Manzini, E., Vezzoli, C.: Design for Environmental Sustainability. Springer, London (2008)
13. Idem, p. 65
14. Manzini, E.: A matéria da invenção. Coleção Design Tecnologia e Gestão. Centro Português de Design, Lisbon (1993)
15. Chapman, J.: Emotionally Durable Design – Objects, Experiences and Empathy. Routledge, London (2005)
16. Bhamra, T., Lofthouse, V.: Design for Sustainability: a Practical Approach (Design for Social Responsibility). Gower Publishing, Hampshire (2007)
17. Kasarda, M.A., Terpenny, J., Inman, D., Precoda, K., Jelesko, J., Sahin, A., Park, J.: Design for adaptability – a new concept for achieving sustainable design. Robot. Comput. Integr. Manuf. **23**, 727–734 (2007)
18. Sherwin, C., Bhamra, T.: Ecodesign innovation: Present concepts, current practice and future directions for design environment. In: Design Society Conference, University of Huddersfield (1998)
19. Cross, N.: Designerly Ways of Knowing. Springer, London (2006)
20. Dorst, K.: The core of design thinking and its application. Des. Stud. **32**(6), 521–532 (2011)

Exploration of Bio-Based Materials and Sustainable Product Design: A Case Study of BioPlastic Preparation and Design

Jiajia Song[1,2](✉)

[1] Academy of Arts and Design, Tsinghua University, Haidian District, Beijing 100084, People's Republic of China
Songjjl9@mails.tsinghua.edu.cn
[2] Beijing Institute of Fashion Technology, No. 2 East Yinghua Road, North End of Heping Street, Chaoyang District, Beijing 100029, People's Republic of China

Abstract. In the contradiction of development, increasing and shortages of sources, we are forced to start considering for the development in the future to seek for more replaceable materials or solution cases. Petrochemical raw materials are not the only way to get the same function as plastic. Rethink people's habits of manufacturing and using materials and try to turn bio-based materials into new materials of products. A new mode and pattern, or even new and alternative materials, are expected to replace the old one. Materials with "plastic" properties can also be obtained from raw materials such as sugar, algae or starch. The thesis sets the preparation of team "Bioplastic" design project and its product designs for example, introducing all kinds of ways of preparation of "Bioplastic" and discussing the possibility of bioplastic applying in product designs in the future. In the view of sustainable designs, it will compare and analyse the function between bioplastic and traditional petrochemical plastic.

Keywords: Bio-based materials · BioPlastic · Sustainability · Product design

1 Introduction

The earth is a closed system-aside from meteorites impacting the planet, there is no way of introducing new matter to the system or of taking it away [1]. So, scientists and environmentalists often emphasize that these limited resources have important value, because nothing in nature can be wasted. "Waste should simply not exist … waste is a resource" [2]. But if we take the reverse of the current situation of our artificial industrial production environment into consideration, which is in sharp contrast to the closed system of the earth or nature, which is one of the global challenges we are facing at present. Up to now, the main production patterns are: to transport natural or artificially synthesized raw materials to large factories, and transform them to the product by high temperature, high pressure, chemical process treatment; then deliver these products to the world for the consumption. From the beginning of production to the output of products, 96% of garbage and waste are produced in this process, only 4% of them are products [3]. However, we always use these products very easily, and discard them

E. Markopoulos et al. (Eds.): AHFE 2020, AISC 1218, pp. 185–192, 2020.
https://doi.org/10.1007/978-3-030-51626-0_22

if we don't need them, this kind of single-lined "production- discard" pattern has reached the physical limit at present [4].

Amidst this apparently peaceful material world, a book The Limits to Growth [5] appeared some 40 years back showing us by modeling the consequences of unchecked economic and population growth with finite resource supplies. Nevertheless, the concern about the diminishing resources to sustain our material world is of recent origin.

"There are professions more harmful than industrial design, but only a very few… by creating whole new species of permanent garbage to clutter up the landscape, and by choosing materials and processes the pollute the air we breath, designers have become a dangerous breed… In this age of mass production when everything must be planned and designed, design has become the most powerful tool with which man shapes his tools and environments (and, by extension, society himself). This demands high social and moral responsibility from the designer."– Victor Papanek (Papanek 1985).

Such as Meadows and Papanek's worry, the raw materials are not only used up, it will also go with the pollution, rubbish and waste issues. In fact, all the evidence indicates that if we continue to develop in the current speed, we will need a second earth rapidly. How to proceed the sustainable development and how to balance the contradiction between development and pollution become the problem which has to be thought about. He re-thinks people's habits to produce and use materials, discusses about the possibility of material recombination and redesign, which can make for finding the sustainable development direction in the future.

Fortunately, we are living in the era of accelerated development of science and technology and after 2000, a large number of design experiments and innovative design cases of new solutions or alternative solutions have sprung up. On the one hand, through studying how to realize the advanced and complex system and structure by using the least resources and simple and ingenious design or combination, so as to form the interdependent relationship and closed network system with feedback mechanism. On the other hand, human beings also constantly learn to open the design door of transformation, such as, designers attempts to transform "garbage, waste" into "new resources or energy", and scientists also try to use microorganisms to treat sewage, and so on.

2 Bioplastic and Product Design

Over the past two hundred years of the human industrial revolution, we have created an unimaginable material world and have achieved a brilliant success. But at the same time, we are consuming the resources of the earth. Because since the 1950s, the plastic has appeared as a "new material". The man-made and synthetic materials occupy a dominant position in the product design and living materials for half a century. The continuous development and use of materials is like a double-edged sword, which leads us to prosperity and destruction. Nowadays, the widely used plastic not only has a bad influence on environment but also lead to traditional fossil fuels shortage and dying up.

Now we may be on the brink of a materials revolution, reflecting from the perspective of materials and the making basis of products, which may help to rebalance or even reshape the relationship between human beings and the earth, and fundamentally realize the sustainability of future development [2]. Materials research constantly evolves to offer novel, superior materials as better alternatives to convention (e.g., bio-based materials, smart materials, recycled and/or recyclable materials, etc.) [6, 7]. Sustainable development, as defined in the United Nations (UN) report Our Common Future (Brundtland Commission, WCED 1987) is the "development that meets the needs of the present without compromising the ability of future generations to meet their own needs".

2.1 Bioplastic

Common plastics, like fossil fuel plastic is refined from petroleum, these plastics mainly depend on fossil fuel and generate greenhouse gas. But there is an important operation principle in the natural ecology – the diversity principle, that is to say, in an ecosystem, the more diverse the elements are, the more complex the energy and material flow methods are, the more intricate the food cycle composition is, the stronger the automatic adjustment and stability ability of the ecosystem will be; on the contrary, the duller the elements is, the simpler the structure is, the lower the resistance ability to cope with environment changes will be. The principle is not only that operated in the natural ecosystem, it is also applied in the designs of Bio-based materials and sustainable products in the future. Diversity is a kind of elastic strategy, when solving the same challenge or facing with the same problem, response in various of solutions and methods to reach the preset target. Exactly as the design practice project "Bioplastic", prepare the material with the same property as the traditional plastics by the material recombination experiment, petrochemical materials are not the only scheme to meet people's use demand, several different food waste materials can be transferred at present through Bioplastic project practice and constant attempt, then the alternative material like the traditional "plastic" property is obtained in many kinds of different preparation methods and ways.

Bio-based materials such as "Bioplastic" is the alternative plastic material from the regenerative biomass resource, it is not the actual plastic, but a material which has the plastic property, such as from vegetable oil, starch or microbiota, the production methods of bioplastic are various, Bioplastic part of the practice project is trying to produce organic bioplastic by food wastes actively, it can produce the alternative material for the fossil plastic currently and put forward the more and possible resolution and design practice [8]. Bioplastic project mainly pay attention to the edible wastes and scrap materials in life, such as nut shell, all kinds of egg shell and Fruit and vegetable waste, it is enlightened by the molecule cuisine, looks for feasible preparation methods by the different prescriptions and proportion of material experiment, finally the organic bioplastic with the fossil plastic characteristic can be obtained. And it designs to produce the bulk materials, sheet materials and products with the aesthetic value but without visual sense and tactile sense effect.

2.2 Preparation and Properties of Bioplastic

Nut and egg shell: The following three ideal bio-plastics were obtained through material preparation experiment of Bioplastic team project. The main raw materials of the three organic plastics are different, but the preparation process is similar. Bioplastic mainly use nut shell and egg shell as the main raw materials. After the discarded shell and eggshell crushed, were heated and mixed with natural dyeing materials, starch gum, glycerin and other materials, and the Bioplastic mainly composed of shell and eggshell can be obtained after drying. There is not a single method of configuration and the Bioplastics with different hardness can be obtained by changing the materials or the proportion (Figs. 1 and 2).

Fig. 1. Nut shell - bioplasctic. Design by Beijing Institute of Fashion Technology, Innovative Materials Research Centre, 2018

Fig. 2. Egg shell - bioplasctic. Design by Beijing Institute of Fashion Technology, Innovative Materials Research Centre, 2019

Fruit and vegetable waste: There are a large number of fruit and vegetable waste every day in agricultural markets, food processing stations and plantations, which can not only be used for composting, but also the main raw material for the preparation of Bioplastics. Our team project has tried to use orange, banana, sugarcane, hawthorn, pumpkin, potato, apple, carrot and other fruit and vegetable wastes, such as skin, rotten fruits and vegetables, after many attempts, to mix them with water and edible gum, cook, and then dry them to get a kind of recycled material, which has similar characteristics of leather and flexible PVC. What's more, products can be made by sewing (Fig. 3).

Fig. 3. Fruit and vegetable waste - bioplasctic. Design by Beijing Institute of Fashion Technology, Innovative Materials Research Centre, 2018

Alga: Alga contain natural colloids, which can also be used to prepare ideal organic plastics with certain transparency. Alga, as the main raw material, added with water, starch, vinegar and glycerin, after a certain temperature mixing, Alga bio-plastics can be obtained. After drying, an organic material can be obtained, with the characteristics of softness, transparency, lightness and thinness, which can be sealed by hot pressing (Fig. 4).

Fig. 4. Alga - bioplasctic. Design by Beijing Institute of Fashion Technology, Innovative Materials Research Centre, 2019

Bio-plastics obtained by three different preparation methods have their own characteristics. The properties of these materials were compared. The comparison of material properties shall be carried out according to the five-level system by Professor Elvin Karana, [8] from the chart below for the specific comparison (Fig. 5).

Types of BioPlastic / Material Characteristics	Nut and egg shell	Fruit and vegetable waste	Alga
hard & soft	●○○○○	○○●○○	○○○○●
smooth & rough	●○○○○	○●○○○	○○○○●
matte & glossy	●○○○○	●○○○○	○○○●○
cold & warm	○○●○○	○○○●○	○●○○○
not elastic & elastic	●○○○○	○○●○○	○○○●○
opaque & transparent	●○○○○	○○○●○	○○○○●
tough & ductile	○●○○○	○○○●○	○○○○●
strong & weak	○●○○○	○○●○○	○●○○○
light & heavy	○○○○●	○○○●○	○●○○○

Fig. 5. Material characteristics. draw by author, 2020

The main materials of Bioplastic are obtained from food waste, alga and other raw materials, trying to achieve zero waste in the process from raw materials to processing and manufacturing to finished materials and products. The raw materials are obtained from food wastes, and produced to materials or products, which can be corrupted or decomposed naturally in the aerobic or oxygen-free environment after using.

At the same time, these products can be recycled and entered the later life cycle of product after completing the life cycle due to the special natural components, which achieves the process from "cradle to cradle" indeed. The following application and development of BioPlastic materials have advantages and potentials in the fields of household products, bags and suitcases and building materials. The material hardness, abradability and the surface natural pattern, texture and color are suitable for the household products or bags and suitcases; by exploring and studying food waste materials to excavate potential value, not only the waste egg shell, to look for more common materials which can be replaced by the recycling materials or wastes, it is the original intention of design of the whole "BioPlastic" project – to collect the alternative product raw materials from industrial, life wastes and waste yard, "waste should simply not exist… waste is a resource." [2] (Fig. 6).

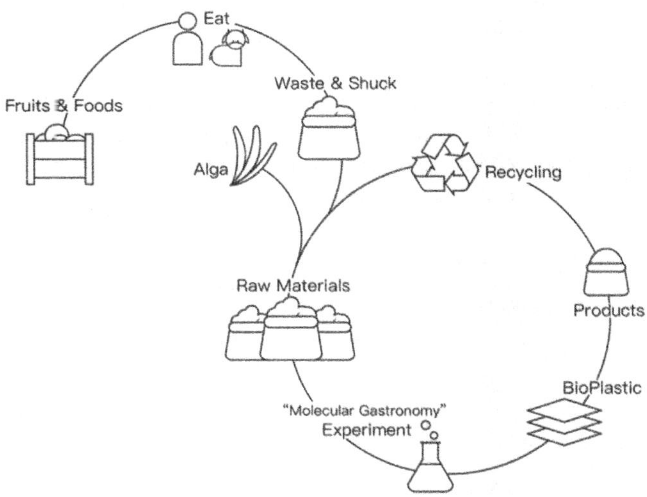

Fig. 6. BioPlastic project cycle chart. draw by author, 2020

2.3 Bio-Based Materials and Sustainable Product Design

Based on biological materials, product design is the organic unity of organic materials and design. Take the Bioplastic project as an example, through the bioplastics, following the design principles of win-win and diversity, designs, harmless to the living environment of human beings and less bad impact on the environment, can be produced. This kind of design will eventually change the production and consumption mode of the public, maximize the utilization of resources, minimize or prevent the generation of garbage, and achieve the goal of effective protection of the environment and human life and health.

Win-win principle: Win-win refers to both the natural environment on which human beings live and the needs to meet human life and daily use can be concerned. The design principle of win-win is to constantly balance the relationship between both of them in the design stage, and finally get the win-win design results. This is also in line with the basic concept of sustainable development, that is, to improve people's quality of life, while gradually reducing energy consumption.

The principle of diversity: The principle of diversity is not only the principle of natural ecosystem operation, but also the design principle of sustainable products based on organic materials. Diversity is a flexible strategy because it allows multiple solutions and ways to solve or respond to the same challenge. A diverse society will be better able to withstand natural disasters, bad growth periods, economic difficulties or cultural crises [9].

In order to meet a human need, the strategy of multi-channel and multi solution can avoid excessive dependence on a certain material or energy. For example, petrochemical raw materials are not the only way to manufacture materials with the same properties as plastics. BioPlastic is made of edible starch, alga, glycerin and other materials, whose hardness and toughness are no less than that of plastic.

3 The References Section

The research of sustainable materials in product design is not only about the simple discussion and attempt of basic materials of creation, but also the practice of products from "cradle to cradle". On the one hand, we the source of new sustainable materials need to be paid attention to, on the other hand, we the sustainability of materials need to be paid attention to after the end of the product manufacturing process and product life cycle. This paper takes our team project "Bioplastic" as an example. It also uses bio-based materials to design the prototype of products and come up with an idea of creative design based on organic materials for sustainable product designs. It aims to lead designers to think about more ideas for continual products in the future when they face the complex nature, ecology and social problems. Materials are the foundation of design creation, and it is a design strategy to make products have sustainable characteristics. Ecological and sustainable product design is not only inseparable from the innovative design of materials, but also inseparable from the interdisciplinary cooperation.

References

1. Kapsali, V.: biomimetics for designers-Applying Nature'-s Processes & Materials in the Real World. Thames & Hudson, Waste eouals resource, pp. 3–25 (2016)
2. Franklin, K., Till, C.: Radical matter-rethinking materials for a sustainable future. Thames & Hudson, 1. Today's Waste, Tomorrow's Raw Material, pp. 6–37 (2018)
3. TED Talk, Janine Benyus, Creating a Bio-Industrial Revolution. https://www.ted.com
4. Dion, C., Jiang, F.: Demain: un Nouveau Monde en Marche. Beijing United Publishing Company, Beijing (2018). (in Chinese)

5. Karana E., Pedgley O., Rognoli V.: Materials experience. Mater. Soci. Sustain., 91 (2014). Butterworth-Heinemann
6. Song, J. (n.d.). Rethinking and reconstituted materials for a sustainable futurt-reconstituting-plan project as an example, Vezzoli, C. (2019). DESIGNING (Vol. 3).1040–1046
7. Smith, T.F., Waterman, M.S.: Identification of common molecular subsequences. J. Mol. Biol. **147**, 195–197 (1981)
8. Karana, E., Hekkert, P., Kandachar, P.: A tool for meaning driven materials selection. Mater. Des. **31**(6), 2932–2941 (2010)
9. Shedroff, N., Liu, X.: Design is the Problem: The Future of Design Must be Sustainable. Tsinghua University Press, Beijing (2011). (in Chinese)

Intuitive Product Design and the Twin Challenges of Innovation and Adoption: An Application

Sandrine Fischer(✉)

San Francisco, USA
sandrine.fischer@ieee.org

Abstract. This paper frames the development of product features in terms of Intuitive Product Design (IPD). Using methods that assess knowledge transfer, IPD identifies which features would benefit from being enriched or restyled and which would benefit from better explanation. IPD is illustrated for a prototype onboard computer. Participants were more successful at using the prototype that was based on IPD, and reported greater intention to recommend it. Preliminary implications regarding technology adoption are discussed.

Keywords: Intuitive Product Design · Innovation · Technology adoption · Kansei analysis · Kano model · Cognitive schemata · Knowledge transfer

1 Introduction

Almost every entrepreneur and startup faces the challenge of taking a product, technology, or service to a market of given maturity. Market maturity is critical to product adoption: basic products may not be differentiated enough to appeal to audiences, while innovative products may fail to resonate or stick within a premature market. A recent analysis of online reviews suggested that the more advanced a mobile application (e.g., personal finance, business analytics), the more traits like intuitiveness prevail in user satisfaction [1]. Of course, early involvement of end-users and continuous integration of feedback are important, but research on intuitive use investigates more principled ways of factoring intuitiveness in product development. This paper discusses what Intuitive Product Design (IPD) has to offer in terms of goals, methods, and potential outcomes for innovation and adoption.

2 The Fundamentals of Intuitive Use

Intuition, or the act of understanding novel situations with little reasoning, is an epiphenomenon of our memory's ability to build *knowledge* or *cognitive schemata* [2]. Schemata represent structure that is common across objects or situations as being fixed,

S. Fischer—Independent Researcher.

E. Markopoulos et al. (Eds.): AHFE 2020, AISC 1218, pp. 193–199, 2020.
https://doi.org/10.1007/978-3-030-51626-0_23

and attributes that differ as being variable [3]. This smoothing out of idiosyncrasies in favor of a common structure holds three interesting properties. First, it makes schemata non-attributable to a particular situation or instance [3]. Second, it lifts constraints on activation, meaning representations are resilient to variations and transferable to never-encountered, yet structurally compatible (aka analogous) situations [4, 5]. Third, structure-based transfer bypasses the analysis of details and renders thinking automatic, effortless, and detail-agnostic [6] - the very characteristics of intuition [7].

One key implication for product strategy is that information compatible with prior schemata is processed through transfer, requires no explanation to be intuitive, and sustains some degree of variation. Information that is incompatible, though, requires costly cognitive operations and will remain prone to effortful thinking and error until new schemata are formed and made available for future encounters [2]. Next, let's see how users' cognitive schemata can be leveraged to design advanced products.

3 Goals and Outcomes of Intuitive Product Design

Advanced products are generally multi-use and likely to present users with information that ranges from already compatible to completely incompatible with their knowledge schemata. Intuitive Product Design proposes to diagnose these cases as a basis for an effective design strategy [8]. Schema-incompatible features impose a trial-and-error behavior, effort, and learning on users that need to be alleviated through design. Product teams should brainstorm and test ways to clarify any schema-incompatible feature their product may have and, in some cases, educate users. Not doing so runs the risk of users getting stuck and rejecting the product, which is why fixing first-time usage performance issues (e.g., task success, error rate) should be the first and foremost focus of IPD [8, 9]. The goal is to ease the understanding and adopting of innovative technology and to properly onboard users.

Once this goal is met, design efforts can turn toward schema-compatible features that already trigger transfer and do not need much explanation to be intuitive. There may not be much room for improving first-time usage performance with these features [6, 8], even by design concepts thought *a priori* to foster intuitive use (e.g., stereotypes, affordances, metaphors, image schemas; [10]). Instead, these features can sustain some degree of variation, meaning a more sensible goal would be to increase their appeal through better differentiation, for example modernizing their style [11], enabling novel interactions [11, 12] or enriching them with additional settings [13].

4 Practical Application

4.1 Prior Work

In 2018, [13] redesigned a prototype onboard computer by applying IPD with schema compatibility assessed via two paradigms, the Screening Method [8, 9] and False Belief Technique [14]. The Screening Method combines a schema induction experiment with effect-size based analytics to sort product features by most compatible with user schemata to least. Because the latter features challenge users the most, the prototype was slightly reorganized, revised with clearer labels, and supplemented with a five-vignette brochure. The False Belief Technique is a delayed memory test for detecting false memories [14, 15], i.e., cases where people remember details not present in a situation but compatible with its schema [6]. In [13], this technique helped to surface features falsely thought to be part of the prototype. Five such features were added to the revised version of the prototype, assuming they would enrich the product without adding any cognitive complexity because of being schema-compatible.

A follow-up test assessed the outcomes of Intuitive Product Design by comparing the original prototype (pre-IPD) with the revised one (post-IPD). The post-IPD prototype included revised labels and structure, along with the brochure and five additional features. Participants in the post-IPD condition were more successful at using the prototype, and reported a higher intention to rent a product with similar or more technology, as well as to recommend the prototype (greater NPS, or Net Promoter Score [16]). However, the prototypes were not rated any differently against seven descriptors derived from AttrakDiff, a subjective instrument for assessing user experience and product attractiveness [17].

For the remainder of this paper, a replication of the above test is presented in order to verify if the prototypes tested in 2018 still yield consistent results and to explore alternative methods for appraising the subjective impact of IPD. To this end, the descriptors based on AttrakDiff were replaced by Kansei descriptors of existing applications in the same category as the prototype. In addition, a Kano questionnaire [18, 19] was administered to get a sense of whether IPD may influence the attractiveness and the expectedness of a product's functionalities.

4.2 Methods

Twenty-seven Amazon Mechanical Turkers (US residents over 18 years old) completed the study in full (about 8 min). The participants started by using either the pre-IPD ($N = 13$) or the post-IPD prototype ($N = 14$), both hosted online [20]. The task scenario involved 1) setting the itinerary to avoid tolls, 2) displaying navigational cues in the rear-view mirror, and 3) activating assistance for safely passing other cars. Log files were recorded to calculate the proportion of participants in each condition who completed the scenario, i.e., correctly found the setting corresponding to each task. The participants then filled out questionnaires meant to qualify their experience in terms of subjective impression of the prototypes (viz. Kansei), intention to rent a product with similar technology, willingness to recommend it (viz. NPS), and perceived attractiveness of 10 functionalities common to both prototypes (viz. Kano questionnaire).

The Kansei questionnaire consisted of descriptors (e.g., *informative*, *intuitive*, *modern*; see left part of Fig. 1 for complete list) that are highly predictive of user ratings for electric vehicle mobile applications. The descriptors were obtained by following a method detailed in [1] from an interaction Kansei catalog and 2 years' worth of online reviews from users of the Tesla app (Android and iOS), My Leaf for Nissan (Android), Scan my Tesla (Android), Tasker plugin for Tesla (Android), and Dashboard for Tesla (Android). The descriptors belonged to five main attributes, which in order of importance were: Easy, Intuitive, Practical, Reliable, and Innovative. Participants' ratings, which ranged from 1 for "Not at all" to 7 for "Absolutely", were later averaged per descriptor and prototype.

One intention question asked whether the participants, when renting a car for a business trip, would "want an onboard computer with more, less, or similar technology". The other question (NPS) was how likely they would "recommend such an onboard computer to a friend or colleague" (1 for "Not at all," 10 for "Absolutely"). The NPS was calculated by subtracting the percentage of detractors, or participants who responded between 1 and 6, from the percentage of promoters, or those who responded 9 or 10. While NPS scores range from −100 (all respondents are detractors) to +100 (all are promoters), a positive NPS is deemed good and one over 50 is considered excellent.

The Kano analysis addressed 10 functionalities of the prototype (e.g., real-time map, customizable navigation voice, parking assistance, gas/energy refill recommendation, see Fig. 2). The usage benefit of each functionality was formulated from the user standpoint in terms of functional and dysfunctional statements. The response choice for the twenty statements were measured on a 5-point scale, in descending order: "I'd like it (5), I'd expect it, I don't care, I'd tolerate it, I'd hate it (1)". Using a technique described by [19], the numerical points for functional/dysfunctional responses were mapped onto a 2D plane. With this mapping, functionalities lying close to the coordinate (0, 4) are considered highly attractive (or delightful), while those close to (4, 0) are considered expected. The closer functionalities lie to the origin (0, 0), the more indifferent participants feel about them; while those closer to (4, 4) yield a satisfaction proportional to how performant they are developed.

4.3 Results

First-use performance registered a near two-fold improvement, from 38.5% of participants correctly completing the entire scenario with the pre-IPD prototype to 69.23% with the post-IPD one (Fig. 1, left). The overall trend is consistent with the 23.5% to 50% improvement for full-scenario completion registered in 2018.

Technologically speaking, the post-IPD prototype was slightly less intimidating, with 2 out of 14 participants saying they would prefer renting a car with less technology, compared to 3 out of 13 for the pre-IPD prototype. This result is in line with, but less pronounced than, the 20% versus 33% recorded in 2018, respectively. The NPS increased from −14 for the pre-IPD prototype to +12 for the post-IPD prototype. This was a stark increase compared to the -0.3 to + 14 change registered in 2018.

The effects of IPD on subjective appraisal (or Kansei descriptors; Fig. 1, right) were more pronounced than those registered in 2018 with the modified AttrakDiff. The post-IPD prototype stood out as less *hard to use* (Easy Kansei attribute), less *complicated* and slightly more *informative* (Intuitive Kansei), less *useless* (Practical Kansei), less *unreliable* (Reliable Kansei), more *modern* and *futuristic* (Innovative Kansei) than the pre-IPD one. Medium effect sizes were observed for the descriptors *complicated* (Cohen's $d = 0.615$), *useless* ($d = 0.562$) and *futuristic* ($d = 0.567$).

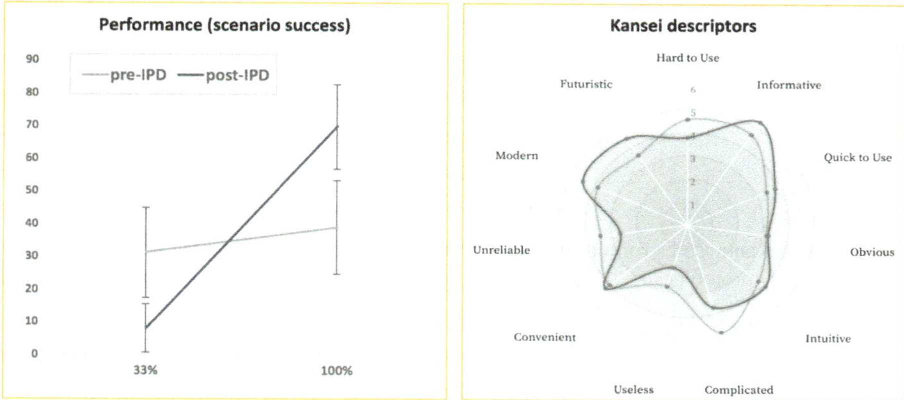

Fig. 1. IPD outcomes in terms of first-usage performance (% task success) and Kansei descriptors (average ratings).

Subjective intuitiveness was broken down into four descriptors: *intuitive* (for direct comparison with 2018), *obvious*, *informative*, and *complicated*. As in 2018, *intuitive* ratings did not differ among the groups. Interestingly, the post-IPD prototype contains a 5-vignette brochure, making it objectively more informative than the pre-IPD one. Participants ratings reflected this only slightly, with a larger difference for *complicated* than for *informative*. All in all, IPD that consisted of adding features and content (viz. brochure) may nonetheless yield an experience "less complicated" over "more informative." This notion would be interesting to explore with higher statistics.

The effect of IPD on attractiveness can be observed by plotting ten of the prototypes' functionalities in a Kano plot (Fig. 2). Although some functionalities did not significantly move between the pre- and post-IPD prototypes, several moved toward the attractive coordinate (0, 4) and away from the expected (4, 0) or the indifferent one (0, 0). Interestingly, the functionalities that did not move were basic ones such as map, date, and calendar.

To summarize the outcomes of IPD, there was a greater chance for first-time users to succeed at using the prototype - a major promise of Intuitive Product Design - and to turn into promoters (people who would recommend the product). While the prototype based on IPD was not rated "more intuitive," it was found to be less complicated, more informative, and easier to use. All these descriptors pertain to Kansei of intuitiveness, per the literature on this construct and similar commercial products [1]. Subjective practicality and innovativeness also improved, while the number of functionalities considered attractive almost doubled, suggesting a better appreciation of the product's technology offering.

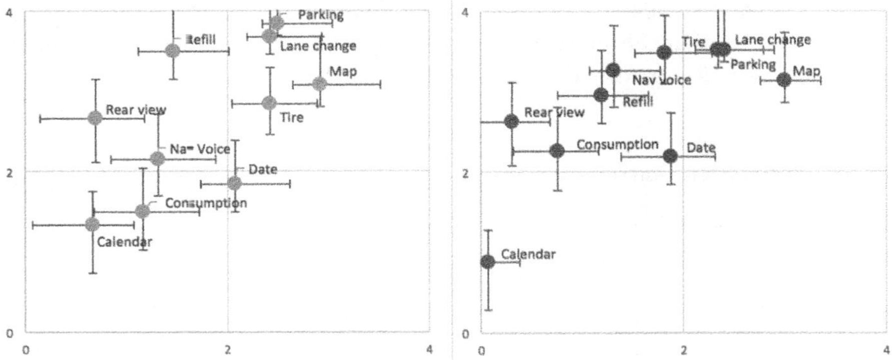

Fig. 2. Mean Kano scores for ten functionalities (with standard error) for the pre-IPD (left) and post-IPD (right) prototypes.

5 Discussion

IPD differentiates product features based on their compatibility with prior schemata. Features already compatible are more understandable, yet potentially less appealing, while features too innovative to be compatible have a steep learning curve likely to hinder understanding of technology. In the latter case, IPD proposes "improving" the product through clarification and/or education, and in the former, through differentiation, enriching, or restyling. The outcome sought is a product *that is and feels intuitive to first-time users*. Two redesign studies conducted to date with IPD registered greater intuitive use (task success) and heightened innovativeness and intuitiveness ratings.

Can such outcomes fuel adoption? Several studies examined survey data for links between perceived intuitiveness and acceptance of utilized [21] or available technology [22–24]. The outcome of IPD on the percentage of task success and subjective intuitiveness is in line with [21], who found correlations among degree of feature use and perceived intuitiveness in their technology acceptance model – see also [24]. Finally, IPD increased subjective innovativeness, which should drive adoption and resonate especially with those people with high personal innovativeness (early adopters) and acceptance towards electric vehicles and ADAS [22, 23].

References

1. Fischer, S.: Examining the incidence of intuitiveness and cognitive-affective states in the Kansei and rating of mobile app products. In: Proceeding AHFE 2020, San Diego, USA (2020)
2. Fischer, S., Itoh, M., Inagaki, T.: Prior schemata transfer as an account for assessing the intuitive use of new technology. Appl. Ergon. **46**, 8–20 (2014)
3. Rumelhart, D.E., Ortony, A.: The representation of knowledge in memory. In: Schooling and the Acquisition of Knowledge, vol. 3, pp. 99–135 (1977)

4. Gick, M.L., Holyoak, K.J.: Schema induction and analogical transfer. Cogn. Psychol. **15**, 1–38 (1983)
5. Gentner, D., Loewenstein, J., Thompson, L.: Learning and transfer: a general role for analogical encoding. J. Educ. Psychol. **95**(2), 393–408 (2003)
6. Brewer, W.F., Treyens, J.C.: Role of schemata in memories for places. Cogn. Psychol. **13**(2), 207–230 (1981)
7. Kahneman, D.: A perspective on judgment and choice: mapping bounded rationality. Am. Psychol. **58**(9), 697–720 (2003)
8. Fischer, S., Itoh, M., Inagaki, T.: Screening prototype features in terms of intuitive use: design considerations and proof of concept. Interact. Comput. **27**(3), 256–270 (2015)
9. Fischer, S., Itoh, M., Inagaki, T.: Identifying the cognitive causes of human error through experimentation. Eur. J. Autom. **48**, 319–338 (2014)
10. Blackler, A., Hurtienne, J.: Towards a unified view of intuitive interaction: definitions, models and tools across the world. MMI-Interaktiv **13**, 36–54 (2007)
11. Hurtienne, J., Klöckner, K., Diefenbach, S., Nass, C., Maier, A.: Designing with image schemas: resolving the tension between innovation, inclusion and intuitive use. Interact. Comput. **27**(3), 235–255 (2015)
12. Macaranas, A., Antle, A.N., Riecke, B.E.: What is intuitive interaction? Balancing users' performance and satisfaction with natural user interfaces. Interact. Comput. **27**(3), 357–370 (2015)
13. Fischer, S.: Designing intuitive products in an agile world. In: Blackler, A. (ed.) Intuitive Interaction Research and Application, vol. 10 (2018)
14. Lee, J., Ryu, H.: I remember/know/guess what I saw: a false belief technique to features selection. In: Extended Abstracts on Human Factors in Computing Systems, CHI 2014, Toronto, Canada, pp. 2593–2598 (2014)
15. Roediger, H.L., McDermott, K.B.: Creating false memories: remembering word not presented in lists. J. Exp. Psychol. Learn. Mem. Cogn. **21**(4), 803 (1995)
16. Reichheld F.: The one number you need to grow. Harvard Bus. Rev. **81**(12), 46–124 (2003)
17. Hassenzahl, M., Burmester, M., Koller, F.: Der user experience (UX) attrakdiff.de. In: Usability Professionals Conference, Stuttgart, Germany, pp. 78–82 (2008)
18. Kano, N., Nobuhiku, S., Fumio, T., Shinichi, T.: Attractive quality and must-be quality. J. Jpn. Soc. Qual. Control **14**(2), 39–48 (1984)
19. DuMouchel, W.: Thoughts on graphical and continuous analysis. Cent. Qual. Manage. J. **2**(4), 20–23 (1993)
20. Fischer, S., Oelkers, B., Fierro, M., Itoh, M., White, E.J.: URU: a platform for prototyping and testing compatibility of multifunction interfaces with user knowledge schemata. In: Proceedings of the HCII Conference 2015, Los Angeles, CA, pp. 151–160 (2015)
21. McAran, D.: Development of the technology acceptance intuitive interaction model. In: Blackler, A. (ed.), Intuitive Interaction: Research and Application, pp. 129–150. CRC Press (2018)
22. Schuitema, G., Anable, J., Skippon, S., Kinnear, N.: The role of instrumental, hedonic and symbolic attributes in the intention to adopt electric vehicles. Transp. Res. Part A Policy Pract. **48**, 39–49 (2013)
23. Keller, K., Carl, K.V., Jöntgen, H., Abdel-Karim, B.M., Mühlhäuser, M., Hinz, O.: "K.I.T. T., where are you?": why smart assistance systems in cars enrich people's lives. UbiComp ISWC 2019, Adjunct, pp. 1120–1132. ACM, New York (2019)
24. Renaud, K., Van Biljon, J.: Predicting technology acceptance and adoption by the elderly: a qualitative study. In: Annual Research Conference of the South African Institute of Computer Scientists and Information Technologists, Wilderness, South Africa, pp. 210–219 (2008)

Life Sciences Product/Service Development Process: Lesson Learned from Thailand Center of Excellence for Life Sciences (Public Organization)

Varit Intrama(✉)

Director of Industry and Investment Strategy Department,
Thailand Center of Excellence for Life Sciences (Public Organization),
SPE Tower Bldg., 9th Fl., 252 Phahonyothin Road, Samsen Nai,
Phaya Thai, Bangkok 10400, Thailand
v_intrama@hotmail.com

Abstract. There is no argument on the benefit of innovation. Much evidence shows how innovation increase the value of the company; surpasses its outcomes, both product and service, and also improves the process's efficiency. Many scholars proposed the model for creating innovation using the term "New Product (Service) Planning and Development Process: NPD process" by employing the inductive methodology, collected case studies and constructs to generate the models such as BAH Model and Stage-Gate Model, However, these models are still not crystalized and incompatible with business requirement, especially in high growth business like life science. There is evidence showing the general NPD process is not fit in the current situation of life sciences product/service development and should be modified to solve the problems. The portfolio management concept is the main principle of the idea screening phase. Ideas that were generated for the previous phase have been prioritized by correlation of market opportunity and operational feasibility and almost categorized as the star, cash cow, question and dog according to Boston Consulting Group: BCG recommendation. But the process cannot be applied with life sciences product/service, because it was selected by the readiness for operation such as good manufacturing practices (GMP) certification by food and drug association (FDA) approval. Although the company has the star category product/service, it cannot produce anything without FDA approval. As the above problem statement, A qualitative method and content analysis were used in this study to find a suitable process for the development of life sciences product/service. The study found a suitable process of development for life science products as pharmaceutical products and cosmetics. Although the results of the study derived from credible resources, quantitative methods are required to confirm the suitability and generalizability of the model in the future.

Keywords: Thailand's life sciences · New product/service development process · Thailand center of excellence for life sciences (public organization) · TCELS

E. Markopoulos et al. (Eds.): AHFE 2020, AISC 1218, pp. 200–205, 2020.
https://doi.org/10.1007/978-3-030-51626-0_24

1 Introduction

Under small and medium enterprises (SMEs) promotion of Thailand's government policy, the government pushes many interventions to many industries of SMEs to increase the number and potentials. According to business start-up factors and innovation strategies research that focus on Hi-Tech SMEs in Thailand [1], the results show that the firm's performances strongly correlate with the score of business start-up factor or strategy execution. Specifically for innovation, entrepreneur(s) who initiates business almost explored their ideas from individual creativity to run the business but not enough to achieve, specific experience in business is required [2].

Although innovations are used as key tools for strategy execution, the success rate of innovation depends on the specific firm environment that resulted in firm's acceptance or rejection in the zero phase of Inno-Kinetic [3]. After innovation's acceptance, their staffs need to improve it as new innovation for better market opportunities. NPD processes, the sequence of steps or activities which an enterprise employs to conceive, design, and commercialize a product [4], are required to use for product/service improvement and launch.

The life sciences (LS) industry in Thailand is growing rapidly. With increased competitiveness and significant expansion of top-notch medical services, Thailand attracts a large number of medical tourists every year. Thailand holds more than 40% share in medical tourist arrivals in Asia. Tectura, a worldwide business consulting company, has projected in its report that Thailand will have the 8th largest pharmaceutical market in the Asia Pacific region by 2016. In the past decade and a half, the number of private-sector biotechnology companies in Thailand has increased at an average annual rate of 11%. With the remarkable paradigm shift envisaged in the ASEAN Plan of Action for Science and Technology 2016–20, the LS industry in Thailand is sure to scale unprecedented heights. Thailand Center of Excellence for Life Sciences (Public Organization), TCELS, was established in 2004 by the government of Thailand with the core intention of providing the central link between innovation and investment and facilitating domestic and international partnerships in LS business in Thailand. It is a public organization and functions under the auspices of the minister of higher education, science, research and innovation [5].

From past experience of TCELS, we found the evidence that may improve NPD processes from the classic model; BAH, Stage-Gate and Ulrich-Eppinger model, to fit with Thailand's LS product development as explained in the following sections.

2 New Product/Service Development Process: NPD Process

According to Cooper [6], NPD key success factors in the project-level compose with 7 elements: intention to outstanding product(service) improvement, real market demand data, development process, speed of responsiveness to product development, planning and organizing resources, quality of implementation, and speed of operation. Moreover, the human factor as a project leader and team member and suitable environment are required for a high-performance NPD project. However, innovation strategy;

strategic planning for innovation, is a critical factor to support targeted oriented NPD project and portfolio management.

Stage-Gate® model was proposed in 2001 to be the general model for NPD [7]. The model categorizes the development step into 2 types which are Stage and Gate. The stage processes for development compose with 5 parts as scoping, building the business case, development process, testing and validation, and launch. Another type is Gate, that means the process of screening and making decision, compose of idea screening, second screening, the decision for development, the decision for testing, and decision for launching.

In 2008, NPD process that focuses on product development was proposed by Ulrich and Eppinger [4]. They constructed model with 6 phases by operation concept, phase 0:planning, 1st phase: concept development, 2nd phase: system-level design, 3rd phase: detail design, 4th phase: testing and refinement, and 5th phase: production ramp-up. Although the screening/decision process was not shown in the model as Stage-Gate®, the process was applied to all phases and feedback mechanisms.

There are many LS companies in Thailand used either of Stage-Gate® or Ulrich and Eppinger model to develop their new products but the results are inappropriate to be acceptable, especially in pharmaceutical products and medical devices. The cause of failure was analyzed by stakeholder and found 4 limitations as follow. The first reason is pharmaceutical product development, the medical device is included, concern on technology-driven more than market-driven. Basically characteristic of them is the lowest human risk product, all compounds were approved both in vitro and in vivo before the clinical trial. International standard operation system certification applies to all operational process again and again as Good Manufacturing Practices: GMPs, Good Laboratory Practices: GLPs or ISO. Although there is much information that easy to assess but asymmetry information phenomena always happen in this business which is considered as the second limitation. The majority of end-user lose decision power to choose the pharmaceutical product or medical device that allocate to medical personnel such as doctors, dentists, pharmacists, and nurses.

The third, there are so many laws and regulation enforcement on pharmaceutical products and medical devices in which affect creativity. The standard initiation phase of NPD process starts with creativity such as idea generation but can not be applied to them. The developer can generate an idea in the scope of law and regulation. In the case of pharmaceutical product, the developer must choose the type of product, new drug, new generic drug, generic drug, or etc., before starting the concept development stage. Every product type has a specific product development protocol and after the developer makes a decision to choose, the specific protocol will be proposed automatically. The last limitation has a very high risk of investment for product development. Numerous cost of investment, both of in cash and in-kind, are required for NPD process in uncertainty return situation. In 2010, the research and development cost of each new molecular entity was about US$1.8 billion [8] with success rate 19.47% from 798 drug discovery project that took place between 1991 and 2015 at 36 academic institutions in the United States [9] while success rate from start as new compound in 15 years (2000–2015) is 1.7–8.3% [10].

With the previous limitation, risk of LS-NPD process, high technology product-based, can reduce with intervention to 4 entrepreneurial factors: opportunity, resources, entrepreneurial team, and context, at the phase of initial [11].

Under the context approach as bio-technology cluster, both of vertical and horizontal development are significant advantage to effect on NPD succession [12]. Merger and acquisition as horizontal integration can build form's capacity passed cost reduction and market expansion at the same time [13]. Although strategic alliance for LS product development, therapeutic area alliance, data mining alliance, technology development, technology transfer, and measuring performance alliance, was proposed as key factors for dynamic response [14] but in the real case, LS product development has to focus on specific resources and firm's core competency from the early stage then almost strategic alliances in LS product tend to apply merger or acquired before starting the NPD project and strategy. Moreover, LS product was recommended to minimize at the first [15].

As mention above, the Conceptual model for LS NPD process development is constructed as Fig. 1.

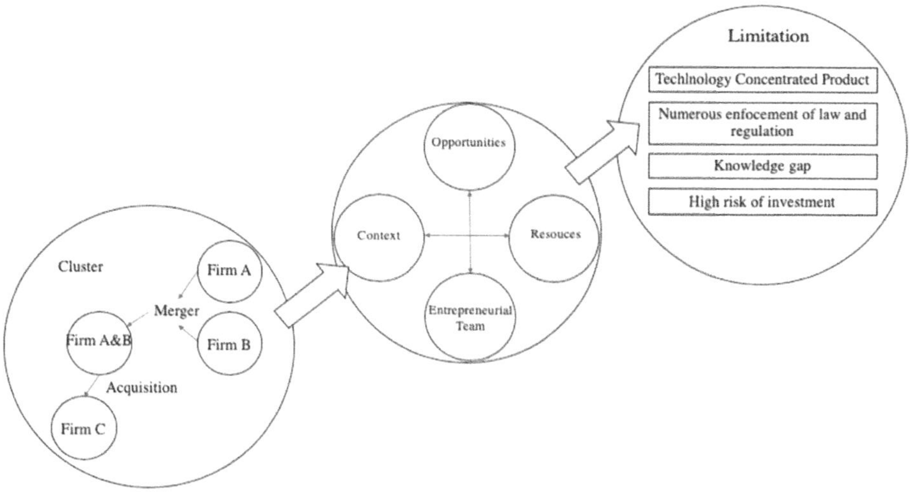

Fig. 1. Conceptual model for LS NPD process development

3 LS NPD Process

According to the development of sustainable bioprocess [16], 3 keys components are required as the principle. Process design and development under regulation is the first requirement. Under the Thai regulatory system of LS product, regulations were classified by the act, medicine act B.E. 2510, cosmetics act B.E. 2535 and medical device act B.E. 2531. Regulation is a critical factor that limits all processes of development especially on the scoping process [17].

After process design is developed, modeling the process and process simulation are required for confirmation of operation feasibility. Reference to Pore Skin® artificial dermis [18], pre-research development and research development are the critical path of NPD show as Fig. 2.

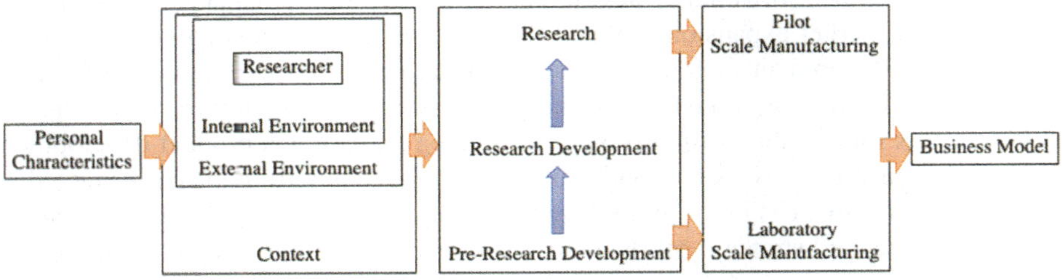

Fig. 2. Pore Skin® artificial dermis's development pathway

The last requirement is the sustainability assessment. The technology roadmap is used to guide new technology development for (or) technology selection in developing new products [19] as future analysis.

4 Lesson Learned from TCELS

More than 10 years of experience in life science business development of TCELS, the general NPD process is not fit in the current situation of life sciences product/service development and should to modified to solve the problems as follow: Life sciences product/service has specific criteria under law and regulation; drug, cosmetic, or medical device. Negotiation or comptonization with product/service standards by law and/or regulation are not allowed, such as, the standard of protocol development, the active compound, plant, operation process, pre-clinical trial and clinical trial, that make a mal-function effect on the phase of idea generation. Then creativity of life sciences product/service has been scoped by the frame of law and/or regulation that prohibited the creativity of the box.

The study found a suitable process of development that composes 5 phases for pharmaceutical product and medical devices; 1st phase: law and regulation assessment, 2nd phase: technology readiness assessment, 3rd phase: process adjustment, 4th phase: FDA approval and 5th phase: run the business. In the case of the cosmetic product, we found only 4 phases in the model, 1st phase: law and regulation assessment, 2nd phase: market assessment, 3rd process FDA approval, and 4th phase: building the business case. Although the results of the study derived from credible resources, quantitative methods are required to confirm the suitability and generalizability of the model in the future.

Acknowledgments. I am grateful to Atthaphon Mumi, Ph.D. who contributed to and moderated this paper and in that line improved the manuscript significantly.

References

1. Intrama, V.: Business start-up factors and innovation strategies, case study: hi-tech SME in Thailand. Int. J. Comput. Internet Manage. **22**(1), 53–61 (2014)
2. Intrama, V., Sookbanjung, S., Santawee, K., Teeraswasdi, P.: Path of creativity in entrepreneur: basic concept of creative economy development. In: The 7th International Conference on Applied Human Factors and Ergonomics (AHFE 2016), Florida (2016)
3. Intrama, V., Uiyamatiti, G., Vadhanasindhu, P.: Inno-kinetic: innovative approach for strategic management. In: The 10th International Conference on Applied Human Factors and Ergonomics (AHFE 2019), Washington, USA (2019)
4. Ulrich, K.T., Epppinger, S.D.: Product Design and Development. Wiley, New Jersey (2008)
5. http://www.tcels.or.th/AboutTCELS/Overview/History. Accessed 15 Jan 2020
6. Cooper, R.G.: Product Leadership: Pathways to Profitable Innovation, New York. Basic Books, NY (2005)
7. Cooper, R.G.: Winning at New Products: Accelerating the Process from Idea to Launch. Perseus Publishing, MA (2001)
8. Paul, M.S., Mytelka, S.D., Dunwiddie, T.C., Persinger, C.C., Munos, H.B., Lindborg, R.S., Schacht, L.A.: How to improve R&D productivity: the pharmaceutical industry's grand challenge. Nat. Rev. **9**(3), 203–214 (2010)
9. Takebe, T., Imai, R., Ono, S.: The current status of drug discovery and development as originated in united states academia: the influence of industrial and academic collaboration on drug discovery and development. Clin. Transl. Sci. **11**(6), 597–606 (2018)
10. Wong, C.H., Siah, K.W., Lo, A.W.: Estimation of clinical trial success rates and related parameters. Biostatistics **2**, 273–286 (2018)
11. Dorf, R.G., Byers, T.H.: Technology Ventures: From Idea to Enterprise. McGraw-Hill, Singapore (2008)
12. Chiesa, V., Chiaroni, D.: Industrial Clusters in Biotechnology : Driving Forces, Development Processes and Management Practices. Imperial College Press, London (2005)
13. Burns, M.H. (ed.): The Business of Healthcare Innovation. Cambridge University Press, New York (2005)
14. Angwin, D., Sammut-Bonnici, T.: Strategic Alliances. In: Wiley Encyclopedia of Management, vol. 12, Wiley (2005)
15. Pfeffer, J., Hatano, T., Santalainen, T.: Producing sustainable competitive advantage through the effective management of people [and executive commentary]. Acad. Manage. Execut. **19** (4), 95–108 (2005)
16. Heinzle, E., Biwer, A.P., Cooney, C.L.: Development of Sustainable Processes: Modeling and Assessment. Wiley, West Sussex (2006)
17. Intrama, V., Karnchanatat, A., Bunaprasert, T., Vadhanasindhu, P.: Critical effect of regulation on Thailand's cosmeceutical development process: human placenta extract. Int. J. Manage. Bus. Stud. **1**(4), 96–99 (2011)
18. Bunaprasert, T.: Development of Pore Skin(R) Artificial Dermis, The First Prototype of Thailand. Bangkok (2008)
19. Belliveau, P., Griffin, A., Somermeyer, S. (eds.): The PDMA Toolbook of New Product Development. Wiley, New York (2002)

A Study on Gamification Product Service System Design of an Intelligent Parking Management System

Jie Xiao, Shuhan Luo, Jue Li, and Long Liu(✉)

College of Design and Innovation, Tongji University, 281 Fuxin Road, Shanghai 200092, China
liulong@tongji.edu.cn

Abstract. The quick development of private car use in China has caused serious problems in parking management. The significant lack in parking space due to unbalanced distribution of urban land use is amplified by the suboptimal utilization of this limited resource. There is no effective parking management system and there is a good opportunity for design of a product and/or service to facilitate the parking. This paper proposed a solution using the gamification approach to design an intelligent parking management system in the framework of the product service system design (PSSD), to provide real time information for the drivers finding a suitable parking spot. The gamification mechanism and the PSSD framework with the key touch points for the drivers to interact with the system are presented in this paper.

Keywords: Gamification design · Parking management · Service system

1 Introduction

China is observing very quick increase of car use in recent years. By 2019, China has the largest fleet of motor vehicles in the world, with 340 million motor vehicles including 250 million cars, of which more than 200 million private cars. Nowadays, 11 Chinese cities have more than 3 million cars and 2 have even more than 5 million [1]. However, as there is delay in urban land use planning to face up the increased car use, the shortage in parking space has become a serious problem in most of the Chinese big cities. The ratio between the number of cars and parking spots is from 1:1.2 to 1:1.5 in the developed countries, while it is only 1:0.6 in China [2]. There is obviously very limited city space for parking, and this situation is further worsened by the unbalanced utilization of the space for parking. The current parking management system is isolated for individual parking lot management and is unable to optimize the utilization of the parking space as a whole [3]. Therefore, a more effective parking management system covering a wider range of parking lots for better utilization of the valuable resource should be developed.

With the quick development of information technology and its applications especially for mobile devices used by most of the Chinese people, it is suggested that a product service system (PSS) for an intelligent parking management system (PMS) to

E. Markopoulos et al. (Eds.): AHFE 2020, AISC 1218, pp. 206–212, 2020.
https://doi.org/10.1007/978-3-030-51626-0_25

reconcile different stakeholders to match the parking demands and supply should be a good solution. This is feasible with the ongoing reconstruction of the infrastructure in many Chinese cities to apply the concept of *City Brain*, which uses artificial intelligence (AI) with big data and computing power to collect and process information in real time to optimize the running of a city efficiently [4].

This paper presents a study on establishing a PSS for parking management to provide drivers with information on parking lots and a gamification mechanism to motivate them to select a suitable parking location based on their balance among time, distance, parking fee, etc. The gamification mechanism, the PSS framework with certain key touch points for the drivers to interact with the system will be presented in the following sections of this paper.

2 Gamification Approach for Parking Management System

Gamification design refers to the application of game methods to non-game situations to guide the users to discover the fun of using the product by stimulating user's engagement, loyalty and commitment. The core of gamification design is to use the game mechanism to enhance the user experience to attract and inspire users to achieve the intended goals [5].

J. McGonigal presented four abstract features of the game as: goals, rules, feedback systems and voluntary participation which encourage users to generate positive emotions during their interaction with the products and services, in order to improve user experience as the key [6]. Other models such as the Fogg's Behavior Model which emphasize the factors like motivation, ability and triggers are also considered in the study to formulate the factors influencing the gamification design [7]:

Challenges and Tasks: Users achieve their goals by overcoming the challenges set by the gamification system;
Bonus points: Users can join the points system to accumulate their credits or compete with others in the system to obtain satisfaction;
User medals: The awards and medals can differentiate user levels and further give users a sense of exclusivity and honor;
Friendly Restrictions: Adding some restrictions to reduce the choices to speed up users' decision and increase the ease in use;
Social Participation: Users can participate in the process with communication with others to promote the long-term use.

These factors can be well integrated into the parking management system to improve the user's experience when accomplishing the task of choosing a suitable parking spot. As parking is essentially a decision on a suitable place for a car, a gamification mechanism can be designed to adjust the relationship between time, distance, parking fee, and other issues, depending on the supply of parking resources, to motivate the drivers to choose a location to optimize the utilization of parking space as a whole. The approaches of product service system design (PSSD) [8] are also applied in the study to define the complex system in a clearer way.

3 Gamification Design of Parking Management System

3.1 Stakeholders and Their Requirements

Parking management involve several parties with different roles and benefits as well as complex relationships between each other. Five types of stakeholders are identified as relevant to parking management, they are: 1. The drivers; 2. The parking lot management party including the parking lot owner and the parking lot management staff; 3. The regulatory organization responsible for planning, construction, and administration of the parking facilities; 4. Support platform for payment, map, navigation, etc.; and 5. Third party service providers of hotels, restaurants, shops, gasoline stations, super markets, etc. in the near. Their relationships are illustrated in Fig. 1.

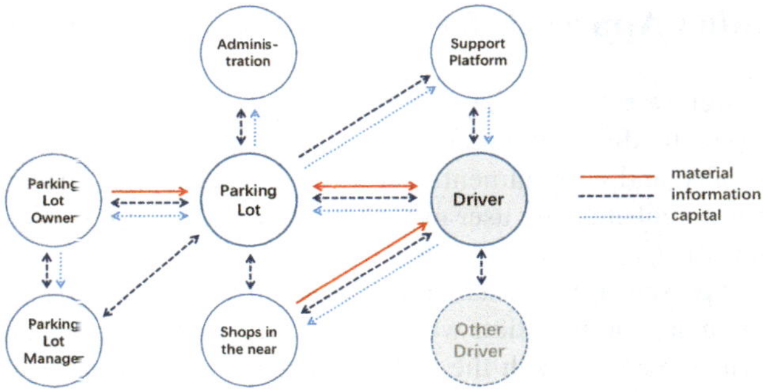

Fig. 1. The relationships of different stakeholders in the parking management system

The role of the stakeholders in the PMS can be further analyzed according to their influencing power on the PMS and their interest in the PMS. As shown in the 2-dimentional diagram in Fig. 2, the driver and the parking lot management are the two most relevant stakeholders in the PMS. The drivers are the direct users and their parking behavior and requirements have significant influence on the effectiveness of the PMS. The parking lot owner and management staff directly run the parking lot and their requirements determine their engagement in the system. Their requirements are then further obtained and analyzed using the questionnaire and interviews.

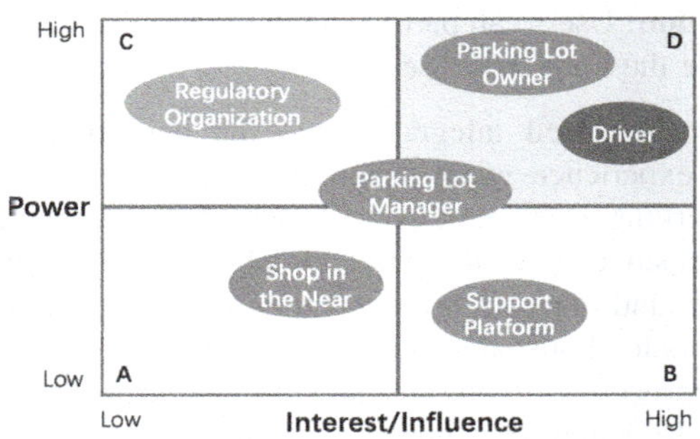

Fig. 2. The role of different stakeholders in the parking management system

The questionnaire investigation of the drivers' requirements was conducted through an online survey with 19 questions regarding the personal information, factors influencing parking decision, requirements on PMS, and previous use experience with mobile applications, etc. Totally 193 questionnaires were collected and the initial results were summarized as the basis for the PMS gamification design. Additional in-depth interviews with 4 drivers were further conducted to reveal their specific requirements on the parking information service provision.

The interviews with the parking lot management staff were conducted with a parking lot manager and a parking management expert. The running of the PMS with its functions and design improvements especially the application of information technologies were extensively discussed.

3.2 Gamification Design Aspects of Parking Management System

It is suggested that the development of an effective PMS should incorporate information technology to reconcile different stakeholders to provide parking management service in a wider scale. In contrast to some existing parking managements systems which focus only on one or two aspects of parking e.g. payment, parking spot lock, car finding, etc., the newly developed PMS should be able to connect more relevant stakeholders to provide parking information in an extensive scale, both in term of the parking lots covered and in term of the diversity of the service provisions. In this sense, the PMS can be integrated as part of the City Brain to specifically optimize the parking issue of a city, with improved interaction with individual drivers to facilitate them finding a suitable parking spot for a specific trip.

The attractiveness of the PMS is largely determined by its interaction with the end-users. As some factors like parking fee, walking distance, queue waiting time, ease of parking, etc. have large influence on the drivers' parking decision and experience, it is possible to adjust the sensibility of these factors through a gamification design to promote the drivers' parking behavior in a way to improve the overall parking space utilization. This idea is the core of an intelligent PMS design and is the basis of the gamification mechanism. For better understanding the relationship between these factors, a small test using the Mental Accounting model proposed by R. Thaler [9] was conducted with 18 drivers. A gamification design framework of the intelligent PMS with 4 general design aspects is then formulated (Fig. 3).

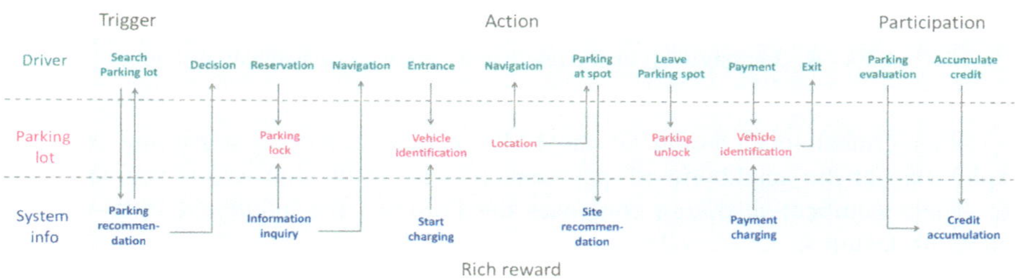

Fig. 3. Gamification design aspects of the intelligent parking management system

Trigger: to recommend the available and suitable parking lots on the mobile phone when the driver search for a nearby parking lot;
Action: to facilitate the driver interacting with the system from searching a parking spot to finding one and finally parking there;
Rich Rewards: to motivate the driver choosing a suitable spot for optimizing the parking utilization by integrating a multiple reward mechanism by adjusting price depending on time and distance, etc. Big data can be used to learn the drivers' parking habits to reduce their decision time and increase satisfaction;
Participation: to encourage more drivers' participation by facilitating the drivers share of their experience and recommendation to others.

4 PSSD for the Intelligent Parking Management System

The customer journey of the drivers interacting with the PMS is illustrated using the PSSD tool (Fig. 4). Critical touch points of the drivers interacting with the system based on the gamification mechanism can be formulated. The most important point is first step of deciding on a suitable parking lot for the trip. It is the trigger of the driver to use the system and is the most important to influence the effectiveness of the system and the user experience. The gamification design factors of challenges, bonus points, user medals, restriction, and social participation should be well designed based on the task requirements on reaching the decision effectively in short time with satisfaction. The availability of restricted amount of parking lots is presented with clear real time information on time and distance as well as discount in fee and bonus, etc. A ranking of different parking lots based on evaluation and recommendation as well as the driver's preference could support the driver's decision in a usable and pleasant way.

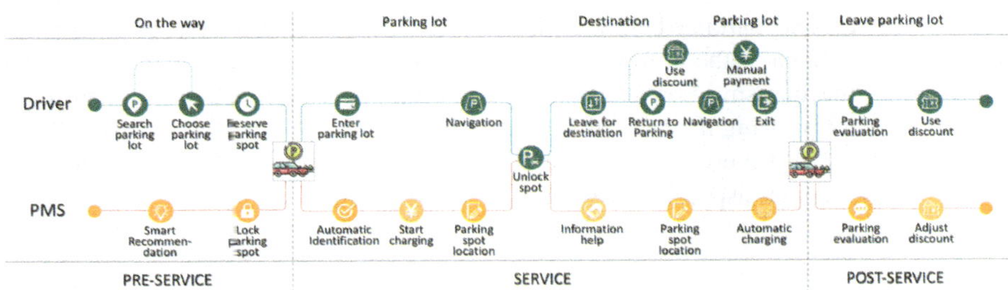

Fig. 4. Journey map of the intelligent parking management system

Other critical touch points of the PMS include: reserving a parking spot, site exploration in the neighborhood, payment, parking evaluation and recommendation, etc. Their gamification design composes the PSSD of the intelligent PMS with the following features:

- To support the drivers' parking decision based on integrated gamification mechanism of achieving certain tasks by balancing the time, distance, parking fee, etc.;
- To encourage the drivers to achieve certain tasks by applying gamification settings of bonus points, user medals, competition, etc. to continuously motivate the drivers to use the PMS for parking;
- To apply the artificial intelligent (AI) technology to ease the interaction between the drivers and the system based on their use habits and to connect the drivers' gamification behavior closely with the overall optimal utilization of the parking space;
- To incorporate more stakeholders in the gamification design framework to provide the drivers with more options to play and in turn to improve the design.

5 Conclusions

The solution presented in the paper of developing an intelligent parking management system covering an increasing number of parking lots with sufficient drivers as active end users provides an effective solution to improve the utilization of overall parking space, and at the same time to improve the parking experience of the user through the gamification design approach. It is clear, with the ongoing reconstruction of the city infrastructure, the intelligent PMS can play a more important role to promote future smart city development. The gamification design of the system can also provide more service options which can be further developed as realistic business applications.

Although the implement of this system still needs much effort to reconcile the benefits of different stakeholders, its promising business prospects could attract more participators to help it come true.

In short, gamification design plays a very important role in the design of the intelligent parking management system with high value to enhance users' behavior change and durable service experience. The study presented in this paper provides a good PSSD framework to support the designer to develop an intelligent parking management system in the future. The approach of applying gamification design to facilitate drivers' interacting with the system also provide good support to the designers to arrange their next design practices.

References

1. The Ministry of Public Security: Report on the national vehicle ownership in the first half of 2019 (2019). http://www.gov.cn/xinwen/2019-07/04/content_5406302.htm. (in Chinese)
2. Yan, K.-F.: Three fundamental measures to solve the parking problem of the cities. Transport Communi. **32**(4), 10–12 (2016). (in Chinese)
3. Wang, H.-L.: Analysis of the problem with parking in big cities. Gansu Sci. Technol. **27**(6), 5–6 (2011). (in Chinese)
4. Hua, X-S.: The city brain: towards real-time search for the real-world. In: SIGIR 2018: The 41st International ACM SIGIR Conference on Research & Development in Information Retrieval, June 2018, pp. 1343–1344 (2018)

5. Zichermann, G., Cunningham, C.: Gamification by design: implementing game mechanics in web and mobile apps. O'Reilly Media, Inc. (2011)
6. McGonigal, J.: Reality is broken: Why games make us better and how they can change the world. Penguin (2011)
7. Fogg, B. J.: BJ Fogg's behavior model. A behavior model for persuasive design (2011)
8. Stickdorn, M., Schneider, J.: This is Service Design Thinking: Basics - Tools – Cases. BIS Publishers, Amsterdam (2011)
9. Thaler, R.: Mental accounting and consumer choice. Market. Sci. **4**(3), 199–214 (1985)

E-Sports Hotel, a Brand New Personalized Hotel in the Age of E-Sports

Lu Xiaoxian(✉)

Department of Industrial Design, School of Mechanical Science and Engineering, Huazhong University of Science and Technology, Wuhan, People's Republic of China
1769268186@qq.com

Abstract. As a brand-new business model, e-sports hotels have no industry norms or general industry labels. As long as the rooms have e-sports facilities and allow consumers to conduct e-sports activities, they can be considered as e-sports hotels. Although many hotels claim to be e-sports hotels, they only move computers into the room mechanically, or turn the room into a mini Internet cafe. This model does not take into account the needs of users, so it cannot bring users a satisfactory experience. So we decided to create a hotel specifically for e-sports.

Keywords: E-sports · Hotel · Innovation · Entrepreneurship · Environmental psychology · User experience

1 Introduction

E-sports is a confrontation sport based on e-sports games, information technology, hardware and software devices. It's played in the virtual environment created by information technology and complied within a unified competition rules. E-sports is a brand new sport."

Affected by the 1997 Asian financial crisis, South Korea adjusted its industrial structure. After the release of StarCraft in 1998, South Korea took the opportunity of "StarCraft",it related TV programs to develop an amazing Korean e-sports industry. In 2000, Korea's e-sports industry Korea International Electronic Marketing Company founded the World Cyber Games, commonly known as WCG, which is one of the most influential e-sports events in the world. It's played a extremely important role for the promotion of e-sports.

In 2004, the annual output value of the South Korean e-sports industry reached about 4 billion US dollars, and the value of the relevant industrial chain even exceeded the South Korean's automobile industry.

In 2017, China's e-sports showed explosive growth, and the overall market scale exceeded 65 billion yuan [1].

The overall industry chain of the e-sports industry includes e-sports game development, e-sports operations, e-sports clubs, live broadcast platforms and other derivatives from top to bottom. In addition, the e-sports industry has also formed and

E. Markopoulos et al. (Eds.): AHFE 2020, AISC 1218, pp. 213–218, 2020.
https://doi.org/10.1007/978-3-030-51626-0_26

integrated with other industries, bringing more business opportunities, spawning many new business models and creating more business value.

Today, as resources become more and more restrictive to economic development, the concept of experiencing economy provides a new way of thinking for economic development. Experience and service have become the core competitiveness of goods. The e-sports industry is naturally a highly engaged and highly experienced industry. Whether it is an e-sports game, an electric competition or an e-sports live broadcast, both required people to participate in it. Although only a very small number of people can participate in events such as the WCG, people's love for this will not fade. E-ports became an indispensable part of the daily life of players. Ball sports enthusiasts can engage in their favorite projects in stadiums, stadiums, etc. E-sports enthusiasts also have their own activities for the venue.

In the early days, when personal computers were not very popular, Internet cafes were a gathering place for e-sports enthusiasts. After the popularity of personal computers, enthusiasts not only limited by computer configuration and other factors, but also can not give up the feeling of face-to-face offline games, so Internet cafes are still the most important e-sports venue. With the improvement of living standards, traditional Internet cafes have not kept up with the needs of the times, and e-sports consumers urgently need a higher quality e-sports sports space.

In recent years, the traditional hotel industry is in a low state. Various types of homestays and apartments have diverted the source of traditional hotels with exquisite decoration and low prices. The fast-developing e-sports industry has brought new possibilities to the traditional hotel industry. When the e-sports industry merged with the traditional hospitality industry, a new personalized hotel, the E-sports hotel, was born.

As a brand new business model, so far, e-sports hotels have not yet formed uniform standards. As long as there is an e-sports facility in the room and allows consumers to conduct e-sports can count as an e-sport hotel. Just because E-sports Hotel is a new business model that has not yet formed a norm, its market demand is huge:

1 E-sports Hotel is In short supply. Overall, taking Wuhan as an example, the number of e-sports hotels is far from meeting market demand. As a populous city with abundant resources, Wuhan has a huge population of young people. However, when I started this research, there were only two E-sports Hotel in the Wuhan area, and there was no e-sports hotel in the Guanggu area.

2 Current E-sports Hotels' quality is not good enough. The existing E-sports hotel just added e-sports equipment to the hotel and it's no longer able to satisfy consumers' growing e-sports experience.

Consumers are in desperate need of hotels designed specifically for e-sports, which can give consumers an excellent e-sports experience.

Through research, we found relevant factors that affect the e-sports experience of e-sports consumers, and created a brand new E-sports Hotel.

2 User Research

2.1 Target Segmentation

Users need to generate user needs, and user goals are specific manifestations of user motivation. Therefore, in order to make the behavioral situation best fit the user's psychological expectations, we must first understand the user's goals.

Because e-sports hotels have functional diversity, the goals of e-sports hotel users are also diverse, overlapping, and overlapping. For example, a couple of lovers staying in an e-sports hotel, one of the goals is naturally for e-sports activities, and the other is to use the e-sports hotel as a dating place. Therefore, the goals of the same and same group of e-sports hotel users are also complex and different. So, we have disassembled and distinguished the main goals as follow: e-sports goals, dating goals, social goals, competition goals, business goals, travel goals, and leisure goals. The goal of e-sports is to entertain e-sports activities in e-sports hotels; the goal of dating is to use e-sports hotels as dating places; the goal of social goals is to use e-sports hotels as places for friends meeting; the goal of the competition is to achieve certain e-sports results, not simply to enjoy and entertain; the business goals are the goals of earning revenue through e-sports; the travel goal is to stay for business or travel; the rest goal is the goal that needs to be rested in the e-sports hotel.

2.2 User Segmentation

We dived the users who stay in the same room as a cluster as follow: dating users, social users, professional users, and other users.

Dating users and social users are consistent with the previous description. It is better to understand.

Professional users refer to users who have clear requirements for e-sports performance and users who use e-sports as a source of revenue. The needs of the competition target users and commercial target users are highly consistent, that is, the level of e-sports. Therefore, the environmental manifestations required by the two target users are basically the same. So the two types of target users can stay in the same type of room. Therefore, we divides these two types of users into the same type.

Other users include: business users, travel users, and users who stay in for other reasons. The number of such users is small and the length of stay in the hotel is also short.

2.3 User Context Segmentation

We have segmented the goals, needs, and types of e-sports hotel users, and segmented e-sports hotel users into dating users, social users, professional users, and other users. The demand points of dating users include: e-sports, dating, and leisure; the demand points of social users include: e-sports, social, and leisure; the demand points of professional users include: e-sports, leisure, computer use and rest. The enjoyment of e-sports and the level of e-sports do not conflict, and they can coexist. While enjoying e-sports activities, only by guaranteeing the user's level of play can users get a satisfactory experience. Therefore, although the starting point of users for e-sports may be different, the design embodiment is the same. So the basic functions that e-sports hotels should have are: e-sports equipment, dating, socializing, and rest.

Therefore, the author divides the demand situation of e-sports hotels into e-sports situations, dating situations, social situations, and rest situations.

3 Design

We embarked on related research on e-sports hotels, and reached a cooperation with Greentree Inn, established Greentree E-sports Hotel, and opened the first Greentree E-sports Hotel in Wuhan, Hubei Province, China.

3.1 Idea

We set the core keywords of the overall space perception of Greentree E-sports Hotel as: passion, youth, leisure and comfort. Hope that users can enjoy passionate e-sports and life in the young age, Greentree E-sports Hotel will provide users with a comfortable staying experience at the same time. We will build the overall environmental space of the hotel, including structure, light, color, texture, and breath to provide users with a passionate, young, casual, and comfortable experience.

3.2 Renderings

Figures 1 and 2.

Fig. 1. Lobby.

Fig. 2. Room.

4 Practice

4.1 Location

After many investigations, we selected No. 01, Block B, located in Yangguang Meishu, Zhuodaoquan East Second Road. It has a well-developed business and a large population, and is a gathering place for young people. Have a good environment and convenient transportation.

4.2 Results Display

Figures 3 and 4.

Fig. 3. Lobby.

Fig. 4. Room.

References

1. Martoncik, M.: e-Sports: playing just for fun or playing to satisfy life goals? Comput. Hum. Behav. **48**, 208–211 (2015)
2. Wibel, H., Donoghue, J.D.F., DeFeo, A., Yao, S.: An osteopathic physician's approach to the esports athlete. Spec. Commun. **119**, 756–762 (2019)
3. Jablonska, J., Trocka-Leszczynska, E., Tarczewski, R.: Impact of New Construction Technologies on Sustainable Hotel Design (2016)
4. The scale of China E-sports market in 2015–2019. iResearch (2019)

The Global Constrains of South African Automotive Industry and a Way Forward

John M. Ikome(✉) and O. T. Laseinde

University of Johannesburg, Johannesburg, South Africa
Ikome20022000@yahoo.com

Abstract. Global competitiveness is a very important part of any nation, particularly the south African automotive industry were the customers demand is constantly changing and radical decisions are needed in order for the organization to stay floating among the big auto players in the world. This justifies the fact that a decision support model is eminent for the success of the South African automotive industry and growth of the country's economy.

Keywords: Competitiveness · South Africa · Automotive industry · Industrialization · Models

1 Introduction

The Automotive industry of South Africa is believed to be one of the leading industries to foster economic growth but due to the complex nature of globalization, it is faced with proliferation pressure in order to stay within the market trend were customers demand are constantly changing. According to a literature review done by John Ikome and Grace Kanakana [16], states that the South African Automotive Industry needs to reduce production costs while maintaining product quality and also decision makers to make the right decision to support and improve the various process in order to be competitive globally. Furthermore, Kehbila, et al. [13], stated that sustainable productivity improvement calls for improvement processes whereby increased productivity as well as the betterment of the well-being for employees is taken into consideration.

Due to the constantly changing demands of customers and intricate nature of markets and competitive strategies today, there is a lot of competition globally for both developing and developed countries. This in general requires a high level of competitiveness, which will include a rational policy from decision makers in order to make the economic of the country competitive and coping with global risks. Ikome and Kanakana [16] further states that, in most industrialized countries, the automotive industry is regarded as the industry of industries and particularly a key pillar to the national economic growth were all type of vehicles that are needed or used in different sectors of the government and nation, includes (tourism, transportation, road construction, agriculture etc.) are manufactured, making it the heart of any nation. Due to the aforementioned, it is clearly justified that any disruption in the automotive industry can deeply affect the economy of the nation in a very native way. According to Alfaro et al. [2], global automotive industry produces approximately 70 million units each

E. Markopoulos et al. (Eds.): AHFE 2020, AISC 1218, pp. 219–226, 2020.
https://doi.org/10.1007/978-3-030-51626-0_27

year, playing a vital role in the world economy and making important contributions to the well-being of societies. This evidently shows the importance of the automotive industries in developed and developing countries and enhancing the competitiveness in these industries is of crucial importance to the economy of any nation [7].

In addition, the success of a specific industry in a country depends strongly on the national competitiveness of that country (Portar, 1990). Consequently, the infrastructure, health care facilities, educational system, the macroeconomic environment and other institutions can greatly affect the competitiveness of a nation's industries [11]. For instance, a well-developed transportation infrastructure and communication network may be a prerequisite for having access to core economic services and activities. Therefore, the infrastructure is highly likely to influence the success of a specific industry. Furthermore, the fact that the macroeconomic environment influences the related microeconomic and firm's operational level conditions, it plays an important role in the success of any industry. Business investment is also critical to productivity.

Therefore, economies with sophisticated financial markets can make capital available for private sector investments [10]. By this logic, it is also possible to demonstrate that the level of the primary and higher education, the labour market efficiency and the financial market development influence the industrial competitiveness in a country. As a result of the aforementioned, there is a strong evidence linking the competitiveness of these industries and the economic wellbeing of a country. This linkage necessitates the development of a competitive model and a framework for decision maker, to analyze and identify policies to support automotive industries that face major foreign competitors.

2 Global Competitiveness of SA

Competing at the nationally level is an important stimulus that helps shapes a firm international competitive position in a country. According to Kao et al. [1], competitiveness at the national level is a measure of the relative ability of a nation to create and maintain an environment for business operation and consequently improve the prosperity level.

Contrast to this, a research work conducted by Baldwin [3], indicates that the chances of South African industries to become globally competitive were still gloomy due to the lack of leadership competitiveness. In the separate incidence, the South African leaders were criticized for failing to attain economic growth levels and made the country to be marginalized in the global economy [4]. Barnes and Kaplinsky [5], argued that for the South African industries to become globally competitive, leadership should be able to embrace both transactional and transformational role.

In addition, Barnes and Morris [6], commented that, given the dramatic change in the South African society, strong transformational leadership to be specific in the manufacturing sector is a requirement. Chopra and Meindl [9], added that South Africa lacked competent and talented leadership that could successfully pioneer industries undergoing transformational phases in this technological era of Industry 4.0. In South Africa, competitive Supply Chain Management (SCM) is one of the important functional areas in the automotive industry and its contribution to the country's economy has been particularly noteworthy [14]. The sector's contribution of 7% of South Africa's gross domestic product (GDP) was R3 251.4 billion in 2012 (AIEC, 2013, p. 7).

However, the industry faces new and pressing competitive supply chain challenges for local manufacturers and internationally uncompetitive (Business Monitor International [8]). Scholars such as Naude and Badenhorst-Weiss [15, p. 70]; Ambe and Badenhorst-Weiss [15, p. 355]; and, as well as publications of Supply Chain Intelligence Report (2009) and the Supply Chain Foresight Report (2010), have identified some of the challenges faced by the automotive industry as: intensified competition in the global market; low levels of collaboration; lack of market sensitivity or reaction to the changing market; pressure from original equipment manufacturers (OEMs) to reduce prices; unreliability of rail transport; rail capacity problems; and the high cost of South African ports.

However, none of these studies indicated the extent to which these challenges are felt by the Automotive Industry or how to overcome them. Considering the profound importance of competitiveness to business success and to the South African (SA) automotive industry [12], it is worth investigating the extent of what is clearly a growing threat (challenge) and how SA automotive industry can best proceed to overcome it.

3 Competing at the National Level

One of the most cited models related to competitive of a nation that has regularly been used is the "Diamond" model of Porter [17]. Conferring to this model, skilled labor or infrastructures and production of a nation are of key important and critical to the competitive advantage of a nation in a given industry.

A number of international organizations, including the World Economic Forum (WEF) and the Institute for Management Development (IMD), have made great efforts to measure national competitiveness. These organizations apply good number of objective and subjective indicators in order to assess the wealth created by the world's nations and subsequently publish rankings of national competitiveness.

The IMD World Competitiveness yearbook has measured 59 countries based on 329 criteria since 1989 (IMD, 2012). A global competitive report of over 100 countries based on 100 criteria was published by WEF, starting from the year 2005, serving as benchmarks to assist policy makers and other interested parties to judge the competitive success of their countries within a global context (WEF, 2012), but still the SA auto industry is still facing challenges in order to use this standards as a blue print to global competitiveness.

4 South African Automotive Industry Globally Competitiveness

According to Zhang and Chen (2006:66), the automotive industry is the union of several companies involved in the Design, Development, Manufacturing, Marketing and selling of motors vehicles. Globalization has further made the industry more complex and cost implication involved in capturing and maintaining a segment in the global market [13] Hence, since the 1990s, South Africa automotive sector is operating by producing and exporting its product in just a few number of countries around the world, using already processed materials from other countries (assembling), this situation gives grounds to the question of how does South Africa fit into this global automotive?

According to an analysis Performed from 2012 to 2015, by South African Trade and industry, shows that SA was ranked, 26th for production of passenger vehicle and 15th for light commercial vehicle production, and above it, are Poland, Iran and Slovakia and immediately behind, Malaysia, Argentina and Hungary. Currently, statistics show that the production of vehicle in SA increased by 1.6% from the 601338 units in 2017 to 610854 unit in 2018, and despite that significant increase, SA has been classified among the countries that producing below one million vehicles a year, which makes the country global vehicle product ranking to remain at 22nd in 2018 with a market share of 0.64%. What is clear is that SA remain the dominant market within the African continent with 54.3% of the total vehicle production. However, as an international global producer, SA still have a lot to do in order to compete with the other countries. Table 1 below shows the yearly contribution of SA automotive industry from 2012 to 2016, while Fig. 1 shows the overall contribution and market trend.

Table 1. South Africa contribution in respect of global production 2012–2016

Year	Contribution vehicles	Passenger	Trucks	Total contribution
2012	0.44%		1.58%	2.02%
2013	0.41%		0.41%	0.82%
2014	0.41%		0.43%	0.84%
2015	0.50%		0.31%	0.81%
2016	0.48%		0.21%	0.69

From Fig. 1, it clearly shows that the SA automotive Industry need a rapid intervention as the total contribution of vehicles declines as the year's go-by. This is evidenced and further justified with the research work done by [9] were they state that SA automotive and Supply Chain is one of the most functional industry of the economics of the country but its contribution has of late particularly noteworthy. According to a report from the (Automotive Industry Export Council {AIEC} 2013:7), this industry contributes about 7% of SA (GDP) which was R3251.4 billion in 2012 but because of intensive competition and the global market trend, requiring shorter product life cycle, the SA automotive industry is facing a huge challenge. Furthermore the Supply Chain Intelligence report (SCIR 2009) also revealed that a huge number of companies within SA do not only operate with a very low collaboration level but also not very sensitive to the market in-terms of strategic reactions during changes in the market trend.

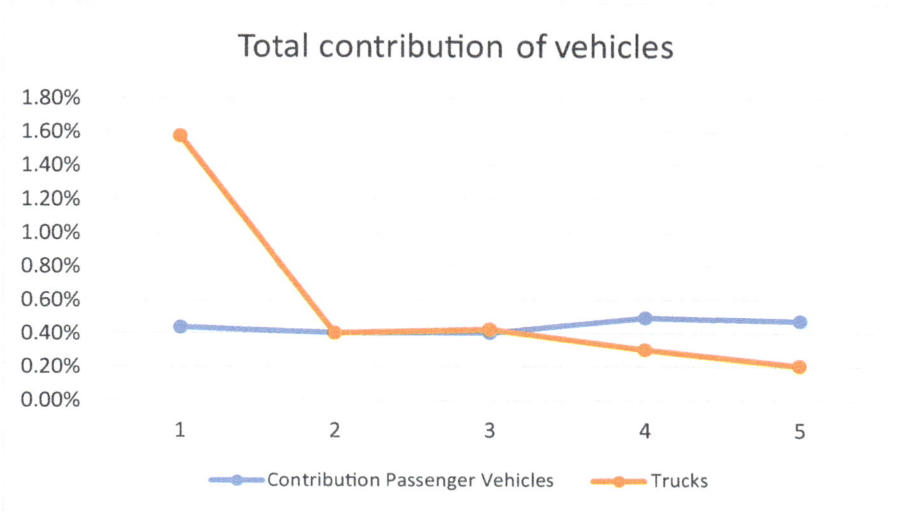

Fig. 1. Total contribution of vehicles, passenger and trucks

The market shares for new vehicles is shown below and it also clealy shows that SA automotive industry needs to develop an aggressive competive strategy in order to stay floating with the big auto players, as it is evidence that the occupy only a small portion of the markert (Fig. 2).

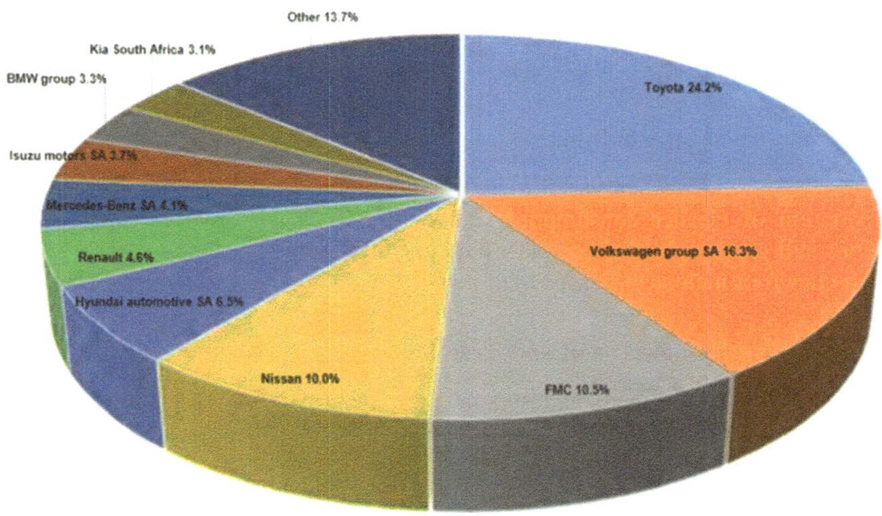

Fig. 2. Overall market shares for new vehicle shares

In an attempt to identify some of these challenges, a conceptualized model was developed as shown on Fig. 3 below. This means if some of these problems can be identify and solutions implemented, SA automotive can manage to compete.

Fig. 3. Challenges of SA auto industry

4.1 Change in Market

The market is evolving on a very basic level, driven by quick innovative improvements in automotive and the development in advanced media applications which were influencing client conduct just as the general automotive industry environment, both discount and retail.

4.2 Infrastructural Challenges

One of the most significant was the introduction of cleaner fuels, which will allow the importation of cleaner consumption and more eco-friendly motors. Most African nations still grant powers to be sold which just meet prerequisites for Euro 2 or three, when numerous nations in Europe as of now expect motors to meet Euro 5 or 6 models.

4.3 Need for Empowerment and Training

The need to make the motor industry far more inclusive through empowerment and training initiatives in both down- and upstream operations.

4.4 Recommendation on How Some of This Challenges Can Be Overcome

Because of the rapid evolution of technology, global automotive manufacturing has changed a lot from the design process, to delivery of the product to the customer.

However, the different challenges currently faced by South Africa automation industry have two main reasons.

The first one is the poor level of commitment to Human resources development. The second reason is the criticality need of South Africa, to advance its own national market. In fact, those two reasons are the key weaknesses of South Africa automotive, which limit the firm to achieve the ability of meeting the international level requirement. Furthermore, to overcome the previous highlighted challenges, SA also need to have skilled work forces (Labour and management), that are committed to the long-term development of that firm. This solution require training people so that they can be aware of new technology and process. In addition, this solution requires a holistic innovation and development start up around the country, in order to bring innovation capacity that will help the firm to expend it services, including an extensive collaboration with foreign companies to develop or extend their activities that will not only help achieving regional expectation but also meet the global expectation and demand.

5 Conclusion

In this paper, it is clear that South African automotive industry is going through a lot of constrains related to global competitiveness and a radical decision need to be taken in order to develop systems in place for continuous improvement; this includes embracing the Fourth Industrial Revolution, research and innovation.

References

1. Kao, C., Wu, W.-Y., Hsieh, W.-J., Wang, T.-Y., Lin, C., Chen, L.-H.: Measuring the national competitiveness of Southeast Asian countries. Euro. J. Oper. Res. **187**(2), 613–628 (2008)
2. Alfaro, A.L., Bizuneh, G., Moore, R., Ueno, S., Wang, R.: South Africa: automotive cluster. Microeconomics of Competitiveness Papers, Harvard Business School, Kennedy School of Government (2012)
3. Baldwin, R.: Trade and Industrialisation after globalisation 2nd unbundling: how building and joining a supply chain are different and why it matters. NBER Working Paper Series, No. 17716, Cambridge (2011)
4. Ballard, R.: Preliminary study on the bovine leather value chain in South Africa. University of KwaZulu Natal, Durban. 16 June 2013 (2001). http://sds.ukzn.ac.za/files/rr_40.pdf
5. Barnes, J., Kaplinsky, R.: Globalization and the death of the local firm? the automobile components sector in South Africa. Regional Stud. **34**(9), 797–812 (2000)
6. Barnes, J., Morris, M.: Staying alive in the global automotive industry: what can developing economies learn from South Africa about linking into global automotive value chains? Euro. J. Dev. Res. **20**(1), 31–55 (2008)
7. Black, A.: Globalisation and restructuring in the South Africa automotive industry. J. Int. Dev. **13**, 779–796 (2011)
8. Business Monitor International: South Africa Autos Report. Business Monitor International, London (2013)
9. Chopra, S., Meindl, P.: Supply Chain Management: Strategy, Planning and Operations, 3rd edn. Pearson, Upper Saddle River (2015)

10. Kgobe, P., Chidi, M., Bwagwan, N.: Research Organisations Skills Plan 2013/14 Update. Centre for Education Policy Development (CEPD). South Africa: Education, Training and Development Practices Sector Education and Training Authority (ETDP SETA) (2012)
11. Gereffi, G., Humphrey, J., Kaplinsky, R., Sturgeon, T.: Introduction: globalisation, value chains and development. IDS Bull. **32**(3), 1–8 (2013)
12. Nag, B., Banerjee, S., Chatterjee, R.: Changing features of the automobile industry in Asia: comparison of production, trade and market structure in selected countries. In: Asia-Pacific Research and Training Network on Trade. (Working paper series, No. 37), July 2007
13. Kehbila, A.G., Ertel, J., Brent, A.C.: Strategic corporate environmental management within the South African automotive industry: motivations, benefits, hurdles. Corporate Soc. Responsib. Environ. Manage. **16**(6), 310–323 (2009)
14. Van der Merwe, J.P.: The state of the automotive sector in South Africa (2009). AIEC (Automotive Industry Export Council). Automotive export manual. NAAMSA, Pretoria, South Africa (2013)
15. Ambe, I.M., Badenhorst-Weiss, J.A.: South African automotive industry: trends and challenges in the supply chain. J. Contemporary Manag. **8**, 337–362 (2011)
16. Ikome, J.M., Kanakana, G.M.: A literature review of South African automotive industry global competitiveness. In: 2018 Portland International Conference on Management of Engineering and Technology (PICMET), Honolulu, HI, 2018, pp. 1–4 (2018)
17. Porter, M.: The Competitive Advantage of Nations. Free Press, New York (1990)

Design of Urban Public Space Through Community Building

Peng Xu(✉)

Tongji University, No. 281 Fuxin Road, Shanghai, China
ben_xp223@tongji.edu.cn

Abstract. This article explores the logic of the production of urban public space and further explains the main misunderstandings. It argues that China's urban public space often follows the rules created by power and capital, neglecting human life. And then it arouses urban residents to create public space from bottom-up in their own way. However, each of the three logics has its advantages and disadvantages. What the key is to build a hybrid community that keeps a dynamic balance between logic of power, capital and life, and promotes organic development. Therefore, it proposes four design models that facilitate community building.

Keywords: Informal · Urban public space · Community building

1 Introduction

Urban public space is the stage for civic life, which reflects the values of freedom, democracy and inclusiveness. The production of urban public space in China follows three layers of logic: power, capital, and life. The logic of power prefers to control and maintain order; the logic of capital pursues profitability and efficiency; the logic of life seeks happiness and well-being. These three logics together determine and influence the performance of urban public space production and management. Urban public spaces in China are more constrained and influenced by the logic of power and capital, which leads to imbalance and alienation of urban public space production [1]. This causes that urban citizens take insurgent actions to produce public space in their way for defending private interests. Unsanctioned, unscripted, and seemingly "undesirable" activities have long appropriated urban spaces in routine and sometimes unexpected ways, bringing new meanings and unforeseen functions to those places [2]. For example, the first-floor residents break the wall to enclose a small garden or run a shop; the café put tables and seats outdoor along the streets, shopkeepers create eye-catching shop signboard to attract people; older people bring their seats to play cards on the streets.

Unfortunately, the local government often views these informal activities as random, disordered, and even illegal actions, and adopts the policy of cutting at one stroke in dealing with these actions. As a result, it falls into the dilemma of *rules to lose* in urban management and law enforcement, which means top-down management would impair bottom-up vitality, while informal activities may bring about messy and chaotic

E. Markopoulos et al. (Eds.): AHFE 2020, AISC 1218, pp. 227–232, 2020.
https://doi.org/10.1007/978-3-030-51626-0_28

without supervision. In the past few years, there have been billions economic loss, millions of SMEs and job opportunities disappeared, owing to the strict law enforcement by the government.

This paper is devoted to explaining the three logic of production of urban public space, and the common misunderstandings of public space might be in China context, and then putting forward a community-building approach.

2 Critiques of Urban Public Space Production

2.1 Power-Oriented Urban Public Space Production

The production of urban public space in China is dominated by the government, from planning, design, construction, to management which reflects a strong power will. What the government pursues are control and order. The advantage is that a monopoly government can rapidly expand cities through implementing related laws, regulations and policies to guarantee the whole system works. However, it is difficult to anticipate and satisfy the need of common people because of the bounded rationality of urban planners. What's more, For the ordinary people, they rarely have the opportunity to express their opinions and preference, and what so-called public participation is often on the surface without actually affecting the decision-making of planning, design, construction and management of urban public. As a result, urban public space has become a space for a certain number of people displaying capital, power, and individual achievements, or become a lost space.

2.2 Capital-Oriented Urban Public Space Production

The capital oriented is that in order to chase profits and efficiency, enterprises exert influence on the planning, design, construction, and operation of urban public space through market mechanisms [3]. The main methods and strategies of capital involved in urban public space production are:

- Buy in the prime locations and advantageous spaces
- Set up an alliance of interests to influence policymaking
- Establish a joint venture company with the government to avoid risks
- Build physical or psychological barriers to isolate from the non-target people

The profit-seeking nature of capital weakens the publicity of public space [4]. Many cities in China, public space has to give way to urban development projects which could bring substantial economic benefits, instead of focusing on space quantity and quality. Those urban spaces with great geographical locations and better natural resources have been developed into real estate, commercial plazas, shopping malls, and government offices, keeping these urban spaces away from the public. In a word, capital oriented public space production invades the public realm and impair public life of urban citizens.

2.3 Life-Oriented Urban Informal Public Space Production

Logic of life is that urban citizens create public space from bottom-up for the need of surviving or pursuing a better life. Because of China's rapid urbanization, a large number of migrant farmers swarmed into the city, due to a lack of knowledge and professional skills, they can only engage in low-end labors. It becomes increasingly difficult to change the status quo of living through their efforts under the squeeze of the capital market. Therefore, many hard-working and diligent workers start to run small informal business, such as a street vendor. Due to insufficient housing supply and expensive rents in the real estate market, numerous private constructions gradually appear in the old cities and urban villages. These informal constructions offer a low-cost place for the disadvantaged people to decrease the living cost.

Besides, some artists and intellectuals use fast, low-cost, and small-scale implements to temporarily intervene the public space in order to challenge the orthodox planning and motivate the common people to participate, such as guerrilla urbanism, tactical urbanism. Although these informal activities are separate from the planning system of official discourse, they reflect flexibility, trial and error, and creativity. It not only makes up for the shortage of top-down supply in cities but also builds and consolidates social relationships in the process, which promotes social stability.

It can be seen that the production of urban informal public space mainly generates from the active resistance of urban citizens to external structural oppression. Power and capital oriented public space production utilize urban planning to regulate and suppress the spontaneous everyday life of urban residents. When biased and unjust urban planning becomes the laws and regulations, urban citizens may take the initiative to resist top-down management for the desire for survival and enjoyable life.

3 Misunderstandings of Urban Public Space Production

3.1 Is Public Space a Public Goods?

Some researchers and citizens have criticized and complained that the public space built by the government is unfriendly and de-humanized, and often hold the view that urban public space is public goods, the government should provide free of charge for citizens because of publicity. In fact, this wrong view directly hinders the benign development of public space.

On the one hand, urban public space is not public goods but private goods. Public goods refer to the items that people use without affecting the use of others, such as storyline, music, sunshine, scientific theorem. Private goods refer to a person using items that no one else can use, including medical services, educational facilities, roads, and parks. Therefore, whether the provider is an individual, a private organization or the government, whether the provider charges or not, the key is whether it will cause competition of use. Public space is a kind of urban space that causes competition. For example, a place cannot be used for playing basketball and dancing at the same time. If users of public space are not charged, the real needs cannot be identified, as it is difficult to evaluate objectively who uses them more efficiently and happily.

On the other hand, the construction of urban public space has a cost. If we view the government as enterprise and urban residents as users, the enterprises provide users with public space and charge a certain fee, which follows the principle of " There's no such thing as a free lunch". Over the past four decades, the China government has obtained large amounts of money from real estate companies to extend cities and build numerous free public spaces for city dwellers through land leases. This model is no longer sustainable until now because cities are entering the stock phase. There are limited lands available for renting, while urban citizens do not pay direct taxes on the use of public spaces. The government, therefore, do not have extra money to pour into upgrade public spaces and provide high-quality public service.

3.2 Can Capital Arbitrarily Intervene Urban Planning?

Urban planning obeys the logic of power in China, where bureaucracy is an open secret. In the capital market, all of the constructions have to measure up the requirements of city regulatory planning strictly. For example, floor area ratio, green landscape ratio and land use, all these aspects above are rigorously controlled. Once breaking the regulation, the constructions are likely forced to face the consequence of being fined or demolished [5]. The capital cannot influence and change urban planning arbitrarily. Therefore, this pre-planned and rigid implement model explicitly or implicitly pushes the capital to rely on cooperating with the government or bribing the government for gaining spatial benefits.

Besides, even if city managers view the informal activities as understandable, they are not entitled to discretion and leniency, because of a lack of negotiation mechanism. In many cases, city rangers have to comply with established laws and regulations to demolish all the informal activities directly.

3.3 Do Bottom-Up Informal Activities Naturally Make Sense?

When moving into a new house, people tend to purchase their favorite furniture and decoration to achieve the highest comfort level; when settling at a new workplace, employees often decorate their workspace with plants and pictures, creating a better working environment. This human nature also applies in public space where people alter their surroundings based on their habits and preferences, turning the cold and aloof physical space into a warm-hearted space filled with humanity. It is safe to conclude that residents are creating mirrored self-image through their active participation. Even armed with all the benefits motioned above, bottom-up approach without proper regulations can cause two serious issues.

One issue lies in the fact that urban public space is a limited resource, which can generate vicious competition. This means that, whenever some people are utilizing the public space, the others automatically lose the opportunity to use yet. It is nearly impossible to define who deserves the public space the best or who can generate the best interest with space. Consequently, if residents exploit the public space competitively based on their self-interest, tragedy of the commons would occur [6]; if public space is only open to certain groups of people, it leads to injustice; if all the public space is prohibited from being occupied, it directly causes waste.

The other issue is that China has experienced rapid urbanization in the past four decades, although it has become the appearance of a modern city in the physical space, the degree of urbanization of human beings is still low. Most migrant workers in urban areas still retain way of life in the countryside, the sense of citizenship should be improved. However, this habitus would not change in the short term. Therefore, a large number of informal activities may affect the benign development of the city and bring about significant challenges for city management and law enforcement, for example, street vendors could impair the traffic, and the unauthorized constructions in city villages may cause a fire hazard.

4 Community Building for Design of Urban Informal Public Space

As explained above, each of the three logics has its advantages and disadvantages. What the key of urban public space production is to keep a dynamic balance between logic of power, capital and life, promote organic development, and facilitate a hybrid community building. As the community is a society of people who share similar values, interests, cultures, identities, norms, and visions [7]. Through adequate communication and negotiation among a community, it may resolve the barriers between government, market, and society. The following four design models of urban public spaces could contribute to community building.

4.1 Design in Everyday Life

IN model refers to the human-oriented design based on everyday life. Designer as a provoker, whose purpose is to shift the design object of urban public space, from grand and romantically squares, parks, and shopping malls to the people within walking distance of everyday life. The designer should follow the human scale as the primary principle, taking into account people's physical, mental, aesthetic, and cultural needs, creating an intimate place that enhances the sense of identity and sense of belonging.

4.2 Design for People

FOR model refers to resilient design. Designer as a facilitator offers people hybrid function places where accommodate diverse activities. Besides, the physical space should also be reserved for the residents to self-adjust and re-design so that they can create themselves space according to their needs in the future. The designer plays a craftsman who is responsible for establishing the stage for the residents, while the residents play the leading role in the show of everyday life.

4.3 Design with People

WITH model refers to participatory design. Designer as a coordinator whose mission is to relieve the tension between residents, government and markets through mild intervention. There are many effective design methods and tools, such as persona, scenario, issue mapping, which can help the public generate better understanding and communication toward design objects. Ideally, residents actively participate in the process of planning, design, construction, management and operation of public space.

4.4 Design by People

BY model generally refers to empowerment design. Designer as an enabler should no longer view urban residents as problem-makers readily to be tackled, as opposed to position them as innovators who have the potential to solve themselves problems. In this case, designers' goals are to empower urban citizens and coordinate among the government, resources, funds, capital markets and human resources, accelerating social innovation.

5 Discussion

To sum up, public space is not public good, and it cannot be evaluated by the utopian, idealized and demanding requirements. It is essentially informal because public space keeps continuously changing with human need. This kind of need is not based on one-sided space production, nor is it to over-satisfy the preferences of certain people. But to make the public space more open and inclusive and create a beautiful urban life for every citizen. The goal of designers is to facilitate a community which keeps a dynamic development between power, capital and life, and eventually achieve co-design, co-construction, co-governance of urban public space.

References

1. Shuisheng, C.: The three logics of urban public space production in china and their balance. ACA Mon. **50**, 101–110 (2018)
2. Hou, J.: Guerrilla urbanism: urban design and the practices of resistance. Urban Des. Int **25**, 117–125 (2020)
3. Carmona, M.: The formal and informal tools of design governance. J. Urban Des. **22**, 1–36 (2016)
4. Amin, A.: Collective culture and urban public space. Cities **12**, 5–24 (2008)
5. Chen, Y.: Informal housing and community in chinese megacities: types. Mech. Responses Urban Des. Int. **34**, 40–46 (2019)
6. Carmona, M., De Magalhaes, C.: Public space management: present and potential. J. Environ. Plan. Manag. **49**, 75–99 (2006)
7. Lou, Y.: Designing interactions to counter threats to human survival. She Ji. **352**, 342–354 (2018)

Communication of Design

A Technique and Method for Mixed Reality Performance with Digital Augments, Gestural Data, and Living Plants

Rewa Wright[1(✉)] and Simon Howden[2]

[1] FASTlab, School of Creative Industries, The University of Newcastle, New South Wales, Australia
rewa.wright@newcastle.edu.au
[2] New South Wales, Australia
simonhowden@hotmail.com

Abstract. We propose a new method, technique, and practices for performing with mixed reality environments and technologies, informed by embodiment and electro acoustics, and critically underscored by new materialism and posthumanism. Through the *Contact* projects, Wright and Howden explore a practice-based approach to live performance in mixed reality, through a novel method that combines gestural controllers, living plants and digital augments. This research investigates potential for multimodal performance in mixed reality (MR), using a novel technical set-up that combines of the HTC Vive with a Leap Motion mount, the MIDI Sprout interface for sensing the bio-electrical signals of plants, transposed to visual data in Touch Designer, and Unity 3D for generating digital augments.

Keywords: Mixed reality · Software assemblages · Plant-human ecology · Augmented reality · Posthuman

1 Introduction

This research develops techniques for affectively co-composing with expressive conjunctions of augmented materials (both digital and organic), including: recursive strategies for modulating digital augments; amplifications of bio-electrical data from plants to produce augmented audio; choreographing hand micro-gestures in tactile and signaletic connections with both augments and plants; and passing augments through the Leap Motion interface in a head mounted configuration. Configured in this way, the Leap Motion is able to transpose hand data (via bespoke software created in Unity), while operating as a head mounted camera, that the performer ‘looks through’ to perceive objects in the world. Digital augments, generated in the Unity 3D engine and tracked to the wearer/performers’ hands in real-time, punctuate the infrared stream and can be manipulated. At the same time, living plants fitted with a MIDI Sprout sensor, emit raw bio-electrical impulses that are captured by sound design software and incorporated by a

Simon Howden—Independent Artist.

E. Markopoulos et al. (Eds.): AHFE 2020, AISC 1218, pp. 235–239, 2020.
https://doi.org/10.1007/978-3-030-51626-0_29

second performer into a bespoke soundscape. The first performer, wearing the HTC Vive/Leap Motion, is able to 'play' the living plants, modulating their signal to shift the pitch of the sound emitted. Tracked to their tactile touch, the digital augments (from Unity 3D) combine with the plant sonics (via Touch Designer) to generate a unique and visually arresting audio-visual experience for the wearer/performer.

This paper discusses two projects in mixed reality performance where new methods and techniques have been investigated: firstly, a live mixed media performance called *Contact Zone*, and secondly, as *Contact/Sense*, an installation and performance that debuted at the SIGGRAPH Asia Art Gallery 2019, Brisbane, Australia. *Contact Zone* was performed at The Black Box, UNSW Art & Design in Sydney, Australia, 23 November 2018. *Contact/Sense*, was performed at the SIGGRAPH Asia Art Gallery, 17-20 November 2019, Brisbane, Australia [1].

At the beginning of the *Contact Zone* performance, a single performer sat at an agave attenuata plant, modulating its bio-electrical impulses that have been transposed to audible sound via MIDI (Fig. 1). In synchronicity with, and activated by the performer's hand gestures, digital augments fill two large screens. On the left screen, the was the live feed from the Head Mounted Display (HMD) worn by the performer, and visible to the audience. On the right screen was a live stream connected to a second Leap Motion interface, controlling custom made software that responded to gestural data. During the performance, the performer would pick up this second Leap Motion, activating the data input. On both screens, digital augments are visible as well on the HMD display of the performer, whose infra-red vision is now punctured with brightly coloured traces of data. Negotiating their new and de-limited MR infrared vision, the performer navigates the responsive media environment.

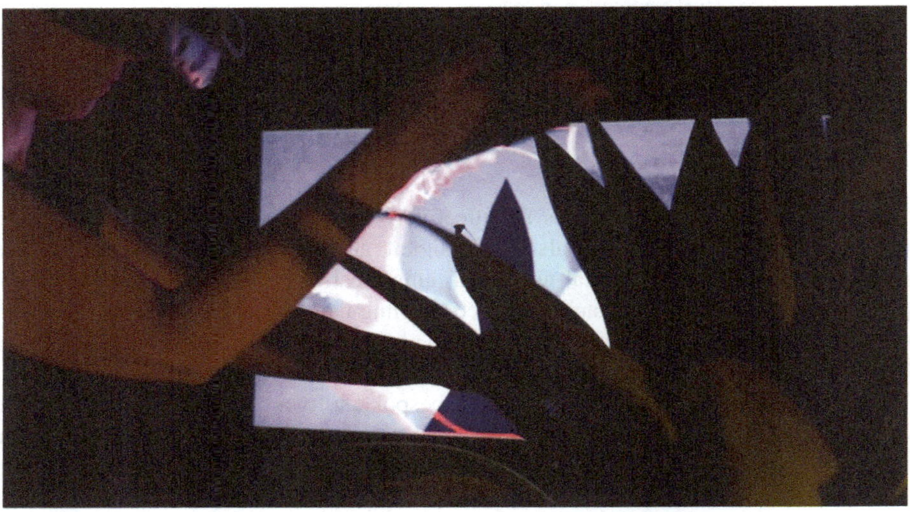

Fig. 1. *Contact/Sense* performance, Sydney, Australia, November 2019.

The technical design of the *Contact Zone* and *Contact/Sense* performances, proceeds from a particular type of interface: the Leap Motion gestural interface worn as a Head Mounted Display (HMD) that doubles as a 'look through' infrared camera. Configured in this way, the Leap Motion is able to transpose hand data (to a bespoke software created in Unity), and operate as a head mounted camera, that the performer uses to perceive objects in the world. With colour removed from natural perception, the grayscale picture plane of the Leap Motion's near field infrared camera becomes the performer's new visual apparatus. Looking through the Leap Motion returns quite a different mode of vision from our human brain-eye combination: instead, the performer sees in the infrared spectrum. Using a Leap Motion gestural interface mounted on an HTC Vive VR headset, digital augments are tracked to the performer's hands (Fig. 2).

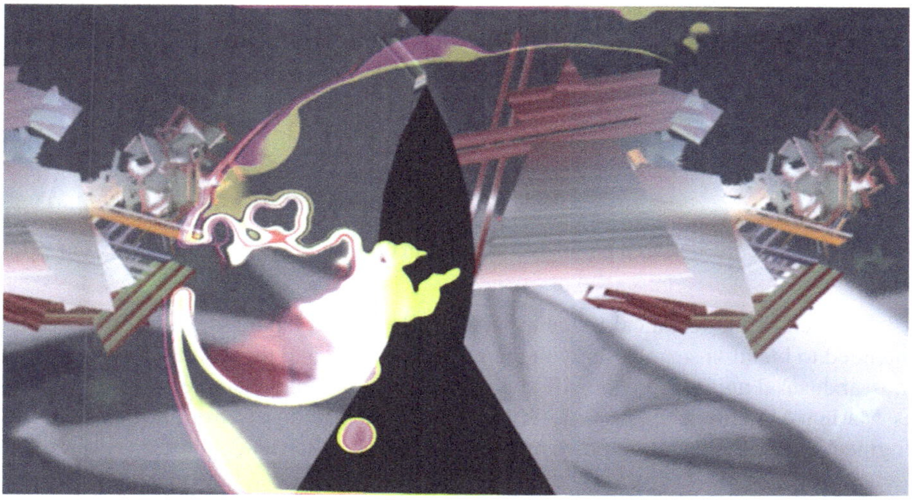

Fig. 2. HMD display of performer in *Contact/Sense,* abstract augments tracked to hands.

The physical effect of this apparatus is that the performer/participant is semi-immersed in an infra-red mode of MR: without colour, vision is de-privileged and tactile gestures and body movements take on greater importance as a technique for sensory inquiry. Using the surface of plants to conjure digital augments into existence, the surface of the plants becomes an interactive surface, where several forces coalesce. The bio-electrical signals emitted by the plants meet the human touch of the performer's hand, thereby adjusting the signal itself. At the same time, digital augments are triggered through the software made in Unity 3D, streamed through Touch Designer, where the augments generated through human performance are blended with visual data created by the bio-electrical impulses of the living plants. Playing alongside and in response to a community of living plants who are emitting bio-electrical impulses, the performer adjusts their actions to elicit differentiation in tonality and harmony. A second performer, a sound designer working in real-time, is able to adjust the timbre and pitch of this combined human-plant sonics. However, one notable factor which the humans are unable to manipulate is rhythm, since this is created solely by the impulses generated by the plants (Fig. 3).

Fig. 3. Performer modulates the sounds coming from the plants by shifting the leaves and stems.

In much of the technical literature about AR/MR, digital augments themselves are intended to blend in seamlessly with the physical world, and this is the desirable goal in most industrial applications of MR as well as recent notable examples in MR art (such as Marina Abramović's *The Life,* 2019). In certain paradigms from commercial and engineering practice, migrated to media art, MR is often encountered as augments viewed through a clear screen display. Understood as both informatic and digital, augments are supplementary content that enhance or extend a human experience of 'reality'. The *Contact Projects* cultivate a contrasting view of augments as emergent via human and nonhuman processes that entangle digital as well as physical spaces, critiquing notions of computational vision as framed by a clear, unproblematized window or screen. In the *Contact Projects*, the performer has their vision profoundly disrupted by a jarring mixture of reality and the virtual. While wearing this apparatus, the performer can only see through infra-red vision, and they must operate in the installation space with this as their primary perceptual apparatus.

The digital augments (created using augmented reality and computer vision techniques), are triggered by the performer's hand tracking data, captured by the Leap Motion interface. Thus, they emerge alongside the tactile hand gestures of the performer, as the performer responds to and plays alongside the bio-electrical emissions from the living plants. The plants become tangible interfaces that elicit sound, and whose tactile surfaces become the trigger for a computer vision system that activates digital augments. The plants are emitting a bio-electrical signal that has been captured through MIDI, and then is manipulated by the performer using the surface of the plants' leaves. While the first performer is directly composing with the bio-electrical signal from the plants, a second performer co-creates with the audio signals generated by the living plants in the space and the acoustic signals produced by the performer wearing the HMD. In terms of the focus on the movement of a performer/participant as they navigate a natural environment populated by plants, this research builds upon earlier work undertaken in the project 'ko maungawhau ki runga/the mountain above', Auckland, Aotearoa/New Zealand, 2013 [3].

For the *Contact Projects*, technical equipment consisted of an HTC Vive, 2 x laptops, webcam, leap motion, a MIDI Sprout interface, various other hardware interfaces and devices, living plants, Unity 3D, Touch Designer and a changing variety of modular display technologies, speakers and screens.

Acknowledgments. Rewa Wright would like to acknowledge the invaluable support of FASTlab and the School of Creative Industries, at the University of Newcastle, Australia, in realising this paper.

References

1. *Contact/Sense*: Association for Computing Machinery, New York, NY, United States SA 2019: SIGGRAPH ASIA Art Gallery/Art Papers, November 2019. https://dl.acm.org/doi/abs/10.1145/3354918.3366368. Article No. 9
2. Wright, R.: Augmented virtuality: remixing the human-art-machine. In: Ferreira, H., Vicente, A. (eds.) Post Screen: Intermittence + Interference. Edicoes Universitarias Lusofonas, Lisbon, pp. 158–166 (2016b)
3. Wright, R.: Exploring the responsive site: ko maungawhau ki runga. In: Kathy, C., Laura, F., Ross, H. (eds.) Proceedings of 19th International Symposium on Electronic Art (ISEA 2013), Sydney, Australia (2013). https://ses.library.usyd.edu.au/bitstream/2123/9700/1/exploringresponsive.pdf

Guidelines for Promoting Thai Television Program Production in Response to the Notion of Creative Economy

Sasithon Yuwakosol(✉)

College of Social Communication Innovation, Srinakharinwirot University, 114 Sukhumvit 23, Wattana, Bangkok 10110, Thailand
Sasithon.y@gmail.com

Abstract. This research investigated three types of television program formats, i.e. news, documentary, and variety show or drama programs, which are suitable for and match the objectives of target audiences and also studied the audiences' satisfaction toward Thai and foreign television programs. The survey research collected data from 548 participants. In addition, data from seven in-depth interviews showed that development of the program production process, funding sources, and promotion from the government and relevant independent organizations are factors influencing the ability of Thai television programs to support the concept of creative economy. Finally, the strategies for producing television content to create cultural products were also suggested in this paper.

Keywords: Creative economy · Television program · News · Documentary · Drama

1 Introduction

The term 'creative economy' was first introduced in a 2001 book by John Howskins, to mean the use of creative imagination to increase an idea's value. Howskins [1] explained that human creativity tended to be successful when such creativity related to their everyday life. Creativity exploits culture, knowledge, and special skills to facilitate changes and drive innovation. That is why the economic growth of advanced developed countries was usually driven by knowledge and creativity. Creative economy adds value to products or services from human ideas and the production that develops into creative economy is called 'creative industry', which means production activities that depend primarily on creativity. For Thailand, the Office of the National Economics and Social Development Board [2] summarized that creative economy focused on propelling the economy on ground of knowledge, education, creativity, and intellectual property in connection with cultural foundation, social knowledge collection, and modern technology/innovation.

Mass media is a format of work that depends significantly on creativity and is an important medium for cultural transmission. Therefore, mass media is considered a group of creative industry. An important example of the countries that succeeded in merging creativity with content and culture is South Korea. The Korean government

E. Markopoulos et al. (Eds.): AHFE 2020, AISC 1218, pp. 240–246, 2020.
https://doi.org/10.1007/978-3-030-51626-0_30

has cultural policies that can convey and present Korean values though a variety of channels such as movies, music, and games, making South Korea one of the leaders in cultural content production [3].

Television is the most accessed media in Thailand. A 2019 survey by Nielsen Consumer Media View showed that 98% of Thai people aged between 12 and 34 had access to television and 99% of those aged 35 had access to television. However, those viewers did not only watch television programs from television devices but also from other platforms such as computers and smartphones, both from live broadcast and recorded broadcast.

The transition of Thailand's television broadcasting system from analogue to digital has resulted in 15 digital television stations in the present. At the same time, media convergence has led to increasing emergence of internet television or online television which contains a wide range of content such as drama series, variety shows, and news programs; those programs can be viewed via mobile phones and notebook computers. Moreover, there are other video content on online platforms as additional options for audiences as well. Hence, competition in Thailand's content production industry is intensifying, not only among Thai media but also with foreign media.

With the advantages of communication infrastructure and meaningful cultures which are great resources of creativity, the integration under the ASEAN Economic Community (AEC) scheme is a great opportunity for Thailand to improve the television industry and generate income for the country to bring about sustainable growth for the society and the economy.

To produce a television program that interests people, the producer should also take the objectives of target audiences into account. Generally, content of television programs is influenced by two factors, which are (1) ideas and past experience of the producers and (2) the sponsors. As a consequence, the content presented may not genuinely meet the objectives of the audiences. Therefore, the television program that wants to expand its audience base in the digital era should give importance to the audiences, the quality production, and the differences from competitors [4].

This research aimed to study the objectives of viewers of news, entertainment programs, and variety shows; the patterns of television programs that meet the objectives of the viewers; factors influencing the ability of Thai television programs to produce content; and strategies for producing television program content. The research also aimed to find the guidelines for promoting Thai television program production in response to the notion of creative economy or creative industry for sustainable development of the country.

2 Methodology

Both quantitative and qualitative research methods were used in this study. The data were collected from a survey research and in-depth interviews. The samples of the study were television viewers and people who worked in the media-related field.

2.1 Participants

The survey samples consisted of 548 television viewers from five regions of Thailand, which were the North, the Central, the North East, the South, and Bangkok and outskirts. The samples consisted of 311 females (56.8%) and 237 males (43.2%). The age ranges of the samples were below 20 years old (15.5%), 20–30 years old (32.8%), 31–40 years old (21.2%), 41–50 years old (12.6%), 51–60 years old (8.4%), and above 60 years old (9.5%). The majority or 63.7% of the samples watched television programs via smartphones, 63.1% via television devices, and 35.9% via computers with internet connection.

The seven samples for the in-depth interviews were television content producers, academics, and government staff whose work is related to media regulations and creative economy policies.

2.2 Instruments

The survey questionnaire consisted of three parts: Part 1 Objectives of watching television programs, Part 2 Expectation from watching news, documentary, and variety show or drama programs, and Part 3 Satisfaction toward Thai and foreign television programs.

The in-depth interviews used a structured interview method to ask the samples about their opinions on supporting factors and obstacles for media to enter creative industry as well as the guidelines and strategies that could help achieve that goal.

3 Findings

3.1 Objectives of Watching Television Programs

Top three objectives of the television viewers were to relax and relieve stress (78.8%), to stay up-to-date with current events and news (75.5%), and for entertainment and happiness (56.8%). Only 13.1% of the samples watched television programs to gain information for persuading others. The majority of the sample group watched Thai television programs more than foreign television programs over the past month before the survey. Most or 39.5% of the viewers watched Thai television programs every day, while 20.1% of them watched Thai television programs three to four days a week. The majority of the sample group watched foreign television programs one to two days a week (23.0%), whereas 11.9% never watched foreign television programs (Table 1).

Table 1. Number and percentage of viewers watching Thai and Foreign television programs in a month before the survey

Frequency of watching	Thai program	Foreign program
Never	7 (1.3%)	65 (11.9%)
Less than once a week	50 (9.1%)	104 (19.0%)
1–2 days a week	88 (16.1%)	126 (23.0%)
3–4 days a week	110 (20.1%)	94 (17.2%)
5–6 days a week	76 (13.9%)	64 (11.7%)
Every day	217 (39.5%)	95 (17.3%)

3.2 Expectation from Watching Television Programs

The research examined what viewers expected from watching three types of television programs, which were news program, documentary program, and variety show or drama programs. The three types of television programs were evaluated in terms of patterns, content, and creativity, with the mean evaluation score ranging from 1.00 to 5.00. The findings were as follows.

News Program
In terms of patterns of news programs, two items that received the highest mean evaluation scores were professional presentation (3.76) and well-selected hosts (3.74). In terms of content, two items that received the highest mean evaluation scores were educational and useful content (3.75) and up-to-date content (3.69). In terms of creativity, two items that received the highest mean evaluation scores were use of new technique or technology in presentation (3.60) and cross channel linking between activities on television screen and activities on other online media, e.g. Facebook live streaming of behind the scene during advertisement break (3.50).

Documentary Program
In terms of patterns of documentary programs, two items that received the highest mean evaluation scores were professional presentation (3.56) and well-selected hosts (3.52). In terms of content, two items that received the highest mean evaluation scores were knowledgeable and interesting content (3.48) and easy-to-understand content (3.46). In terms of creativity, two items that received the highest mean evaluation scores were use of new technique or technology in presentation (3.50) and inspirational effects (3.42).

Variety Show or Drama Program
In terms of patterns of variety show or drama programs, two items that received the highest mean evaluation scores were having extra features to facilitate the disabled such as a subtitle or a sign language interpreter for the hearing impaired or a voice over narration for the visually impaired (3.99) and using quality production technique (3.65). In terms of content, two items that received the highest mean evaluation scores were entertaining and humorous content (3.66) and interesting and captivating story telling (3.54). In terms of creativity, two items that received the highest mean evaluation scores were use of new technique or technology in presentation (3.60) and good taste of entertainment (3.44).

3.3 Satisfaction Toward Thai and Foreign Television Programs

The majority of the samples were more satisfied with foreign television programs than Thai television programs in every aspect. Only the item "The program content matches the objective of watching" showed no significant difference in the level of audiences' satisfaction. On the other hand, the item that showed the most difference in audiences' satisfaction level was "The program was professionally produced", as the mean satisfaction score toward Thai programs was 3.33 and the mean satisfaction score toward foreign programs was 3.83 (Table 2).

Table 2. Comparison of mean satisfaction scores toward Thai and foreign television programs

Satisfaction toward watching	Thai program	Foreign program	t-test	Sig.
The program was produced with quality process	3.31	3.46	−2.81	.01*
Elements e.g. scene setting, costume, and original soundtrack of the program are fascinating	3.33	3.40	−2.38	.02*
The program content matches the objective of watching	3.32	3.44	−1.31	.18
The program content is useful in some ways	3.19	3.62	−4.53	.00*
The program widened your visions and horizons	3.21	3.55	−7.00	.00*
The program is entertaining and not boring	3.30	3.51	−3.98	.00*
The program is enthusiastic	3.24	3.57	−4.59	.00*
The program content gave knowledge and new experience	3.23	3.58	−6.13	.00*
The program content is about local or national wisdom and culture	3.22	3.47	−3.82	.00*
The program can encourage creativity	3.25	3.58	−5.82	.00*
The program is inspiring	3.26	3.60	−6.11	.00*
The program gave different experience from other television programs	3.23	3.65	−7.39	.000*
The program was professionally produced	3.33	3.83	−8.23	.00*

* Statistical significance level is 0.05.

3.4 Guidelines for Promoting and Improving Television Program Production

According to the in-depth interviews with television content producers, academics, and government staff whose work is related to media regulations and creative economy policies, three important factors television producers needed in order to survive and join international competition included (1) development of the production process, which includes new technology that rivals foreign production, new marketing channels, and more audience-oriented content, (2) appropriate promotions and regulations from government organizations that allow media producers to present content freely without unfair intervention, and (3) business partners and sponsors as important funding sources.

For Thailand, to develop Thai television programs that can capture the heart of the audiences both locally and internationally required the following strategies.

1) **Audience-oriented production** – As competition in the media industry is intensifying against the backdrop of a vast array of media choices for consumers, television producers should improve the content of the program, the quality of the program, and the creativity of the program to best suit the demand of the audiences.
2) **Engagement and experience creation** – This strategy is to keep the audiences engaged with the program and to build its popularity. Moreover, creating audience experience will make the program part of the audience's lifestyle and everyday life.
3) **Power of online social media** – Social media is a hugely powerful media at present. Knowing how to use those media to create viral marketing could increase the program's presence and reach.
4) **Characteristics of content of each types of program** – Television programs that are popular among general Thai viewers usually have light, amusing, and easy-to-understand content. They can be own-produced local programs, franchise programs from overseas.
5) **Research and development to build production knowledge** – Viewers usually have different objective of watching each type of program. Conducting research will help television producers understand the insight of viewers' objectives and the expected formats of each type of programs.

4 Discussion

Creative industry is the cycle of creation, production, and distribution of goods and services that use creativity and intellectual capital as primary inputs. Creative industry also constitutes knowledge-based activities that are not only limited to arts and those activities can generate revenues from trade and intellectual property rights [5]. For these reasons, television programs are considered a type of creative industry. This research discussed and introduced the guidelines for promoting Thai television program production in response to the notion of creative economy.

The findings showed that Thai viewers were more satisfied with foreign television programs than Thai television programs. However, those viewers had the similar level of satisfaction from watching Thai and foreign programs that match their objectives. This suggests that if Thai television producers have an insight about the objectives of Thai viewers and create the content accordingly, they have the potential to become popular among Thai viewers given their existing familiarity with local background.

Speaking of the audiences' expectation from three types of television programs, in terms of patterns and presentation the viewers would like to see professionalism from news programs and documentary programs because both types of programs need to emphasize on accuracy of information and professional work. For variety shows or drama programs, the viewers expected the programs to provide extra features to facilitate the disabled such as a subtitle or a sign language interpreter for the hearing impaired or a voice over narration for the visually impaired. This is possibly because those programs were produced to entertain people and everyone should have the equal access to such entertainment. Television producers, therefore, should take this issue into consideration and produce a program that can reach every group of people. In terms of content, the findings showed that viewers expected different content from different types of programs. For example, viewers would like to see educational and useful content from news programs, while they expected knowledgeable and interesting content from documentary programs and entertaining and humorous content from variety show and drama programs. Accordingly, television producers should produce the content that matches those expectations of the audiences. In terms of creativity, what viewers expected from all types of television programs were use of new technique or technology in presentation, so television producers should improve their production standard and apply new technology in their program production.

In conclusion, important strategies to create cultural goods from television program content are (1) producing content with audience-oriented approach, (2) creating audience's engagement and experience, (3) using the power of online social media as a marketing tool, and (4) producing content that is suitable for each type of programs. All these strategies will improve the production quality of television programs and ensure their survival in the highly competitive red ocean market.

Acknowledgement. This research received funding support from College of Social Communication Innovation, Srinakharinwirot University, Thailand.

References

1. Howkins, J.: The Creative Economy. Penguin, London (2001)
2. Office of the National Economics and Social Development Board: The Creative Economy Report (2009). http://www.thaiembassy.org/seoul/contents/files/business-20130905-090645-658148.pdf
3. Chatsakulpairach, V.: The effects of Korean T.V. dramas on Korean values assimilation and culture imitation of the Thai Youth. Panyapiwat J. **6**(1), 175–185 (2014)
4. Kuljitjuerwong, S.: TV program in the digital age. Panyapiwat J. **7**(2), 245–257 (2015)
5. The United Nations: Creative Economy Report 2010 (2010). https://unctad.org/en/Docs/ditctab20103_en.pdf

Wayfinding Systems: Digital Versus Analogical Media

Maria Luisa Costa[1,2](✉)

[1] CIAUD, Faculdade de Arquitetura da Universidade de Lisboa, Rua Sá Nogueira, Pólo Universitário, Alto da Ajuda, 1349-063 Lisbon, Portugal
design.luisacosta@gmail.com
[2] Instituto Superior Miguel Torga, Largo da Cruz de Celas nº1, Coimbra, Portugal

Abstract. In today's society space is overloaded with visual information, thus digital platforms are being increasingly used in design and could contribute to hinder this trend, assuming digital media and mobile platforms as an alternative to analogical media (paper support). In what concerns wayfinding systems, they need to allow more then to reach a point, starting from another, they should allow the experience and the knowledge of space, by allowing several layers of information, that people can use or not, depending on their interests and ambitions. This paper presents a study about the preferences and difficulties of people over 50 years old in the use of analogical and digital maps and discusses the possibility of the use of wayfinding systems in digital means for this age group, regarding the Design principles and inclusivity.

Keywords: Design · Wayfinding systems · Digital · Analog · Inclusivity

1 Introduction

Nowadays, accessing the city and different social or cultural spaces is an arduous task due to increasing territorial complexity, population growth, proliferation of visual information that "inhabits" the territory, and above all the fact that the city to assume, according to Acher [1], as an immense palimpsest that does not change and successively receives different writings.

Cities are growing and, according to UN data UN [2], it appears that since 2008, more than half of the world population has lived in cities. In parallel with the growth of cities and according to data from the National Statistics Institute [3] in Portugal, there is a progressive aging of the population, and it is expected that in 2031, of the 7.5 million inhabitants, 2.8 million will be elderly. This situation leads Designers to consider the need to design while taking care of the needs of an aging population.

The need to use wayfinding systems increases in direct proportion to the growth of cities and in technological potential. Wayfinding systems should respond to the needs of the population, in some cases with objective, clear and unambiguous information on the fastest way to reach a certain point, as is the case with transport interfaces, garages, airports, hospitals, etc. In other cases, when the movement is driven by leisure, and particularly in the context of cultural itineraries, these systems must allow individuals

E. Markopoulos et al. (Eds.): AHFE 2020, AISC 1218, pp. 247–254, 2020.
https://doi.org/10.1007/978-3-030-51626-0_31

to enjoy the space more fully, create their own experiences and, through access to multiple information, expand knowledge. The design task is often to build tools and systems through which others can create their own experiences [4].

This paper reflects about individuals who are above 50 age and have not grown in contact with technologies, which today are an integral part of our daily lives. In this sense, and due to the importance, that maps assume in the context of wayfinding systems, and the increasing use of digital maps, we developed a questionnaire where we evaluated the use of digital and paper maps by individuals above 50. We questioned the difficulties and preferences and assessed the existence of differences between genders and academic levels.

2 Designing the City

Designers actively participate in the construction of the visual universe of cities, through informational messages such as signs and toponymy but also through advertising. According to the Design Manifesto [5], competence and creativity, as well as the activity of designers, have been directed towards promoting consumption, and designers use their time and energy in the creation and development of products that, after all, are superfluous. This practice needs to be reversed, and designers must direct projects towards the common good, bettering the lives of citizens. In the present case, we are concerned with improving accessibility in the context of the city, in order to make them more inclusive and adapted to contemporary social realities.

According to UN data [2], the number of individuals living in cities is expected to double from 3,300 million in 2007 to 6,400 million in 2050. In addition to this, a high number of individuals living outside the city travel there to access goods and services and, also, tourist flows supported by the current ease of travel, by the improvement of social conditions and the growing cultural interest which the widespread access to information enhances, as well as the free time older generations have to dedicate themselves to leisure and knowledge.

The forms of urbanity produced by design projects and developed in the city space are elements of communication that promote meaning and sense [6].

In order to read the city according to Lynch [7], the clues and signs that enable recognition and orientation are decisive and the greater this legibility is, the greater the comfort and well-being of citizens. However, there is an increase in territorial complexity, where natural images, such as buildings, streets and landscapes, which would assist in the reading of the city, overlap multiple images produced, resulting mainly from advertising. Both live together and accumulate in multiple layers in the public space. From this growing complexity and multiplication of images, reading the city's structural elements becomes very difficult, making it necessary to expand the auxiliary means of reading it, especially when citizens move outside their "comfort space" in the residence or work area.

The city can be interpreted as a picture of mental and symbolic maps that reflect a different way of appropriating the same space depending on the presence of individuals and the relationships between them [8].

Moving around and accessing different places with tranquility and security, without feeling disoriented is very important for all citizens, which will contribute to establishing a harmonious relationship between you and the outside world. This is the reverse of the fear that derives from disorientation; it means that the sweet sense of home is the strongest when the home is not only familiar, but also distinctive [7]. In this sense, the wayfinding systems can contribute to an easier circulation around the city, promoting a sense of security and well-being, especially among the elderly.

3 Inclusive Design and Aging

In view of the aging process, individuals lose their physical, sensory and/or cognitive capacities, considering that the increase in life expectancy that results from this demographic implies that individuals, once released from family and professional commitments, have more time to devote to hobbies, to move and to travel. We know that the complex, confusing and crowded city is full of barriers that limit accessibility and these are not exclusively physical, many derive from the impossibility of reading the city and the recognition of geographical space and make safe travel impossible. Most of the barriers that prevent many citizens from engaging in democratic life, and limit the full exercise of active citizenship, are designed and built by men [9].

Inclusive design, also known as Universal design is an approach to creating environments and products that are usable by all people to the greatest extent possible [10], placing emphasis on identifying as many needs as possible, across all potential users. Considering the physical, cognitive and emotional changes, as well as the social changes that individuals experience throughout life [10].

The need to safeguard accessibility and inclusiveness has long been an objective of design and the focus on older groups is becoming increasingly necessary in the face of demographic aging. As in many other countries, in Portugal the population is expected to decline, from 10.3 million to 7.5 million in 2031, while the number of elderly people will increase from 2.1 to 2.8 million [3].

The WHO defines that the age groups of the elderly should be conceptualized according to the socio-cultural and economic-political characteristics of the countries [11]. In the European Union, it is consensual to consider elderly people over 65 years of age, in accordance with retirement standards. The United Nations argues that older people are all people over 60 years of age. In Africa it is consensual to define that the elderly are people over 50 years of age [12].

In fact, the classification of elderly people is defined according to what is indicated by the WHO, but we must consider that the aging process is not homogeneous and chronological age is not the best indicator of the individual process [13]. The reality is the entire multicellular organism undergoes physiological changes over time, with aging being characterized by a decline in the functional capacity of the organism [14].

However, it is a mistake to think of the older user as a wheelchair user or as severely disable. Older users are that vast number of people who, in advancing age have a little discernible impairment, but have a strong drive to remain independent and to contribute to the community, but are hampered by inappropriate design. Better design can play a crucial role in enabling older people to remain active [15].

Only above 75, the visual acuity becomes a problem, namely the difficulty in reading signs, they are more dependent on high contrast between sign backgrounds and lettering; a reduced capacity for sensory processing or problem solving can increase difficulties in older adults when negotiating unfamiliar environments; the reduced manual dexterity and coordination can affect their ability to operate common mechanisms such as phones [16].

According to Boch and Gharaveis [17], we defend that humans privilege independence, whether they move around the house or take a long trip. The lack of confidence in their navigational skills can reduce their desire to travel, so wayfinding projects aimed at an older population should take care of these needs and constraints.

4 Wayfinding Systems

Wayfinding systems facilitate access to places that individuals are not familiar with, allowing a set of information that, depending on the space in question, may include landmarks, signage and maps.

The wayfinder is an individual who travels on foot or by bicycle with the aim of finding a specific place [18]. According to this definition, we are all wayfinders. Assuming navigation issues are an essential activity in our lives. Recently, many people use smartphones for outdoor navigation in the city, or indoor in hospitals, museums and others. With the success of smartphones, outdoor and indoor navigation systems for pedestrians have reached the end consumer market [19].

The trend of population aging in contemporary societies determines measures implemented regarding inclusiveness [8]. The main means used in wayfinding systems usually consist of maps and signal systems, symbols, colors and typography that indicate the direction of individuals. Progressively these means integrate mobile applications and automatic identification systems, paving the way for new approaches in wayfinding systems [6]. Currently, maps continue to be an integral and necessary part of wayfinding systems, with digital maps becoming increasingly important, but when designing a wayfinding system, we can use these means without risk of exclusion of individuals over 50. This is the answer to which we seek to respond.

5 Method

This paper aims to analyze how people over 50 use printed and digital maps. We were interested in this age group because they did not grow up with technology and we want to understand if despite this, they improved their knowledge in this matter and if they are able to use digital platforms without important restraints.

For this, the first part of the survey creates a small socio-demographic framework, in order to obtain information on participants' gender, age, level of education and work. The second part, surveys the subjects on their preferences and difficulties using digital and analogical media.

The questionnaire consists of multiple choice answers, short answers and a set of open ended questions that allow us to obtain more information that we didn't have thought yet, none of the questions was mandatory.

The Google Forms platform was used to carry out the questionnaire and was disseminated through email lists, and Facebook social network.

This study involved 94 participants, women (n = 64) and men (n = 30), the ages of respondents ranged from 51 to 80 (M = 61.3).

Regarding academic level, 76 respondents have higher education, 18 middle or professional education, and no responses from individuals with basic education.

6 Results and Discussion

With regard to the total number of people moving to unknown places, 62 claims to use paper maps (66%) and 79 digital maps (93%) (Tables 1, 2 and 3).

Table 1. Use of paper and digital maps for frequent and rare trips to unknown places.

Frequent trips	Paper maps	Digital maps
Use	48 (67%)	62 (94%)
Do not use	24	4
Rare trips	Paper maps	Digital maps
Use	14 (64%)	17 (89%)
Do not use	8	2

Table 2. Use of paper and digital maps – by gender

Use of maps – by gender	Paper maps		Digital maps	
	Use	Do not use	Use	Do not use
Masculine	24 (80%)	6	23 (92%)	2
Feminine	38 (59%)	26	56 (93%)	4

Table 3. Use of paper and digital maps – by academic level

Use of maps – by academic level	Paper maps		Digital maps	
	Use	Do not use	Use	Do not use
Middle or professional school	14 (78%)	4	12 (86%)	2
University	48 (63%)	28	67 (94%)	4

Regarding the use of paper maps, of the 32 respondents who stated they no longer use paper maps, 26 justify replacing them with digital maps due to their accessibility and reliability of information; 2 justify it due to the fact that paper maps become outdated very quickly, 1 answer refers to problems with the handling of these maps.

When asked about the reasons for continuing to use paper maps, 10 responses mention there is no need for an internet connection, nor dependence on the battery of electronic equipment; 11 respondents report that paper maps provide more comprehensive or global view of space; 2 respondents highlight the complementary use of analog and digital maps; while another 2 answers mention the advantage of being able to write and take notes on these maps.

Regard to digital maps, 8 respondents indicate that these maps are very practical, accurate and up-to-date; 2 responses refer to the importance of digital maps in reducing the use of paper and printing ink; the possibility of obtaining geographic information at various scales and varying them during the consultation, was 1 of the responses obtained; 3 responses mention the fact that it is possible to access information on exact distances (meters or kms), information on travel time and directions to follow; there is also 1 answer that refers the possibility of exploring the route until the destination point.

Regarding the answers given to specific questions previously identified by us and classified on a scale from 0 (without difficulty) to 5 (maximum difficulty), with regard to paper maps, the difficulties related to legibility are particularly relevant, due to the size of letter (M = 3.3) and use of color or chromatic contrast (M = 2.9). As positive aspects (0 nothing to 5 very much), affection and personal taste for paper maps (M = 3.0), as well as the fact that printed maps are part of the memories that individuals like to bring (M = 3.0).

Regarding the use of digital maps, internet access difficulties are referred (M = 3.0), about this aspect, it is worth noting that 14% of respondents only use free wi-fi. Regarding the positive aspects in the use of digital maps, it refers to the functionality within the scope of the trip preparation (M = 3.6), as well as the possibility of accessing more information in addition to the exclusively geographical (M = 3.8).

7 Conclusions

The results of this study reveal the use and difficulties experienced by individuals over 50 in the use of analog and digital maps. We consider that it may be a limitation of this survey the lack of answers given by individuals with basic education. The sample presents results based on individuals with above-average academic training in Portugal, which may mean that the results cannot be applied in all contexts.

The data obtained allow us to verify that, in the age groups above 50, both analog and digital maps are used simultaneously and in complementarity, 66% of individuals claim to use paper maps and 93% use digital maps, a percentage that allows us to conclude that the use of digital maps is not an obstacle for these age groups, having been mentioned by 26, of the 32 individuals who stopped using paper maps, that replaced them by digital ones because they are more reliable and are updated. Advantages are also mentioned in the use of digital maps in the preparation of the trip and exploration of the routes.

Regarding the difficulties in the use of digital maps, issues related to the capabilities of the equipment (mobile phones and batteries) and internet access are mentioned, considering the respondents that the platforms are very intuitive and easy to use.

With regard to paper maps, handling and legibility are indicated as the main difficulties, with benefits being the possibility of a broader view of space, as well as factors related to an emotional connection to maps as objects that one likes and wants to bring as a souvenir.

About the analysis of the use of digital maps by gender, there are no significant differences between the female and male universe, whereas with regard to the use of paper maps there is a greater use by men (80%) than by women (59%).

Concerning academic levels, it can be verified a larger percentage of digital maps use by individual with university education (94%), than by those with middle school level (86%). The situation is reversed with paper maps: less use by individuals with university education (63%) than by those with middle school education (78%).

By crossing results obtained, revised literature and direct observation, we can conclude that is possible and desirable the integration of digital maps in wayfinding systems, namely those for cultural itineraries, without excluding people over 50, making it possible to use and enjoy all the possibilities that technologies offer, by implementing different playful and informative layers. However, we consider that it should not be overlooked the importance that paper maps still have for these age groups, and they should present a macro vision of the territory and fulfill their role as a 'trophy' and souvenir in the context of a trip.

References

1. Archer, F.: Novos princípios do urbanismo, seguido de novos compromissos urbanos – um léxico. Livros Horizonte, Lisboa (2010)
2. United Nations: World Population Prospects: The 2015 Revision, Key Findings and Advance Tables (2015)
3. National Institute of Statistics (2017). http://www.ine.pt
4. AIGA Design Futures. https://www.aiga.org/aiga-design-futures/complex-problems
5. First Things First: A Design Manifesto. AIGA J. (2000). http://www.xs4all.nl/~maxb/ftf2000.htm
6. Costa, M.L., Amaral, I.: Communication design and space narratives. In: Raposo, D., Neves, J., Silva, J. (eds.) Perspective on Design. Springer Series in Design and Innovation, vol. 1. Springer, Cham (2020)
7. Lynch, K.: A Imagem da Cidade. Edições 70, Lisboa (1960)
8. Costa, M.L., Amaral, I., Daniel, F.: The contribution of information design to age-friendly cities: a case study on Coimbra and public transport. In: Rebelo, F., Soares, M.M. (eds.) AHFE 2018, AISC 777, pp. 485–496. Springer International Publishing AG, Berlin (2018)
9. Simões, J.F., Bispo, R.: Design Inclusivo: acessibilidade e usabilidade em produtos, serviços e ambientes. Centro Português de Design (2006)
10. Beneficial Designs: Designing sidewalks and trails for access, Part II - Best Practices Design Guide. Department of Transportation. Federal Highway Administration, USA (1999)
11. World Health Organization: Active Ageing: A policy Framework. World Health Organization, Geneva (2012)
12. Costa, M.L., Amaral, I.: Cidades e Sistemas Wayfinding. Convergências - Revista de Investigação e Ensino das Artes, vol. XI, no. 21 (2018)
13. Marques, S.: Discriminação da Terceira Idade. FFMS and Relógio D'Água Editores, Lisboa (2011)

14. Hoffmann, M.E.: Bases biológicas do envelhecimento. Idade Ativa–Revista Eletrônica da Terceira Idade (2003)
15. Huppert, F.: Designing for older users. In: Coleman, R., Lebon, C., Clarkson, J., Keates, S. (eds.) Inclusive Design: Design for the Whole Population, pp. 30–49. Springer, London (2003)
16. Beneficial Designs: Designing sidewalks and trails for access, Part I - Review of Existing Guidelines and Practices. Department of Transportation, Federal Highway Administration, USA (1999)
17. Boch, S.J., Gharaveis, A.: Flying solo: a review of the literature on wayfinding for older adults experiencing visual or cognitive decline. Applied Ergonomics **58**, 327–333 (2017)
18. Mollerup, P.: Wayshowing: A Guide to Environmental Signage. Lars Muller Publishers, Baden (2005)
19. Rodriguez-Sanchez, M.C., Martinez-Romo, J.: GAWA-manager for accessibility wayfinding apps. Int. J. Inf. Manag. **37**, 505–519 (2017)

Practical Application of Bullet Point Systems for Constructing an Information Hierarchy

Amic G. Ho(✉)

Department of Creative Arts, School of Arts and Social Sciences, The Open University of Hong Kong, Hong Kong, People's Republic of China
amicgh@gmail.com

Abstract. Bullet point systems are widely employed in digital communication. However, very few people are concerned with how bullet points work for communication purposes. There is a need to explore how bullet points could influence the attention of readers. Some aspects of typographic design are arranged as various data models. However, there is a limited understanding of how this data is transformed into information and knowledge. This study aims to explore ordinary people's understanding of bullet points and their various functions in digital applications. In order to achieve this, a pilot study was conducted. The relationship between people's attended information and elicited emotional changes was investigated. The relationship between emotional concerns and communication design those implied in bullet point systems application would be understood. The insights put forward in this study will suggest approaches that will enable information engineers to prepare the foundations of a well-organised document.

Keywords: Bullet point systems · Information hierarchy · Usability · Typographic design

1 Introduction

Many people employ bullet point systems in their digital communication. As most readers tend to skim-read instead of concentrating deeply on the text, bullet points are the most useful tool to present information effectively and hold readers' attention, as well as stimulating emotional changes in readers. However, few people are concerned with how bullet points work for communication purposes; it is rare for people to recognise the bullet points and their applications in a report or article. Instead, the bullet point system is understood as an essential visual element that is applied to typographic design. There is a necessity to explore how bullet points influence the attention of readers from the point of view of typographic design.

E. Markopoulos et al. (Eds.): AHFE 2020, AISC 1218, pp. 255–261, 2020.
https://doi.org/10.1007/978-3-030-51626-0_32

2 Information Hierarchy

Influenced by the computerised media, readers changed their reading habits. In the past, most readers are deep readers, and they had plenty of time to read. However, in the era of the World Wide Web, technological development is employed to organise information that is more complex. The amount of information bloom as well. Readers thus are leaded to read the information they received in minimal time. Scholars investigated how the classes of information were organised. Rohrer and Swing [1] developed several "web-based information visualisation prototypes". The research team applied these prototypes into several famous information visualisation techniques on web platforms. In their study, the research team provided some detailed background knowledge of online information and analysis of information visualisation. There were three significant levels of information, included: data, documents and published information. Daum [2] conducted a further study. He pointed out that there were limited tools for planning and managing documents. Technology provided the advent of storing documents and information. However, it also created the needs for managing information. He carried out the importance of making communication policy as well as systematic and scientific control of information. Information engineers should be aware, in both paper and web-based media, information of individual documents had to be successfully retrieved and identify.

3 Bullet Point System

In order to present the information effectively for holding readers' attention as well as stimulate their emotional changes, bullet point system mostly are employed as tools for organising information according to a certain logical sequence. It is working as tools in the concept development of the information hierarchy; scholars suggested various hierarchies. Some comment shapes such as circular, dot, arrow, and square were adopted in some important word processing applications. Summarised the research result of Bly [3] and previous studies conducted by Blokzijl's research team [4], Ho [5] identified the categories of bullet point systems. They include 'regular symbols', 'punctuation', 'word dividers', 'general typography', 'intellectual property', 'sequence symbols', and 'graphical dividers' (Fig. 1). They included common symbols, icons, as well as graphics. These systems were employed in text-driven documents.

In some cases, the bullet point systems were drawn in handwriting, and thus various styles would be employed. Among the categories, the 'index symbol' which is the oldest style, were employed in classic and formal document frequently. Several aspects including but not limited to wisdom hierarchy, knowledge hierarchy, information hierarchy. Also, some of the categories are arranged as various data models. These concepts refer to sets of the system for representing structural and functional relationships between data, information as well as knowledge. However, there is limited understanding of how data transferred to information and knowledge. Under the concepts of emotion and design study, the close relationship between emotion and communication design were explored. The theories of communication design may provide possible hints for modifying the application of bullet point systems. Their effectiveness of constructing information hierarchy would be enhanced.

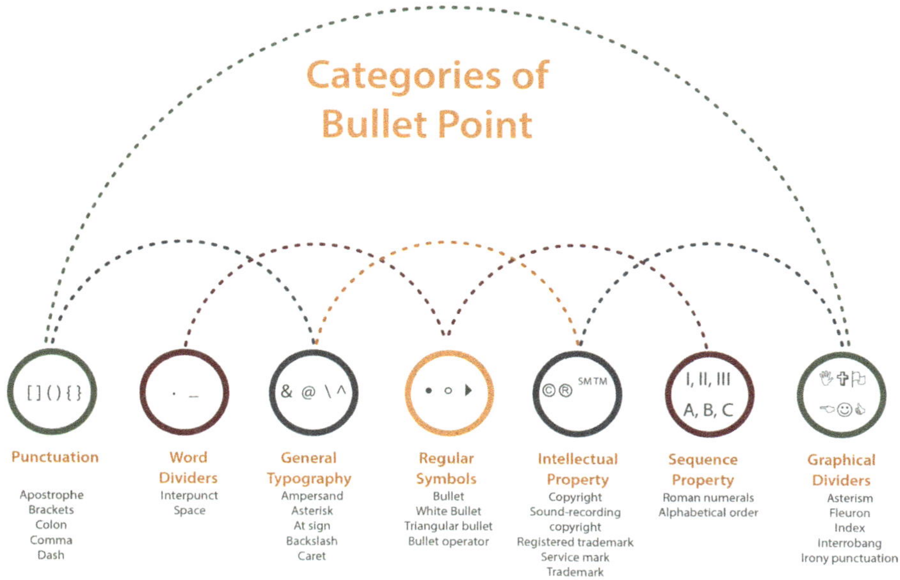

Fig. 1. Model of bullet point system in information hierarchy

4 Model of Bullet Point System in Information Hierarchy

In order to present information effectively, a bullet point system is employed for organising information according to a certain logical sequence [6]. After reading the previous studies, a model of how data is transformed into information and knowledge was developed (Fig. 2). Within the concept of the bullet point system in an information hierarchy, three core elements were included: component, type, and system. The component refers to all elements included in organising information; the type refers to a saved and named variation of a family, which is often based on standard or recognisable characteristics; and the system refers to all information that has been defined. In the process of constructing the information hierarchy, components of information were categorised into various types, and these types were compiled into a system.

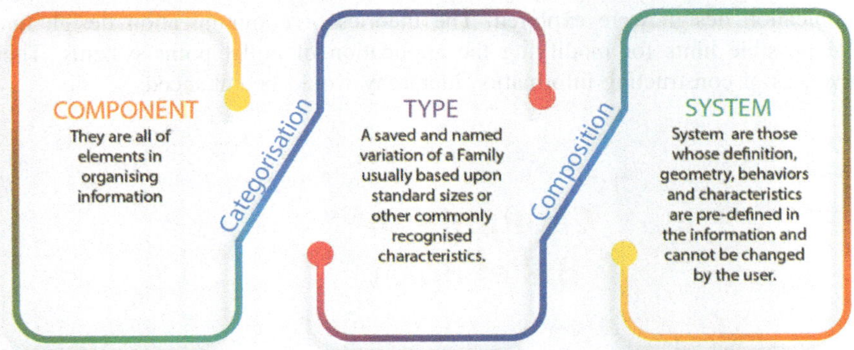

Fig. 2. Model of bullet point system in information hierarchy

5 A Pilot Study for Understanding the Effectiveness of Bullet Point System in Information Hierarchy

This study aimed to explore the understanding of ordinary people on bullet point and their applications in digital applications. In order to achieve this, a pilot study was conducted for collecting both quantitative and qualitative data for analysis.

5.1 Research Methods

In the first phase of this pilot study, all thirty participants were invited to answer a questionnaire about their using habit of bullet point systems without any professional knowledge of typographic design. The objective of this stage is to find out the recognised types of bullet point and application of them from readers' perspectives. The current problem of applying bullet point systems would be explored. In the second phase, ten participants (who took part in the first phase) were invited to share their feedback about how they feel when they are reading as well as employing bullet point systems in their document. All participants were randomly invited through online platform. They were all ordinary readers who did not obtain any professional knowledge of typographic design. Thus, they provided appropriated feedbacks and knowledge about how ordinary readers rationally read the information.

In the questionnaire about their using habit of bullet point systems, ten sets of bullet point were shown (Fig. 3). Participants were asked to identify the types of the bullet point system. They also asked to compare which type of bullet point system would provide the most effective information hierarchy. In the second phase interview, some difficulties of applying bullet point were reflected by the feedback provided by the participants.

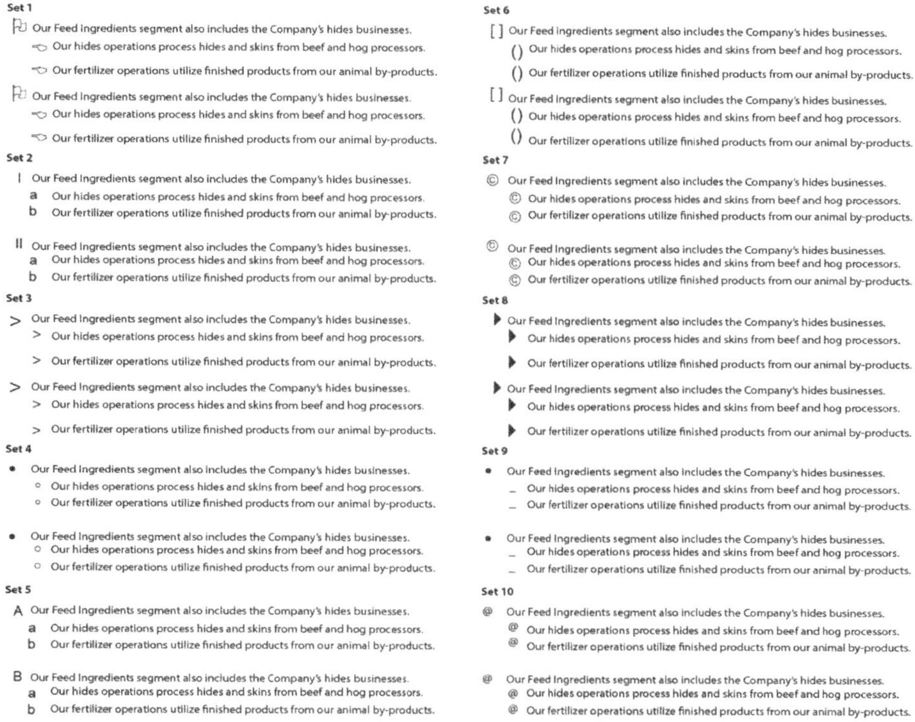

Fig. 3. Ten sets of bullet point were shown for asking participants to identify the types of the bullet point system.

5.2 Research Results

The relationships between their attended information and elicited emotional changes were investigated. The relationships between emotional concern and communication design those implied in bullet point systems application were briefly understood. The insight proposed in this study suggested the factors of information organisation to the information engineers (Figs. 4 and 5).

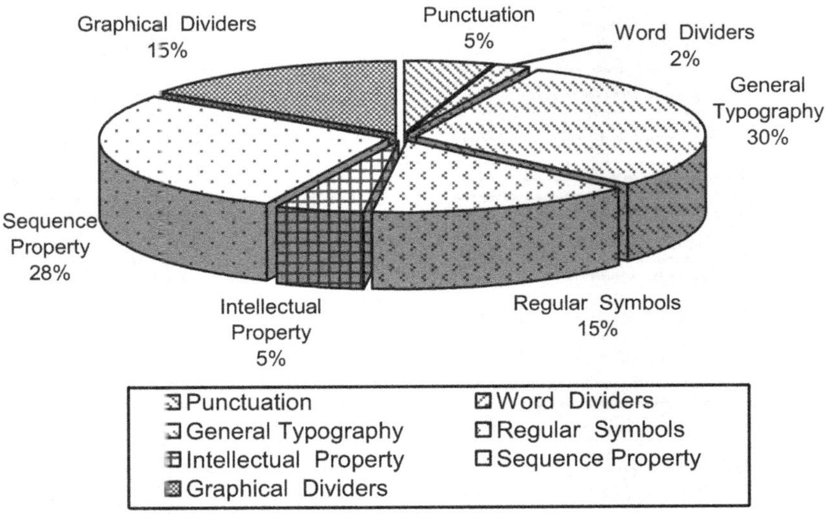

Fig. 4. Pie-chart that shown the most identifiable type of bullet point system.

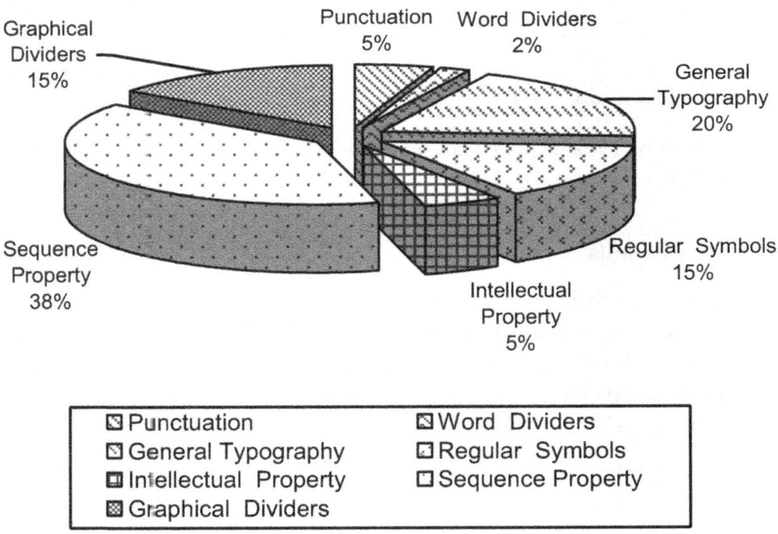

Fig. 5. Pie-chart that shown the most effective type of bullet point system in constructing the information hierarchy.

6 Conclusion

The objective of this study was to explore the recognised bullet point types, such as ‘punctuation’, ‘word dividers’, ‘regular symbols’, ‘intellectual property’. Some applications of them from readers’ perspectives were reflected through the questionnaire. The current problem of applying bullet point systems were explored. In the second phase, some participants were invited to answer a questionnaire about how they feel when they are reading as well as employing bullet point systems in their document without any professional knowledge of typographic design. The relationships between their attended information and elicited emotional changes would be investigated as well. The relationships between emotional concern and communication design those implied in bullet point systems application would be understood. More investigations on approaches for information engineers preparing the foundation for the well-organised document were needed.

References

1. Bly, R.W.: The case against PowerPoint. Success. Meet. **50**(12), 50–52 (2001)
2. Blokzijl, W., Andeweg, B.: The effects of text slide format and presentational quality on learning in college lectures. In: Proceedings of International Professional Communication Conference, pp. 288–299 (2005)
3. Rohrer, R.M., Swing, E.: Web-based information visualisation. IEEE Comput. Graph. Appl. **17**(4), 52–59 (1997)
4. Daum, P.B.: Technology and the four-level information hierarchy. Inf. Manag. **31**(4), 8 (1997)
5. Ho, A.G.: Bullet points: applying emotional symbols in information management. In: Ho, A. (ed.) Advances in Human Factors in Communication of Design. AHFE 2019. Advances in Intelligent Systems and Computing, vol. 974. Springer, Cham (2020)
6. Mosenthal, P.B., Kirsch, I.S.: Toward an explanatory model of document literacy. Discourse Process. **14**(2), 147–180 (1991). https://doi.org/10.1080/01638539109544780

The Invisible Art of Storytelling and Media Production

David Kei-man Yip(✉)

OUHK, Hong Kong, People's Republic of China
dkmyip@ouhk.edu.hk

Abstract. As the early filmmakers were magicians, cinema has developed from the early days of visible 'tricks' to invisible 'effects' [1]. As narrative cinema is to engage the audience with the story, many of the cinematic effects have been made to look invisible to the untrained eyes and ears. Film is both magical and make-believe. Like a magician who manages to hide his or her trick in a magic act, a filmmaker often hides his or her magic tricks from the audience. The invisible art of hiding cinematic effects is one of the important skills in media production. In storytelling, the hidden subtext can be as important as the visual text. This comprehensive article discusses the invisible art of storytelling and filmmaking.

Keywords: Art of storytelling · Cinematic arts · Creative arts · Film art · Content design · Narrative content · Media production · Invisible editing

1 The Invisible Story Elements

A film can be structurally and conceptually divided into three basic story elements: pre-text, text and subtext. A pre-text of the film is a precursor of a text. It inspires the creation of a text in the form of concept or theme. The film text consists of visual images of characters and locations put together through the cinematic vision and artistic treatment of the director and his or her key crew. Subtext is the different meanings a film can be read beneath the text. Except for the film text, all the elements in the pre-text and subtext can be considered as the invisible contributions of the film.

The writer of the film is responsible for developing all the pre-texts of a film. It is a long creative process where the writer or writers work repeatedly to refine their concepts and ideas that lay the foundation or creation of the text. Pre-text cannot be directly told or shown to the audience. It can only be visualized to the audience through the production of the text expressed in the visual or cinematic style of the film director. Pre-text often provides guiding principles to what and how the rest of the film text should look. Design sketches of the characters are made to turn abstract concepts into concrete visual materials for the different stages of pre-production. Meantime, the writer or writers are refining the script to turn different levels of character conflicts [2] into words. The visual text gives ways to the different interpretations of the pre-text and subtext. Ernst Hemingway's famous short story illustrated a textual relation between the text and the hidden subtext in storytelling. The famous author entered a short story contest with his fellow authors as the pre-text of his short story. He wrote six words to

E. Markopoulos et al. (Eds.): AHFE 2020, AISC 1218, pp. 262–266, 2020.
https://doi.org/10.1007/978-3-030-51626-0_33

form the text of the story, which was later known to be the shortest story ever told. His famous six-words were '*For sale: Baby shoes. Never worn.*' Without telling what happened directly, the text simply implied a tragic story behind the meanings of the text. This example illustrated less can be more through the powerful implication of the unseen and the unsaid.

2 Inner-Conflict as Pre-text

One of the pre-texts in the conceptual stage is conflict design. Dramatic conflict and tension make a story interesting. In a nutshell, conflict and tension can be described metaphorically as if two opposite forces were pulling a string from two sides with increasing strength and tension until the string breaks. To describe it in human term, drama is often about a person up against all odds with increasing internal and external conflicts and tensions imposing on him or her until he or she eventually makes it or not. According to Robert Mckee [2], character conflict has three levels: inner, personal and extra-personal. The latter two are respectively about one's relation with others and society and therefore these conflicts are physical and visual in nature. Inner conflict is internal and almost always hidden from the surface. Designing inner conflict is one of the main tasks of a screenwriter. Filmmaker must visualize the inner conflict of a character externally through film languages such as mise-en-scéne, which includes staging and the use of motif or symbol [3].

3 Three-Acts with Rising Action as Text

The classic Three-Act structure has been used to describe the arc of storytelling with rising action. The First Act, typically the first twenty to third minutes of a feature movie, is used to set up the story premise that establishes the major conflict of the drama through a crisis or a series of crisis. The Second Act, the large middle part of the story, is used to complicate or intensify the conflicts with rising tension as the complication of the key conflict makes the drama. Increasing the stake of the protagonist is one common way to increase the tension of a story. The Third Act of the story reaches the climax of this tension building followed by the final confrontation and resolution of the whole journey of the hero or protagonist. Establishing internal and external conflict of the Act I is followed by intensifying these conflicts with increasing tension as rising action in the Act II.

Alfred Hitchcock [4] the Master of Suspense described tension or suspense building with his famous analogy of a ticking time bomb planted underneath the table of the characters without them knowing about it. As time passed and the 'bomb' got closer to explosion by the second, level of tension and suspense would rise and build up. Since the ticking time bomb was an analogy to describe a plot point or situation that resembled the function of a ticking time bomb for building tension on the screen, its presence in the story was not supposed to be taken visually and literally. Hitchcock's time bomb analogy illustrated a direct relation between time and tension building when these two invisible elements worked together to create dramatic effects.

4 Invisible Editing

Cutting in film can create sense of rhythm like music. Alfred Hitchcock [4] compared film shots as music notes played by different music instruments. When different sizes of shot are put together in a montage sequence, it has a structure similar to music. Invisible editing can be achieved by the smooth transition of shots that move the story forward with consistent match of shot composition, action continuity, consistent flow of screen movement, direction and/or eyeline match to hide the cuts with good rhythm. Oscar-winning film editor and editing theorist Walter Murch [5] offered his insight about why cuts worked. His famous Rules of Six in editing stated the six important criteria in making a good cut, which include the same criteria of making invisible cut mentioned above. What made Murch's rules special was emphasis on cutting on emotion. Walter Murch believed cutting on emotion was the most important criteria in achieving seamless cuts in editing. He used blinking of the eye as both a metaphor of an emotional cue for making a seamless cut as if a cut was '*blinking for the audience*'. On the contrarily, the cinema of the French New Wave countered this invisible approach to filmmaking and editing. It treated cinema as an expressive art form with a very different philosophy of using discontinuous editing such as jump cut as a cinematic style. Unlike invisible editing of continuity, jump cut discontinuity operated a different sense of rhythm outside the Murch's Rules of Six but by the subjective freestyle impulse of the editor. Jump cut discontinuity was used to deliberately remind the audience to be conscious about film as an expressive art form. It asked the audience to detach themselves emotionally from the film and to engage with it intellectually.

The ending shot of the movie *Eternal Sunshine of the Spotless Mind* had an interesting and subtle moment with a hidden message that could be unnoticeable to typical audience. There was a series of repeated jump cut looping the ending wide shot where the two lovers were walking and fading away from the scene. Such an editing style was a rare choice in film, especially in Hollywood mainstream film. As discussed earlier, jump cut could disrupt space time continuity and is considered a violation of invisible editing. However, the repeated jump cut loop in this ending shot was not out of the film context. Quite the opposite, this editing choice fitted well with the context, pre-text and the subtext of the film. The pre-text of the film was about eternal love and fate. The editing implicitly stated that no matter how many times the lovers tried to re-set or erase their memories in their life, they would still end up together as lovers and in the same spot holding hands over and over again. It was a cinematic way to express and embrace the theme of eternal love of life. This example also illustrated how the invisible pre-text and subtext were expressed through the editing style of the visible film text.

5 On-screen Monster vs. Off-Screen Monster

Some movie fans of sci-fi monster genre have been debating whether *Godzilla* can beat the nameless monster of another monster movie *Cloverfield*. The sizes of both monsters are mega huge but the one big difference of the two monsters is that the nameless monster in the latter hardly appeared in the movie. Such a visual treatment of the invisible 'presence' can be comparable to the famous killer jaw in the original classic movie *Jaws*, in which the killer jaw also hardly appeared in the movie. This classic movie has put the then young director Steven Spielberg on the road to become one of the great film masters in film history. The absence of the giant *Cloverfield* monster or the killer jaw has created an invisible level of cinematic suspense and tension in film as if they could be anywhere and just pop up anytime in the movie. Such a cinematic treatment of the invisible 'presence' also illustrates the power of the unseen and the meaning of less is more.

6 Invisible Visual Effects

Long before there were green or blue screen shooting for compositing effects, generations of filmmakers have used their camera's perspective and matte painting technique to create cinematic effects and managed to hide them in a shot from the audience. Matte painting, developed from double exposure in photography, has long been a common special effect in film to replace the background of a shot with hand-drawn painting to give an illusion of a different location [6]. Charlie Chapman's famous breathtaking scene in *The Modern Times* was a work of matte painting to give an illusion that he was roller-skating dangerously at the edge of a high floor when the background of the lower floors was just a painting matted on the shot. Hand-drawn matte painting technique was still being used in the early Star Wars films in the 70s before George Lucas started his digital transformation in film effects. Since then, hand-drawn matte painting has gradually been replaced by digital compositing through blue or green screen studio shooting. Compositing is made by removing the color of the background with image of other location to create an illusion of another location. To make the illusion seamless, the lighting and other conditions of the added background must match consistently with the studio conditions of the shooting set. Nowadays, new emerging digital technology such as real-time rendering has eliminated this problem and made compositing effects even more incredibly seamless than before. It has taken advantage of the latest real-time immersive virtual reality technology that enables the composite technology goes beyond concept of layering and allowing actors to immerse themselves in the 3D virtual studio with hyper-realistic quality. Actors still need to perform in front of the green screen but instead of compositing the image by layering, a 3D virtual environment will replace the green screen altogether combined with real-time automatic camera tracking [7] to match the motion and do the compositing instantaneously. Within this 3D virtual set, the actors' own shadow and reflection can be stimulated digitally in real-time with hyper-realistic picture quality as if the actors were being instantly teleported to another three-dimensional environment. Instead of the conventional way of changing the background and matching it with the foreground, the new technology allows the actors to be immersed into a new 3D virtual environment. From hand-drawn matte painting to hyper-realistic virtual studio, new digital technology has made the compositing effects become invisible to the typical audience. AI automation has been used in recent years to reduce the tedious and time-consuming in-between animation works required in key frame animation. Instead of spending a great deal of time to draw all the in-between frames, animators can now use AI technology to generate and clean up this uncreative production process, down to individual strokes with superb quality.

7 Conclusion

Digital technology has gone from modem dialing speed to real-time lighting speed, which has delivered us endless supply of visual content instantaneously and interactively. Living in the visual world with the abundance of high-resolution big screens and media platforms, the audience is constantly tempted to thirst for more visual stimulus and spectacles. It's an endless cycle of feeding and wanting more. Arguably, our visual senses might have already been desensitized or even become addicted by the constant bombardment of digital images. With so much attention being focused on the visual, the intangible contributions of the invisible art of storytelling and media production are not to be ignored. In storytelling and media production, the invisible is as important as the visible. The visible design attracts eyeballs but it is the invisible design in storytelling and filmmaking that truly capture the audience's imagination by making the magic appear to be real and the trick of making the magic real invisible. This article pays tributes to all the creative and innovative minds behind the scene that continue to make cinematic magic invisible. It calls for more attention on the hidden contributions of the invisible art of storytelling.

References

1. Christie, I.: The visible and the invisible: from 'tricks' to 'effects'. Early Pop. Vis. Cult. **13**(2), 106–112 (2015)
2. McKee, R.: Story: Style, Structure, Substance, and the Principles of Screenwriting. Harper Collins, New York (1997)
3. Mamet, D., Block, B.: On Directing Film. Penguin, London (1991)
4. Hitchcock, A.: Hitchcock on Hitchcock, Volume 2: Selected Writings and Interviews. University of California Press, California (2014)
5. Murch, W.: In the Blink of an Eye, vol. 995. Silman-James Press, Los Angeles (2001)
6. Vaz, M.C., Barron, C.: The Invisible Art: The Legends of Movie Matte Painting. Chronicle Books, San Francisco (2002)
7. Dobbert, T.: Matchmoving: The Invisible Art of Camera Tracking. Wiley, Hoboken (2006)

The Impact of Website Design on Users' Trust Perceptions

Kairi Fimberg(✉) and Sonia Sousa

School of Digital Technologies, Tallinn University,
25 Narva Rd, 10120 Tallinn, Estonia
kairi.fimberg@gmail.com, scs@tlu.ee

Abstract. With rapid digitalization, trust has become a critical issue in designing and maintaining e-commerce platforms – without trust, no transaction takes place. Companies that design for trust have a strategic advantage over competitors. Although trust is a crucial factor in e-commerce, designing a trustworthy website can be challenging for companies that make most or part of their profits online. The study builds on prior research to propose a comprehensive and up-do-date checklist. Trust components are divided into three dimensions of website design; their impact on users' trust perceptions is studied in an online experiment with two websites. The results present demonstrable evidence that website design has a powerful impact on users' trust perceptions. Professional design, primarily visual aspects, is responsible for creating a positive user first impression. Furthermore, additional trust is built through different dimensions of website design, increasing the likeliness of buying from the site.

Keywords: Website design · User trust · Trust components · Trust design · Visual design · Content design · Social-cue design · Trust assessment

1 Introduction

A good part of communication previously carried out between humans now relies on technology and human-computer interaction. With Internet technologies and infrastructures to support e-commerce mostly established – although always evolving –, focusing on the psychological factors that affect e-commerce acceptance by online users has been on the rise. One such factor, playing a significant role in the success of e-commerce, is trust [17].

For a company, a website is often the first point of contact for potential customers, responsible for first impressions and generating revenue. Companies that design for trust have a strategic advantage over competitors. Without trust, visitors leave the website immediately. The probability of leaving is higher during the first seconds. While most companies know they need to have trust components on a website, they are often overlooked because sometimes, it does not provide measurable value [3].

Research problem and significance. It has been shown that web browsing exhibits a significant "negative aging" phenomenon, meaning that some initial screening has to be passed before a page is examined in detail, giving rise to the browsing behavior called

E. Markopoulos et al. (Eds.): AHFE 2020, AISC 1218, pp. 267–274, 2020.
https://doi.org/10.1007/978-3-030-51626-0_34

"screen-and-glean" [12]. To gain only several minutes of users' attention, a website must clearly communicate its value proposition within 10 s [13]. A study by Bentley University Design and Usability Center demonstrates that when people first navigate to a website, businesses have about 6 s to create a positive impression with users [1]. This is an extremely small window of time to convince users that the website is one that they can trust making a purchase from.

Although trust is a crucial factor in e-commerce and the concept has been widely studied by several authors, a comprehensive, easy-to-follow checklist of components that companies could utilize when designing their website was either missing or outdated. Nevertheless, almost every company struggles with website credibility, as visitors are immediately skeptical [14].

Goal of the research. In line with the aforementioned shortcoming, the study aims to understand the components in website design needed to build a trustworthy website; and to propose an up-do-date design checklist. The following research questions were formulated to reach this goal.

RQ1: How important is visual design as an initiator of trust?
RQ2: How is trust influenced by different dimensions of website design?
RQ3: How important is trust-inducing design for the purchase decision?

2 Background

Trust is a common and essential concept in different domains, the term has been defined in different ways, and there is no widely-accepted definition [6, 15]. Trust and trust relationships in the offline world have been a topic of research in various disciplines, such as philosophy, sociology, psychology, management, marketing, etc.; and each of these disciplines has produced its own concepts, definitions, and findings [7].

As a social being, trust is incorporated into every aspect of human life. Trust is essential for all kinds of personal relationships, "the loom on which is woven the social fabric of society" [4]. Without trust, social life breaks down, no business transactions take place, new technology is not adopted, and even political legitimacy collapses [11]. Trust is the basis for decision making in many contexts, and the motivation for maintaining long-term relationships based on cooperation and collaboration [6].

As emphasized by several authors, trust is the key to the success of e-commerce [11, 16, 17], and a prerequisite for actions involving another agent in which one may suffer physical, financial, or psychological harm [2]. Lack of trust has been identified as one of the most formidable barriers for engaging in e-commerce, involving transactions in which financial and personal information is shared [8, 18, 19].

Establishing trust relationships in a digital environment involves more aspects than in the social world because communications in the computing network rely on not only relevant human beings and their relationships but also digital components [19].

The model of trust for e-commerce (MoTEC) by Egger provides a framework of six components, regrouped into three more significant categories: pre-interactional filters (pre-purchase knowledge), interface properties, and informational content [8]. To render the design process more coherent, the model components have been redistributed into three qualitatively different types of requirements: appeal (graphic design), usability (structure and navigation design), and trustworthiness (content design). Cheskin Research focuses on website interface cues and presents a model of six primary components that play a major role in communicating trustworthiness [5]. The building blocks of trust are seals of approval, brand, technological sophistication, navigation, presentation, and fulfillment. These blocks, in turn, can be divided into a total of 28 components that can be used to communicate functional trustworthiness. Patel lists over 40 factors that influence website credibility [14], using the four types of credibility by Fogg and Tseng [9]. *Presumed credibility*: general assumptions in the perceiver's mind; *reputed credibility*: what third parties have reported; *surface credibility*: based on simple inspection; *earned (experienced) credibility*: first-hand experience, reputation built over time. According to Patel, "The goal with each of these credibility factors is to stack the deck in your favor" [14].

From the perspective of the current study, grouping trust components within website design dimensions seems to be the most fitting framework. Elaborating on the framework by Wang [17], that allocates trust components in three design dimensions (visual, content, and social-cue design), and considering the components of a trustworthy website suggested by other cited authors, the author of the current study proposes a refined checklist of trust-inducing design components (shown in Appendix). The last dimension, the social-cue design, is amended to incorporate both social presence and social proof components.

3 The Study

To answer the research questions and test the checklist of trust components proposed, an experiment was carried out online, focusing on the quantitative results collected from two different websites of Estonian furniture manufacturers. The A/B test featured a between-subjects study design. The questionnaire guided participants through four separate parts: first impression, design assessment, trust assessment, and final comments. A 7-point Likert scale was used for all questions. To evaluate participants' trust perception, the trust assessment model by Gulati et al. [10] was put into practice. Lookback.io, an online user experience and screen recording platform, was used to gain additional insight into users' browsing behavior.

A pilot study was carried out on a small group of participants to evaluate the time and statistical effect to predict the appropriate sample size, plus test the experiment protocol. The sampling technique used was convenience sampling, which is often used in business studies to gain initial primary data regarding specific issues like the perception of an image of a particular brand or opinions of a new design.

A total of 50 participants were recruited for the study. Participants were randomly assigned into two groups, 25 in each. In Group A, there were 17 female and 8 male participants. The majority of them fell in two age groups, 25–35 (13 participants) and 35–44 (11 participants); 1 participant was older, aged 55–64. In Group B, the gender was slightly more equal; there were 14 female and 11 male participants. The majority of them fell in the same two age groups, 25–35 (9 participants) and 35–44 (15 participants); again, 1 participant was older, aged 55–64. Most of the participants in Group A and B were Estonians, with 2 and 1 Russian respectively. In both groups, 23 out of 25 participants had higher (tertiary) education, 2 had secondary education. A majority of participants shopped online regularly.

4 Results and Discussion

The results of the study present demonstrable evidence that website design has a powerful impact on users' trust perceptions. The data revealed that the website with an attractive and contemporary design (the perceived average quality of design in Website A was 5.71; first impression 5.67) implicated considerably higher trust than the website with a dull and outdated design (the perceived average quality of design in Website B was 3.54; first impression 3.16) (Fig. 1).

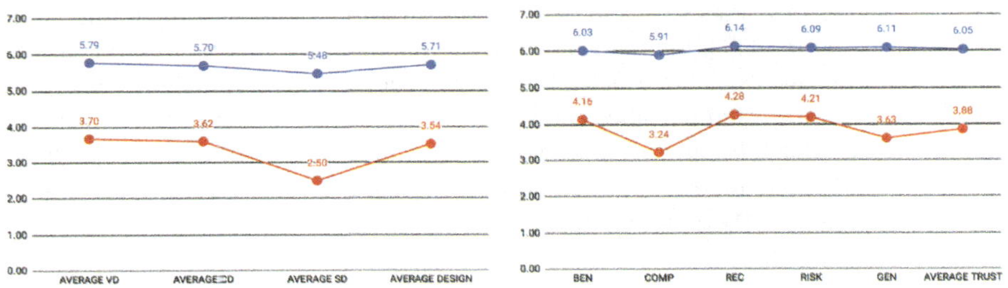

Fig. 1. Perceived quality of website design *(left)* and perceived trust *(right)* in Website A *(blue)* and Website B *(red)*. The three design dimensions are visual design (VD), content design (CD), and social-cue design (SD). Altogether 21 trust-inducing design components were evaluated.

In order to measure users' trust perceptions, participants were asked about benevolence (BEN), competency (COMP), reciprocity (REC), risk (RISK), and general trust (GEN) — the constructs coined by Gulati et al. [10]. The overall trust level of Website A was 6.05 (86.41%), while the overall trust level of Website B was 3.88 (55.39%). Based on the trust model, scores between 80–90% are considered as a high trust level, while scores between 50–60% are considered as a very low trust level.

To find answers to the research questions, correlation analysis was done (see Table 1).

RQ1 → strong positive correlation. Participants' browsing behavior confirmed that when visitors first come to a website, they develop their first impression in a matter of seconds. The average time spent on Website A was 1 min 16 s, during which, on average, 7.81 clicks were made by 87.5% of participants. In contrast, the average time spent on Website B was only 32 s, during which, on average, 2.17 clicks were made by 25% of participants

Professional design, primarily visual aspects, is responsible for creating a positive first impression, which, in turn, is strongly correlated with trust, leading us to believe that visual design is a vital initiator of trust.

RQ2 → strong or moderate positive correlation (with some exceptions). The influence on users' trust perceptions is active within all three design dimensions (visual, content, and social-cue design), indicating the importance of them all when creating or redesigning a website. There was no significant correlation between some of the design components and trust. This, however, does not mean that website visitors think of these components as unimportant. It tells us that trust perceptions are not based on one component only but rather on a collection of them. Lacking in quality of some components does not significantly decrease the overall trust level, as having a few key components does not increase users' trust to a sufficient level.

RQ3 → strong positive correlation. Similarly, the study affirmed the importance of trust-inducing website design for the visitor's purchase decision, i.e., whether the website succeeds in converting a visitor into a customer. The process chain here is the following: strong design fosters higher trust; higher trust makes it more likely that a visitor engages in a purchase decision. However, as with the first impression and interest, visitors' purchase decisions also depend on other factors, like their actual need for the product or service, and whether they can afford it.

Table 1. Correlation analysis.

Correlations (Pearson)					
	Visual design	Trust		Design	Likeliness of buying
First impression	.846**	.782**	Trust	.898**	.760**
Visual design	1	.845**	Design	1	.763**
Correlations (Pearson)					
		Trust			Trust
VD1 [Design]		.815**	CD1 [Brand information]		.727**
VD2 [Color]		.840**	CD2 [Company information]		.238
VD3 [Font]		.811**	CD3 [Contact information]		.415**
VD4 [Images]		.803**	CD4 [Content]		.718**
VD5 [Search]		.405**	CD5 [Blog/news page]		.822**
VD6 [Navigation]		.725**	CD6 [Grammar]		.731**
VD7 [Links, buttons, forms]		.619**	CD7 [Product information]		.653**
VD8 [Technical functioning]		.231	CD8 [Price information]		.481**
		Trust	CD9 [Policies]		.775**
SD1 [Customer service]		.631**	CD10 [Guarantees/warranties]		.748**
SD2 [Social presence]		.753**	CD11 [Important questions]		.541**

** Correlation is significant at the 0.01 level (2-tailed). N = 50.
VD = visual design, CD = content design, SD = social-cue design.

The independent samples t-test was used to determine whether there was statistical evidence that the means of two groups were significantly different. All results (first impression, design, trust, likeliness of buying) came back as significantly different.

The trust assessment model by Gulati et al. [10] proved to be a reliable tool for measuring trust in a website, with the Cronbach's alpha coefficient of 0.969.

5 Limitations and Further Research

The main limitation was the stimuli. The two websites used in the study differed in terms of product design and style. The product itself can influence the first impression, trust, and the likeliness of buying. Although the results of Website A were significantly different from those of Website B, ideally, we should have an identical product that is unknown to all participants, with the only difference being the website design, if we wanted to measure solely the impact of website design on users' trust.

While the setup of this study was able to list the trust-inducing components and provide quantitative results to confirm the relationship between design and trust, further research is required to investigate the importance of these components in different situations (their effects based on users' gender and age, for example, or a company's field of business), and how to strategically place them into the website.

A more detailed discussion and suggestions can be found in the author's full thesis.

Acknowledgements. This project has received funding from the European Union's Horizon 2020 research and innovation program under the NGI_TRUST grant agreement no. 825618.

Appendix: The Design Checklist

Dimensions	Explanations	Trust-inducing design components
Visual design	Defines the graphical design aspect and the structural organization of displayed information on the website	✓ Professional design ✓ Color scheme to suit the product/service ✓ Nice and legible fonts ✓ High-quality (and authentic) images and visuals ✓ Good on-site search ✓ Easy-to-use navigation ✓ Clear anchor text and microcopy ✓ No technical problems (broken links, missing pictures or pages)
Content design	Refers to the informational components that can be included in the website, be they textual, graphical, etc.	✓ Brand-promoting information (logo, slogan) ✓ Company information ("About" page, facts & figures) ✓ Contact information ✓ Physical address ✓ Useful (expert-level) content ✓ Good grammar, minimized jargon

(*continued*)

(*continued*)

Dimensions	Explanations	Trust-inducing design components
		✓ External links (sources) ✓ Up-to-date blog/news page ✓ Clients (client logos) ✓ Client case studies ✓ Product information ✓ Price information ✓ Order information (transaction reports) ✓ Clear policies (privacy, return) ✓ Guarantees and warranties ✓ Helpful FAQs ✓ Trust seals ✓ Awards
Social-cue design	Relates to embedding social and interpersonal cues, such as social proof, social presence and face-to-face interaction, into the website via different (communication) media	✓ Staff photos and bios ✓ Easy access to customer service (e.g., contact form) ✓ Instant messaging/chat option ✓ Social presence (social media) ✓ Testimonials ✓ Reviews ✓ Reviews from influencers and notable customers ✓ Professional product reviews (from review sites, bloggers, customers) ✓ Press articles (media logos)

References

1. Albert, B.: How quick are we to judge? A case study of trust and web site design, 6 June 2012. https://www.slideshare.net/. Accessed 9 Jan 2019
2. Bickmore, T., Cassell, J.: Relational agents: a model and implementation of building user trust. In: Proceedings of the SIGCHI Conference on Human Factors in Computing Systems, vol. 3(1) (2001)
3. Blue Fountain Media (n.d.): 8 trust factors to help build a website and convert users [Blog post]. https://www.bluefountainmedia.com/. Accessed 10 Jan 2019
4. Cassell, J., Bickmore, T.: External manifestations of trustworthiness in the interface. Commun. ACM **43**(12), 50–56 (2000)
5. Cheskin Research: Trust in the wired americas, July 2000. http://saywhatcr.com/. Accessed 15 Jan 2019
6. Cho, J.-H., Chan, K., Adali, S.: A survey on trust modeling. ACM Comput. Surv. (CSUR) **48**(2), 1–40 (2015)

7. Corritore, C.L., Kracher, B., Wiedenbeck, S.: On-line trust: concepts, evolving themes, a model. Int. J. Hum.-Comput. Stud. **58**(6), 737–758 (2003)
8. Egger, F.N.: Trust me, I'm an online vendor: towards a model of trust for e-commerce system design. In: Extended Abstracts on Human Factors in Computing Systems (2000)
9. Fogg, B.J., Tseng, H.: The elements of computer credibility. In: Proceedings of the SIGCHI Conference on Human Factors in Computing Systems (1999)
10. Gulati, S., Sousa, S., Lamas, D.: Modelling trust in human-like technologies. In: Proceedings of the 9th Indian Conference on Human Computer Interaction (2018)
11. Hasan, Z., Gope, R.C., Uddin, M.N.: Do aesthetics matter in long-established trust? Int. J. Comput. Appl. **77**(13), 31–36 (2013)
12. Liu, C., White, R.W., Dumais, S.: Understanding web browsing behaviors through Weibull analysis of dwell time. In: Proceedings of the 33rd International ACM SIGIR Conference on Research and Development in Information Retrieval (2010)
13. Nielsen, J. How long do users stay on web pages? September 12, 2011. https://www.nngroup.com/. Accessed 15 Jan 2019
14. Patel, N. (n.d.): 41 Factors that influence your website's credibility [blog post]. https://neilpatel.com/. Accessed 15 Jan 2019
15. Rousseau, D.M. Sitkin, S.B., Burt, R.S., Camerer, C.: Not so different after all: a cross-discipline view of trust. Acad. Manag. Rev. **23**(3), 393–404 (1998)
16. Tan, F.B., Sutherland, P.: Online consumer trust: a multi-dimensional model. J. Electron. Comm. Organ. (JECO) **2**(3), 40–58 (2004)
17. Wang, Y.D.: Trust in B2C E-commerce interface. In: Encyclopedia of Information Science and Technology, 1st edn. (2005)
18. Wang, Y.D., Emurian, H.H.: An overview of online trust: concepts, elements, and implications. Comput. Hum. Behav. **21**(1), 105–125 (2005)
19. Yan, Z., Holtmanns, S.: Trust modeling and management: from social trust to digital trust. In: Computer Security, Privacy and Politics: Current Issues, Challenges and Solutions (2008)

Research on the Influencing Factors of the Communication Effect of Tik Tok Short Videos About Intangible Cultural Heritage

Siqi Wang and Rongrong Fu(✉)

East China University of Science and Technology, No. 130, Meilong Road, Xuhui District, Shanghai, People's Republic of China
elkwang@126.com, muxin789@126.com

Abstract. Tik Tok, as the top app of short video socialization in China, has also become a front for the spread of intangible cultural heritage in new media. However, only a few short videos on intangible cultural heritage have achieved effective transmission. Most of them still have problems such as inability to capture audience's attention and low transmission effectiveness. In order to solve these problems, we construct a theoretical model of the effect of the Tik Tok intangible cultural heritage short video transmission and a qualitative analysis based on NVivo to explore the influencing factors of the communication effect, then we propose an improving path for the creation and dissemination of short videos on intangible cultural heritage, which provides an effective method.

Keywords: Mobile short videos · Communication effect · Intangible cultural heritage · Tik Tok

1 Introduction

Nowadays, short video is the most popular field in the mobile internet industry. Since its launch, Tik Tok has sweep around the world in a short period of time and turned into the flagship product in the short video social interaction. Nowadays, it is the largest platform for the spreading of knowledge and art in China. Tik Tok has also become a front for spreading intangible cultural heritage in new media. 93% of the 1,372 national intangible cultural heritage projects have utilized the platform of Tik Tok, gaining 3.33 billion likes. However, only a few intangible cultural heritages have realized effective communication through short videos. Most of them still face problems such as inability to catch the attention of audience, low degree of users' participation, poor audience loyalty and low communication effectiveness.

When it comes to the existing research on the communication of Tik Tok short videos, no effective solutions have been proposed for the poor transmission effect of most intangible cultural heritage videos and no scientific and effective influencing factor model has been constructed yet. This study cross-applies two research methods with regards to the following three aspects: (a) Factors affecting the transmission of short videos of intangible cultural heritage on the platform of Tik Tok; (b) Construct the influencing factor model based on the Layer-of-Effects Model; (c) Qualitative

E. Markopoulos et al. (Eds.): AHFE 2020, AISC 1218, pp. 275–282, 2020.
https://doi.org/10.1007/978-3-030-51626-0_35

analysis of communication effects based on grounded theory. Based on the analyses of these influencing factors, a path strategy is proposed for short videos of intangible cultural heritage to provide an effective reference for content production and dissemination.

2 Theoretical Model Construction of Communication Effect

2.1 Theoretical Basis and Research Hypothesis

Informative Profit, Cognitive Effect and Emotional Effect. Porter believed that virtual community members could search and find valuable information in the community, which could help them learn and solve problems or make decisions [1]. It can be said that information acquisition is one of the motivations for users to use mobile social media products. The short videos that carry valuable information for users and generate the willingness to buy products can be regarded as the need of both users and video creators. Therefore, the informedness of video content can arouse the rational attraction of users. At the same time, judging the usefulness of video information can help users decide whether to participate in the communication process and whether benefits and self-efficacy can be achieved. Accordingly, this study proposes the following hypothesis:

H1: Informative profit has a positive effect on the cognitive effect of the audience.
H2: Informative profit has a positive effect on the emotional effect of the audience.

Entertaining Profit, Cognitive Effect and Emotional Effect. The recreational benefits brought by video content can make users feel happy, which shows users' pursuit of self-satisfaction [2]. Ducoffe proposed that entertainment, irritation and information were the main determinants of online advertising attitudes when studying online advertising models [3]. Tik Tok users are used to getting addictive pleasant sensation in a very short period of time, which attracts their attention and encourages them to generate a positive emotional attitude. Therefore, this study proposes the following hypothesis:

H3: Entertaining profit has a positive effect on the cognitive effect of the audience.
H4: Entertaining profit has a positive effect on the emotional effect of the audience.

Cognitive Effect, Emotional Effect and Behavioral Effect. According to Lavidge & Steiner's layer-of-effects Model, the communication effect of Tik Tok short videos at the audience level can be divided into cognitive effect, emotional effect and behavioral effect. After audiences are exposed to the short videos of intangible cultural heritage, they form a basic attitude through the cognitive and emotional effects, and finally externalize it into the behavioral effect. Consequently, this study proposes the following hypothesis:

H5: In the process of short video dissemination, the effect of audience's cognitive psychology has a positive impact on the behavioral effect.

H6: In the process of short video dissemination, the effect of audience's emotional experience has a positive impact on the behavioral effect.

As to the analysis above, this study analyzes the factors affecting the communication effect at the audience level on the basis of new media effect theory, technology acceptance model (TAM) and advertising psychological effect model. This study takes the informative and entertaining profit brought by video content as independent variables, the audience's cognitive and emotional effects as intervening variables, and the audience's behavioral effect as dependent variable to construct a theoretical model (Fig. 1):

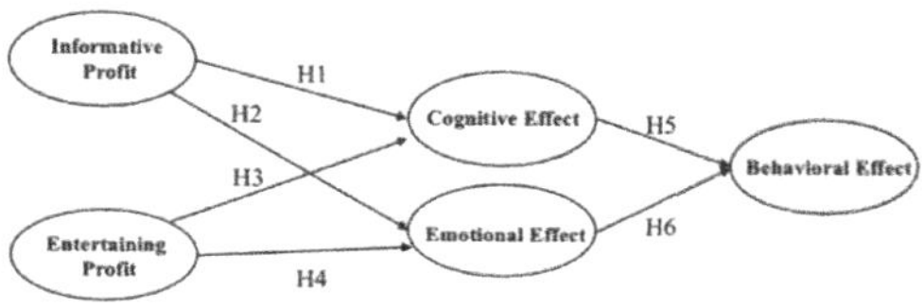

Fig. 1. Theoretical model

2.2 Research Methods and Data Collection

This study uses a questionnaire to conduct an analysis of the proposed research model and hypotheses. The questionnaire is set in the form of Likert 5 scale with reference to relevant researches at home and abroad. After modifying the items according to the actual situation, this research conducts a pre-investigation, and reviewed with experts based on the results. The final questionnaire was determined through repeated verification and modification.

A total of 400 questionnaires were distributed in this study, and 369 were actually recovered. The number of questionnaires that met the specifications was 282. The subjects were all Tik Tok users and their gender and age distribution accorded with the characteristics of Tik Tok user groups, which can be used in this research analysis.

In the choice of data processing method, the structural equation modeling (SEM) was used, which applies mathematical models and language to present the objective state of subjects. Moreover, it can effectively analyze the multivariate data relationship among potential variables, and intuitively reflects in the form of a road map.

2.3 Data Analysis and Hypothesis Testing

Reliability and Validity Analysis. With AMOS software, a confirmatory factor analysis was performed on the variables in the constructed influencing factor model, as shown in Table 1. The factor loadings are all above 0.6 and all dimensions of the model's confirmatory factor analysis are up to standard.

From the analysis of the mean and standard deviation of each variable, it can be found that the informative profit reaches a fairly high level, and the audience still has high expectations for entertainment profit. The emotional effect layer is up to the maximum value, which indicates that the formation of perceptual attitudes and the audience's emotional experience play an important role in the process of intangible cultural heritage transmission.

Table 1. Confirmatory factor analysis

Variable	Mean	S.E	Cronbach's α	CR	AVE
Standard			**>0.7**	**>0.6**	**>0.5**
Informative profit	3.697	0.048	0.887	0.886	0.568
Entertaining profit	3.270	0.052	0.939	0.938	0.720
Cognitive effect	3.862	0.044	0.859	0.859	0.551
Emotional effect	3.986	0.047	0.847	0.847	0.581
Behavioral effect	3.657	0.051	0.886	0.881	0.598

Table 2. Structural Equation Modeling path analysis

Hypothesis path	Standard path loading	S.E	C.R.	Result
H1: IP → CE	0.177^{*}	0.076	2.585	Valid
H2: IP → EE	0.107	0.081	1.581	Invalid
H3: EP → CE	0.026	0.055	0.405	Invalid
H4: EP → EE	0.150^{*}	0.061	2.268	Valid
H5: CE → BE	0.171^{**}	0.075	2.645	Valid
H6: EE → BE	0.378^{***}	0.075	5.444	Valid

P.S.: *indicates $p < 0.05$, **indicates $p < 0.01$, ***indicates $p < 0.001$

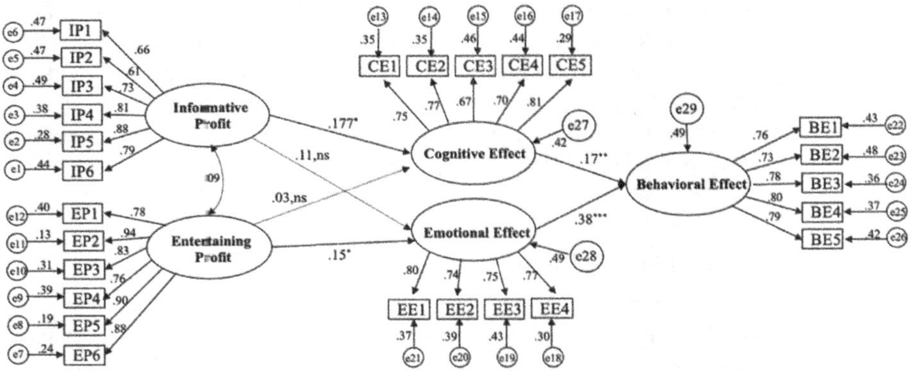

Fig. 2. Structural Equation Modeling path map

Fit Index. The fit index of the Structural Equation Model: $x^2/df = 2.323$, RMSEA = 0.069, GFI = 0.858, AGFI = 0.830, IFI = 0.914, TLI = 0.903, CFI = 0.913, except for the GFI index, all other fitting indices reach the corresponding standards. The fit index indicates that the data fits the model well and the model is acceptable.

Hypothesis Testing Results. Structural Equation Modeling path analysis is shown in Table 2, and the path is shown in Fig. 2. In the dissemination of intangible cultural heritage short videos, the audience's informational profit has an important impact on their cognitive effect, while the entertainment profit makes a difference in the emotional effect. As a mediating variable, the emotional effect has a more significant influence on the audience's final behavioral effect than the cognitive effect.

3 Qualitative Analysis of Communication Effect Based on Nvivo

This study also introduces qualitative analysis method, taking the comment data of phenomenal Tik Tok short videos about intangible cultural heritage as samples. The ground analysis theory is applied to encode the text data by the aid of Nvivo12 software. The study focuses on the explanatory understanding of communication effect from the behavior and meaning construction of successful cases.

3.1 Sample Selection and Text Data Acquisition

In this study, the video content of the intangible cultural heritage phenomenal KOL Li Ziqi, who has 34.58 million fans on the Tik Tok platform, is used as sample. Li Ziqi has created the Oriental Intangible Cultural Heritage Video Collection since 2018. As of January 8, 2020, the video with the highest number of likes and view counts is the *Paper-making* of the *"Four Treasures of the Study"* series (58.18 million view counts, 2.11 million likes). Tik Tok has a huge number of users and a high degree of activity that users can post instant feelings and thoughts at any time. Therefore, video comments are highly original and authentic. In this research, a Python data crawler was used to collect the comments of the *Paper-making* video. The text data was utilized as the qualitative analysis object after removing the part that is unrelated to the video content.

3.2 Text Data Encoding

Viewing short videos is an acceptance process. Video content can cause cognitive, attitude, and behavioral changes to the audience. The research attaches importance to the role of the audience as the interpreter and secondary disseminator, considering their acceptance psychology, perception, understanding, attitude changes, etc. Then text analysis is introduced to encode the comment data. Through induction and deduction, the huge text data can be abstracted into concepts.

After sorting and screening the data, open coding is the first step. Nodes were adjusted and improved according to related literature, and finally 60 free nodes were obtained. Secondly, by the way of axial coding, the free nodes are in-depth explored and conceptually integrated to 11 child nodes. Finally, selective coding is performed, and the child nodes are further integrated. The following parent nodes are obtained, as shown in Table 3.

Table 3. Free nodes and tree nodes

Parent node	Child node	Free node	Reference point
Audience cognitive layer	Knowledge memory	Gain knowledge, knowledge popularization, know for the first time, learned	416
	Information perception	Intangible cultural heritage, paper making technology discussion, video content teasing, open eyes to rural life, tools	1055
	Value perception	Cultural self-confidence, cultural transmission, intangible cultural heritage communication, elegance of culture, promotion of traditional culture, multi-regional culture, chinese cultural admiration, hard-earned paper	1087
	Impression formation	Excellent video quality, meaningful video, enjoyable viewing experience	279
Audience emotional layer	Immersion	Get stuck, addicted, immersed	122
	Association and empathy	Long for home, childhood recall, extend families, historical association	245
	Psychological feelings and appeals	Anticipation, asking for updates, thanks, moving, self-deprecating, yearning for a beautiful life, dissatisfaction with real life, psychological self-healing, suggestions	801
	Personal charisma identity	Support, admiration, worship, love dearly, affection, pride, adoration, hard-working and dedication, great to marry li ziqi, all-around, traverse from ancient times	3788
	Negative attitude	Video team packaging, doubt authenticity	166
Audience behavioral layer	User retention	Watch from the beginning to the end, watch all the videos, the series, watch for a few hours, never tired	208
	Sharing/purchasing	Purchase, share, amway, recommend	355
	Social/interactive	Like, follow, fan, ask	581

3.3 Discussion and Analysis

The video of *Paper-making* lasts 54 s, mainly showing the process of using the ancient method to make paper by Li Ziqi, as well as the plain and beautiful scenery in the countryside. The information of the video comments can be regarded as the audiences' feedback on the effective communication of the content. Through coding analysis of the text information, it can be found that in the feedback given by the audience, the emotional level of the audience accounts for 56.27%, the cognitive level accounts for 31.56%, and the behavioral level accounts for 12.57%. Audiences mainly express their perceptual knowledge of video content. We can say the video of *Paper-making* has successfully realized the construction of audience's emotional attitude towards intangible cultural heritage.

At the emotional level, 73.96% of the audience has expressed their acceptance of Li Ziqi's personal charm and style. 15.64% of the audience has expressed their feelings of self-ridicule, self-healing and longing for a beautiful life. At the same time, it cannot be ignored that some audiences express the empathy of video content, which triggers the associations with their own experiences. At the cognition level, information perception and value perception account for 37.19% and 38.32% respectively. Even though there are no text and voice-over in the video of *Paper-making*, the value delivery of intangible cultural heritage is effectively achieved. At the behavioral level, 50.79% of social contact and interaction also reflect the high viscosity between the audience and the communicator.

4 Conclusion

With the support of the above two research methods, we can draw some convincing conclusions: The formation of the audience's emotional attitude is of cardinal significance in the effective communication of short videos about intangible cultural heritage; Based on the features of the platform, the audience has a strong desire to socialize and interact with each other in the communication process, which is the most obvious manifestation of the behavioral effect; Value perception and information perception are of significance in cognition level of the audience. According to the conclusions of the research, this paper advances the following suggestions:

(a) Improve the recreational gains of the audience; give them emotional energy to stimulate interaction in the information level. Combining auditory, visual and other immersive experience elements and presenting a fashion style is more in line with the need of young audiences. The emotional cognition and experience can make the audience take the initiative to obtain information and realize spiritual and mental participation, which will undoubtedly improve audience's acceptance of intangible cultural heritage.
(b) Attach importance to the audience's need to participate in social sharing and set motivations to arouse interpersonal interaction. The unique intangible cultural heritage suddenly appears in the life of the ordinary audience, satisfying their curiosity. The values it conveys can strike a chord with the audience. In this way,

the membership symbol of intangible cultural heritage fans is formed, thus establishing the emotional bond.

(c) Build a complete value chain with micro narrative content; satisfy the real-time pleasant sensation and open the window for audience's self-integration. A single short video of intangible cultural heritage needs to convey accurate information in a very short time and generate a complete narrative. A complete value chain of intangible cultural heritage should be built between contents. Only if the instant pleasure is linked through the value chain can the audience build a profound positive attitude towards intangible cultural heritage from the series of videos.

By constructing the theoretical model and the qualitative analysis, this study puts forward suggestions on the content production of short videos of intangible cultural heritage. It is aimed to promote the in-depth and effective transmission of intangible cultural heritage on new media platforms thus cultivating young audiences. Simultaneously, this study also provides a new method for the research on the communication effect of short videos on new media platforms.

References

1. Porter, C.E., Donthu, N., MacElroy, W.H., et al.: How to foster and sustain engagement in virtual communities. Calif. Manag. Rev. **53**(4), 80–110 (2011)
2. Van der Heijden, H.: User acceptance of hedonic information systems. MIS Q. **28**(2), 695–704 (2004)
3. Ducoffe, R.H.: Advertising value and advertising on the web. J. Advert. Res. **36**(5), 21–35 (1996)
4. Zeithaml, Valarie A.: Consumer perceptions of price, quality, and value: a means-end model and synthesis of evidence. J. Mark. **52**(3), 2–22 (1988)
5. Carlson, J., Rahman, M.M., Taylor, A., et al.: Feel the VIBE: examining value-in-the-brand-page-experience and its impact on satisfaction and customer engagement behaviors in mobile social media. J. Retail. Consum. Serv. **46**, 149–162 (2017)
6. Hsiao, C.H., Chang, J.J., Tang, K.Y.: Exploring the influential factors in continuance usage of mobile social apps: satisfaction, habit, and customer value perspectives. Telematics Inform. **33**(2), 342–355 (2016)

Visual Elements and Design Principles in Media Production

David Kei-man Yip(✉)

OUHK, Ho Man Tin, Hong Kong
dkmyip@ouhk.edu.hk

Abstract. Visual elements such as dot, line, plane and color are the basic components of an image. By applying design principles to these visual elements in visual expression, the creative possibilities are infinite. Cinematic style is created by the filmmaker's control over the medium. Many visual styles of media projects are the application of the design principles to the visual elements. This systematic approach is not only confined to a shot composition but also to the whole visual structure of the work involving the development of both story and visual elements. This article analyzes this design approach to visual styles by discussing the works of several film masters as examples of how media project can use this approach to design and structure visual content.

Keywords: Visual elements · Design principles · Cinematic arts · Film art · Creative arts · Content design · Art of storytelling · Media production · Visual structure · Color structure

1 Introduction

Many moviemakers believe that moviemaking is about creating a dream world for their audience. Film Master David Lynch had this to say about movie and dream: 'When you sleep, you don't control your dream. I like to dive into a dream world that I've made, a world I chose and that I have complete control over.' [1] According to this great cinematic stylist whose movies had been known to be visually masterful, his way of creating great visual styles for his movies was to have complete control over the medium. To control the visual medium is to control its visual elements. Built on the foundations of visual communication theories [2] and visual literacy in the design of digital media [3], this article discusses visual styles with film examples. It analyzes how these visual styles are created in the context of the narrative content. It aims to discuss ways filmmakers used to control visual elements in their works and how this approach can also be applicable to other forms of media production.

2 Visual Elements and Design Principles

The smallest visual element is a dot or point. Others are line, shape, planes, textures and sizes. If it is a moving image, visual elements include direction as well. Visual elements and design principles work together to form visual contrast, harmony,

E. Markopoulos et al. (Eds.): AHFE 2020, AISC 1218, pp. 283–288, 2020.
https://doi.org/10.1007/978-3-030-51626-0_36

balance, repetition, dominance, graduation and unity as a whole [4, 5]. Lines and shapes can be used to create various design purposes or effects such as to visualize and communicate senses of chaos or unity through applying the design principles. The degree of chaos is subject to the level of contrast created by the visual elements. More lines juxtaposing in opposing directions can create more senses of conflicts visually. On the other hand, the different degrees of unity or harmony can also be created by the levels of visual consistency or affinity, which is the opposite of contrast. Basic visual elements and visual design principles work and apply together to achieve consolidated effects.

Pointillism is an art and technique of drawing with dots or points to form visual patterns. Expressionism and Cubism are art styles created by different textures and patterns of geometric lines, shapes, planes with the use of colors for the expression of the artist's feelings and emotions. Kandinsky's *Composition VIII* was created in the styles of Russian Suprematism and Constructivism of the early 20th century that focused on basic visual elements of geometric forms and shapes such as circles, squares, rectangles, straight and curved lines with a limited range of colors. On the other hand, Jackson Pollock's abstract paintings composed of a chaotic form of visual texture and pattern expressed and communicated the artist's impulse visually. Unlike Kandinsky's *Composition VIII*, which represented a more structural approach to the design principles, Pollock's unique visual pattern and style through his dynamic and random strokes of repeated conflict and visual dominance illustrated a different impulsive freestyle approach to the same design principles with the basic visual elements with very different effects. Kandinsky's *Composition VIII* and Pollock's *Number 31* represent two opposite approaches to the application of the same design principles with the basic visual elements. Both of these examples of abstract paintings show how the artist's abstract feelings and emotions can be expressed to the viewers not in a literal way but through the artist's sensual and impulsive treatments of the visual elements. Although abstract painting and film are different visual mediums, they can also illustrate the different approaches or applications of the visual elements.

3 Visual Contrast and Affinity

Visual contrast or affinity is a systematic way to control visual elements in storytelling, of which its plot should have twist and turn or up and down. Contrast attracts attention by signifying a visual change in content. On the contrarily, visual affinity with similar elements or components maintains certain degree of visual consistency or pattern. Visual contrast or affinity can exist within a shot or between shots. Film Master Bernardo Bertolucci was famous for his masterful cinematic imagery in his films. His use of contrast in cinematic space in his first film *The Conformist* illustrated the relation between a totalitarian state and its people within a shot.

Fig. 1. Bernardo Bertolucci's The Conformist 1970 [6]

As shown in Fig. 1, the wide shot was used to visualize the ideology of the totalitarian authority overpowering an individual visualized here as small and insignificant carrying the symbolic heavy political artifact. This visual contrast between the small and big showed that insignificance of an individual as compared to the overwhelming power of the political ideology under the totalitarian state captured in a wide shot. A change in plot point in a story could also be signified by a visual contrast or change in visual affinity in a montage sequence to enhance its dramatic effects. The famous opening sequence of David Lynch's *Blue Velvet* showed a sharp contrast from images of a peaceful innocent town to a dark and filthy underground world full of insects. This sharp contrast created a dramatic interest by visually foretelling the theme of the story. These film examples above illustrated the different applications of design principles with the basic visual elements of line, shape and space in film and how these visual elements also worked structurally with the narrative content for a more consolidated effect.

4 Visual Structure of a Film

Visual style is not confined to a shot but to the whole visual structure of the story. Film Master Roman Polanski's *Repulsion* was known to be full of rich and original visual styles that cinematically enhanced its story progression. Long before there were computer visual effects, one of the director's ingenious mise-en-scène used in the film was to externalize and visualize the inner turmoil of the female lead was through changes of space [7] as his way to signify her changes to the audience. To keep these changes subtle or less noticeable, the director moved his film set to a different location with deeper space but decorated it with the same furniture setting to make it look as if it was the same location but in a different spatial dimension. Shot in a wide shot and in a somewhat distorted space and view, this spatial change aimed to visualize cinematically and externally the inner deterioration and turmoil of the female lead and keep the audience stay emotionally engaged and focused on the story.

Fig. 2. Roman Polanski's mise-en-scène in *Repulsion* 1965 [8]

Repulsion also illustrated another aspect of visual design principles through the use of lighting. In a nutshell, lighting in media production is to create and control contrast between light and shadow. Textbook definition of lighting technique in media production begins with the classic three-point lighting with key, fill and background lights. In practice, there are always many alternative lighting styles but the lighting contrast, which determines atmosphere, is always controlled by the ratio between the intensity of key and fill lights. High key lighting effects can be achieved by the low ratio between the key and fill lights showing little contrast. Contrarily, low key lighting, as illustrated above, means high contrast between key and fill lights. As shown in Fig. 2, the low key lighting visualized that the main character was in her most torn and trouble state of mind towards the end of the film. This lighting contrast worked with the distorted spatial dimension as application of the visual elements. Contrarily when the same character appeared to be a normal girl in the beginning of the film, the visual was shot in normal lighting conditions.

Film Master Stanley Kubrick's signature visual style of his films was his consistent use of symmetric deep shot with one-point perspective in all his films. The important aspect of Stanley Kubrick's visual style of storytelling [9] was not only confined to individual shot composition but to the overall visual structure in relation to the story development for the rest of the film. Shot design in the beginning of the classic horror film *The Shining* were some typical and relatively flat shot but as the story progressed with the arc of the story picking up its intensity, many shots would gradually get deeper in perspective and shot composition. The visual development through changes in shot dynamic worked alongside with the story development of the film to consolidate a more subtle psychological and emotional effect on the audience. Arguably, whether or not it was the visual images that drove the story forward or the plot of the story that motivated the visual development, it was for sure that the visual form and the content worked collectively to engage with the audience emotionally.

5 Color Structure of a Film

Color as an important part of visual element is both art and science. The science of color is beyond the scope of this article. Instead, this article focuses color as visual elements and its application in media production with the same design principles of contrast and affinity. Color generally ranges from cold to warm in a spectrum of color temperature normally presented in the form of a two-dimensional color wheel, in which the cold and warm colors are opposite to each other as contrast color. Contrarily, the nearby colors in the color wheel are analogous colors that can create a more consistent look. However, with more color configurations such as the various levels of hues, saturation, tints, tones and values of color considered, the fuller extent of colors should not be confined to a two-dimensional wheel or chart but in a three-dimensional color sphere. When combinations of colors are used to create various visual effects and perceptions by controlling the levels of contrast or affinity and the possibilities are endless. Bernardo Bertolucci's *The Last Emperor* was best known for its use of different colors, from warm to cold, to illustrate the different stages and the downfall of the Puyi's life. The ancient Kung Fu action movie *Hero* directed by Zhang Yimou stylistically used colors to identify the different tales of the heroes through the use of analogous colors that separated each plot in the story.

6 Conclusion

Visual style in film is a form of film literacy distinctive and unique to each filmmaker. This article aims to use the application of visual elements and design principles as theoretical framework to analyze how a visual style is created in the context of film content. It illustrates a systematic and structural approach to the applications of visual elements in visual storytelling that can be applicable to a creative process normally regarded as abstract and full of unpredictability [10]. The applications of these visual design principles have endless possibilities and they are not only limited to filmmaking or media production but in many other visual applications such as website and game design.

References

1. Libquotes. https://libquotes.com/david-lynch/quote/lbu6m9r
2. Berger, A.A.: Seeing Is Believing: An Introduction to Visual Communication. Mayfield Publishing Company, Mountain View (1989)
3. Roth, S.: Visual literacy and the design of digital media. ACM SIGGRAPH Comput. Graph. **29**(4), 45–47 (1995)
4. Gatto, J.A., Porter, A.W., Selleck, J.: Exploring Visual Design: The Elements and Principles. Davis Publications, Worcester (2000)
5. Digital Web Magazine. http://www.digital-web.com/articles/principles_of_design/
6. https://dovilemeilute.wordpress.com/2015/12/16/mise-en-scene-in-the-conformist/

7. Craig, J. R.: On Visual Tropes as Expressions of the Psyche in Roman Polanski's "The Tenant". J. Fantast. Arts **9**(2(34)), 131–141 (1998)
8. http://kafkasfilm blogspot.com/2012/11/repulsion-roman-polanski.html
9. Falsetto, M.: Stanley Kubrick: A Narrative and Stylistic Analysis. Greenwood Publishing Group, Westport (2001)
10. Kaufman, J.C., Sternberg, R.J. (eds.) The Cambridge Handbook of Creativity. Cambridge University Press. New York (2010)

Reviewing the Usability of Mosaic Artwork Simulation Systems

Chi-Fu Lai(✉) and Jeff K. T. Tang

The Open University of Hong Kong, Jockey Club Campus, Ho Man Tin, Kowloon, Hong Kong
{wcflai, jkttang}@ouhk.edu.hk

Abstract. Mosaic is a form of art created by assembling many small colored tiles. There are numbers of previous research works proposed for simulating various types of mosaic artworks. Most of them focus on generating virtual mimics of physical mosaic artworks that compute the shape of tiles and their arrangement. Though automated mosaics design could be treated as a packing optimization problem, the product could be aesthetically challenged and tedious for mosaicking. In this paper, we discuss existing computational mosaic simulation systems and suggest improvements to cope with mosaic artists.

Keywords: Mosaics · Algorithm · Human-computer interaction

1 Introduction

Mosaic artworks are created by arranging a large number of small colored stones, tiles, glasses, etc. (tesserae), to form patterns and pictures. The history of mosaic arts can be chased back to the 3rd millennium BC in Mesopotamian temple, where tesserae in form of ivory, seashells, and colored stones of irregular shapes were used. Roman Mosaics were a common feature of ancient Roman homes and buildings, which reveal the activities at that time such as sports and agriculture [1]. In the 4th century two-dimensional repetitive mosaic patterns had been widely adopted decorations such as floor design [1].

1.1 Types of Mosaics

Throughout thousands of years of art, culture, and technology development, artists have invented many variations of mosaics. We introduce here some of the most common types of mosaic artworks. Some of them are special due to the materials used, and some are focusing on using tesserae of particular shapes.

Ancient Mosaics. Tiny stones are crafted into nearly cubic or elliptical shapes to be used as tesserae. The tesserae are assembled according to colors or alignments to form patterns and figures. The Roman mosaics described in [1] belong to this type. Ancient mosaics can be further subdivided into subgroup according to different techniques where the tesserae are cut and placed. For example, opus vermiculatum emphasizes the alignment of tesserae to form the contour of an object.

E. Markopoulos et al. (Eds.): AHFE 2020, AISC 1218, pp. 289–295, 2020.
https://doi.org/10.1007/978-3-030-51626-0_37

Stained Glass Mosaics. Stained glasses are normally manufactured to be relatively large pieces; and can be easily cut in smaller pieces of any shape. Some artists would purposely cut stained glass with similar texture into similar sizes and shapes. Those similar tesserae can be grouped together to produce the clustering effect in order to present a shape feature or to mimics an object. An example is shown in Fig. 1 (right), where the mountains, the sky, and the ground are clearly segmented by stained glasses of similar textures, sizes, and shapes. This type of mosaics can be commonly found as windows of a church (Fig. 1 (middle)).

Object Mosaics are created by assembling objects of various shapes, colors, and sizes. They can be assembled on a flat surface to mimic a figure. The objects used could be recycled items from daily life (like Fig. 2 (left)), or remains from dead creatures like ivory, seashell, and bones (like Fig. 2 (middle)). This type of mosaics has a blur differentiation with collage, which does not have a base.

Fig. 1. Left: A computer generated crystallization mosaic [2]; middle: A stained glass mosaics church window by Sunny Days Glass Art studio; right: A hand-cut stained glass mosaics by Kasia I. Polkowska mimicking the mountain spring of Colorado.

Fig. 2. Left: One of the artwork creation collection: "Plastic Classics" by Jane Perkins, remaking the famous Mona Lisa Painting by Leonardo da Vinci; middle: A Commemorative Panel in the Shell Grotto of Margate.

Photographic Mosaics (or Photomosaics) is a special placement of massive images and photographs that views to be one figure or photo at low magnifications. The idea is to regard one photograph (a tile) as one superpixel replacement of the target figure. This type of mosaics is mainly created digitally and printed out as physical images if needed. In this paper, we focus on studying handcrafted mosaics artworks; thus, we are not going to discuss further on Photomosaics.

1.2 Mosaic Crafting

[3] summarized the history of mosaic artworks; and reviewed the techniques of crafting mosaics - two primary mosaicking techniques: (i) the direct method; and (ii) the indirect method. The main difference between the two methods is that: In direct method, tesserae are fixed directly onto the base; whereas in indirect method, tesserae are only temporarily fixed onto a removable backing with the front face downwards, then fixed onto the base after the design (or a module of the design) is completed.

1.3 Overview

The beauty of various mosaic artworks admires not only artists, but also computer scientists in the past decades. Computational algorithms and frameworks have been proposed to simulate different forms of mosaics. However, to the best of our knowledge, there are no research works done on reviewing how these computational methods can cope with the traditional mosaicking techniques. In this paper, we are going to review representative computational algorithms that simulate different types of mosaic artworks; and suggest how it could fit into mosaic artists design practice.

2 Computational Mosaic Simulation

In the past decades, computer scientists have been simulating mosaics of different forms. Digital mosaics are generated by non-photorealistic rendering techniques [4, 5] best classified those algorithms proposed into four main kinds of mosaics simulations: Crystallization Mosaics, Ancient Mosaics, Photomosaics, and Puzzle Image Mosaics. In the following subsections, we review and discuss each family of computer simulated mosaics, but not Photomosaics as explained in Sect. 1.1.

Crystallization Mosaics. This type of computational generated mosaic simulates stained glass mosaics. The synthesized tesserae are irregular but convex shapes that mimics stained glass or crystal materials, for example [2] (Fig. 2 (left)) attempts to simulate the rendering style of stained glass mosaics typically used as church glass windows like Fig. 2 (right). More examples could be found in [6] (see Fig. 3(b)); and [7]. In general, the algorithms we studied for generating crystallization mosaics are all based on Voronoi diagrams [8] with additional image feature extraction and energy optimization. The simulated results are aesthetically challenged and hard to reproduce because: (i) the size and shape of the Voronoi cells are similar which is not the case for physical stained glass tesserae (see Fig. 2 (middle & right)); and (ii) it is tedious for artists to cut stained glass to match shapes and colors of the synthesized tesserae.

Fig. 3. Mosaic images of Lenna generated by different non-photorealistic algorithms: (a) Original image of Lenna; (b) Voronoi diagrams method [6]; (c) Centroidal Voronoi diagrams method [9]; and (d) Statistical Region Merging method [11].

Ancient Mosaics. There are works aim at synthesizing an image by breaking down the original image into regular square tiles of different color, orientation and size [4, 9–13]. Their computing pipelines can be generalized into three steps: (1) extract image features and segment it into smaller regions, (2) compute the alignment of tiles and the image feature, and (3) place the tile. An example workflow using [10] is shown and explained in Fig. 4; and example simulation results are shown in Figs. 3(c and d), and 4(d). Other than tiny tiles, [14] attempted to simulate ancient mosaics with pebble tesserae. Those works are conducted without the consideration on the physical feasibility of mosaicking and tesserae. The following assumptions on tesserae are made: (i) freely synthesized color, texture, and shape [11–14]; (ii) non-rigid body tesserae [10, 11]; and (iii) zero volume (i.e. 2D) tesserae. Thus, it is tedious for mosaics artists to reproduce such synthetic mosaics. There are works attempted to synthesize solid tesserae in 3D [15–17]; however, tesserae can either be fixed square size or synthesized. Beside virtual simulations, a robotic system has developed to automatically place physical square tiles on a planar surface in grid [18]. To the best of our knowledge, there are no known works on automatically crafting irregular tesserae nor non-grid placement.

Fig. 4. (a) Parallel curves to a feature curve (in grey) can be calculated as offsets of the given curve; (b) an image of a dinosaur taken at the Utah dinosaur museum in Ogden; (c) the feature curves (in white) used to place the tiles; and (d) final mosaic result [10].

Puzzle Mosaics. Puzzle mosaics are virtual simulations of object mosaics. Some works formulated as an image generation problem [19, 20], where the object tesserae are only considered as 2D images. Real-world objects tesserae can be captured as photographs and used to synthesize 2D tesserae and mosaicking on 3D surface [21]. However, the tesserae obey assumptions of zero volume non-rigid body and cloned. On the contrary, [22] considers rigid body 3D volume as tiles; however, they are solving a modeling problem on collage problem instead of a mosaicking problem.

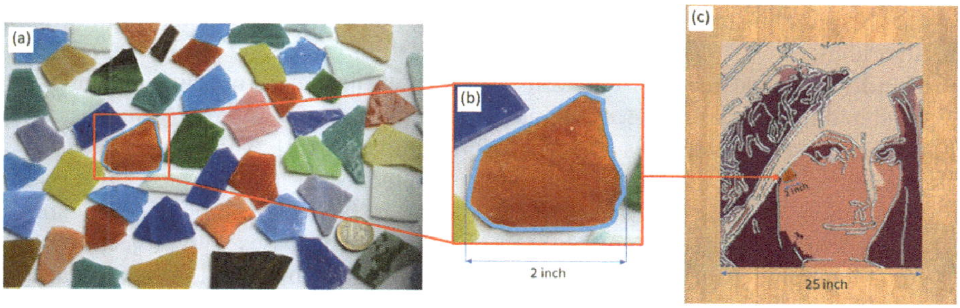

Fig. 5. (a) State-of-the-art object recognition and 3D scanning algorithm can easily recognize and extract the contour of a tessera; (b) and is measured digitally; (c) with placement suggestion visually overlaid with design guides on the physical base. Note the color coding visualization of the tessera. Image of (a) retrieved from [23].

3 Suggestions

Fully automating the mosaicking process is expensive and aesthetically challenging; instead, we suggest a user interactive mosaicking system to tickle the above problems. Given an input target digital image, a set of physical tesserae, and a user defined working base, the mosaicking system can suggest a subset of tesserae and their placement configuration on the base. In detail, a mosaicking system should incorporate the following features:

1. **Tesserae Recognition, Identification, and Tracking.** Given input RGBA video of the physical tesserae, the system should be able to recognize, 3D reconstructed, measure (related to the base), and extract the contour of each piece of tesserae, as illustrated in Fig. 5(a–c). Recognized tesserae will be identified and tracked. Unlike [21], each tesserae should be unique, non-reusable, and as 3D volume. The pool of available tesserae should be interactively updated for any used or new tesserae.
2. **Direct/Indirect Method Mode.** When the indirect method is used, the system should be able to compute a mirrored solution, where tesserae are placed with their backface upwards. Thus, all recognized tesserae have to be computed mirrored.
3. **Design Guides Suggestion.** Design guides can be computed from the target figure using image processing toolkits; and can serve as references for mosaic artists to follow. The design guide can be visualized by overlaying on the working base (like Fig. 5(c)) using Augmented Reality (AR) display technology.

4. **Interactive Suggestion Feedback.** The system should provide tesserae placement suggestions as an interactive feedback to the user (i.e. mosaic artist) in various situations. First, when the mosaic artist picks up a tesserae, the system can feedback by hinting a suggested placement configuration on the base. Second, when the mosaic artist needs assistance on further tesserae placement, the system can hint multiple tesserae placement on the base. Third, if the mosaic artist placed a tessera of his/her own preference, the system should immediately update all the hints. The hints can be displayed by overlaying a color-coded tesserae contour on both the physical tessera and the base as illustrated in Fig. 5(a and c). The suggestion system can be built by training a state-of-the-art neural network model or optimization algorithms.
5. **Hand-free Hardware.** Mosaic artists need both hands for mosaicking; thus, any handheld hardware or devices are not feasible for such a system. We suggest using head mounted AR devices like Microsoft HoloLens [24]. This is due to the consideration of all the computational requirements aforementioned: i) tesserae recognition, updates, and measurement require instant image input from a camera; ii) overlaying visualization on physical objects require glasses with projection prism for holographic display; and iii) a powerful computing unit for all the computational calculation.

4 Conclusion

In this work, we reviewed the common types of mosaics, and the two typical mosaicking techniques: the direct method, and the indirect method. We also reviewed computational algorithms on mosaics simulation. We compared the synthesized mosaics with handcrafted mosaics; and discussed how those algorithms are not feasible for mosaic artists to use for mosaicking. We proposed five essential features for building an interactive mosaicking system for mosaic artists.

References

1. Cartwright, M.: Roman Mosaics, https://www.ancient.eu/article/498/roman-mosaics/. Accessed 12 Feb 2020
2. Dobashi, Y., Haga, T., Johan, H., Nishita, T.: A method for creating mosaic images using Voronoi diagrams. In: Eurographics (Short Papers). Eurographics Association (2002)
3. King, S.: Mosaic Techniques & Traditions: Projects & Designs from Around the World. Sterling Publishing Company, Inc. (2003)
4. Battiato, S., Di Blasi, G., Farinella, G.M., Gallo, G.: A novel technique for opus vermiculatum mosaic rendering. In: 14th International Conference in Central Europe on Computer Graphics, Visualization and Computer Vision (WSCG 2006). Václav Skala-UNION Agency (2006)
5. Battiato, S., Blasi, G.D., Farinella, G.M., Gallo, G.: A survey of digital mosaic techniques. In: Eurographics Italian Chapter Conference. The Eurographics Association (2006)

6. Haeberli, P.: Paint by numbers: Abstract image representations. In: Proceedings of the 17th Annual Conference on Computer Graphics and Interactive Techniques, pp. 207–214 (1990)
7. Faustino, G.M., de Figueiredo, L.H.: Simple adaptive mosaic effects. In: XVIII Brazilian Symposium on Computer Graphics and Image Processing (SIBGRAPI 2005), pp. 315–322 (2005)
8. Preparata, F.P., Shamos, M.I.: Computational Geometry: An Introduction. Springer, New York (2012)
9. Hausner, A.: Simulating decorative mosaics. In: Proceedings of the 28th Annual Conference on Computer Graphics and Interactive Techniques, pp. 573–580 (2001)
10. Elber, G., Wolberg, G.: Rendering traditional mosaics. Vis. Comput. **19**, 67–78 (2003)
11. Di Blasi, G., Gallo, G.: Artificial mosaics. Vis. Comput. **21**, 373–383 (2005)
12. Battiato, S., Milone, A., Puglisi, G.: Artificial mosaic generation with gradient vector flow and tile cutting. J. Electr. Comput. Eng. (2013)
13. Jetchev, N., Bergmann, U., Seward, C.: GANosaic: Mosaic Creation with Generative Texture Manifolds (2017). http://arxiv.org/abs/1712.00269
14. Doyle, L., Anderson, F., Choy, E., Mould, D.: Automated pebble mosaic stylization of images. Comput. Vis. Media **5**(1), 33–44 (2019)
15. Lai, Y.-K., Hu, S.-M., Martin, R.R.: Surface mosaics. Vis. Comput. **22**, 604–611 (2006)
16. Dos Passos, V.A., Walter, M.: 3D mosaics with variable-sized tiles. Vis. Comput. **24**, 617–623 (2008)
17. Dos Passos, V.A., Walter, M.: 3D virtual mosaics: Opus Palladium and mixed styles. Vis. Comput. **25**, 939–946 (2009)
18. Peganant, P., Koomsap, P.: Improving decision making in product flow-based tiling automation for custom mosaic design. Assembly Automat. **37**, 391–399 (2017)
19. Kim, J., Pellacini, F.: Jigsaw image mosaics. ACM Trans. Graph. **21**, 657–664 (2002)
20. Di Blasi, G., Gallo, G., Petralia, M.: Puzzle image mosaic. In: Proceedings of the IASTED/VIIP, pp. 33–37 (2005)
21. Hu, W., Chen, Z., Pan, H., Yu, Y., Grinspun, E., Wang, W.: Surface mosaic synthesis with irregular tiles. IEEE Trans. Vis. Comput. Graph. **22**, 1302–1313 (2016)
22. Chen, W., Ma, Y., Lefebvre, S., Xin, S., Martínez, J., Wang, W.: Fabricable tile decors. ACM Trans. Graph. **36**, 1–15 (2017)
23. Tiffany Broken Glass Mosaic, untreated.: Amazon.in: Garden & Outdoors, https://www.amazon.in/Approx-5060pieces-Multicoloured-Tiffany-untreated/dp/B01MDTGIVV. Accessed 14 Feb 2020
24. Blokdyk, G.: Microsoft HoloLens A Complete Guide - 2020 Edition. 5starcooks (2019)

Documentation and Scientific Archiving: Digital Repository

Priscila Verdugo[1,2](✉), Catalina Astudillo-Rodriguez[1,2], Jackelín Verdugo[1,2], Juan-Fernando Lima[1,2], and Santiago Cedillo[2]

[1] LIDI, Universidad del Azuay, Cuenca, Ecuador
{pverdugo, cvastudillo, jverdugo, flima}@uazuay.edu.ec
[2] Escuela de Ingeniería en Ciencias de la Computación, Universidad del Azuay, Cuenca, Ecuador
scedillo@es.uazuay.edu.ec

Abstract. Documentation and archiving tools were developed for using with descriptive linguistics and then became a discipline called Linguistics of Documentation in 1998. The aim of this discipline is to preserve documents and primary data of linguistic, anthropological, archaeological and humanistic research. In 2018 the University of Azuay signed a Framework Agreement for Scientific and Technological Cooperation with several partners: CONICET Argentina, the Peruvian Amazon Research Institute, the University of Chile, the National University of Formosa, the Center for Anthropological Studies of the Catholic University of Paraguay, the National University of San Juan, Argentina; and the Pontifical Catholic University of Peru to contribute to the documentation, preservation and dissemination of linguistic and cultural heritage. The main objective of this research is to create a website called 'Documentation and scientific Archiving' that facilitates the use and exchange access to a corpus of resources from various disciplines for scientific and educational purposes. The premise of the methodology used for the development of the website is an orientation to various disciplines. The synthesis focused on the review of the characteristics of sites of similar purpose. The study considered three major stages for the synthesis of the web application: 1) the detailed review of 11 websites allowed identification of the entities and attributes used as the basis for development; 2) the quantification of entities and the attributes found in those pages were the key to deciding that the website covered multiple disciplines or lines of research i.e. the key attributes for handling multimedia files were evidenced; 3) the identification of user requirements. The "Documentation and scientific Archiving" website is expected to be functional for both researchers and users to help maintain and preserve the country's linguistic and cultural heritage. In addition to UDA researchers, other local and national universities can file their research inventories on the site where they can be preserved, reviewed, exchanged and used by other Latin American researchers. Finally, today, technology has become a fundamental aid for the development of research projects, building bridges between linguistic, anthropological, archaeological, humanistic research and the development and creation of applications and websites.

Keywords: Documentation · Archiving research · Linguistics data

E. Markopoulos et al. (Eds.): AHFE 2020, AISC 1218, pp. 296–302, 2020.
https://doi.org/10.1007/978-3-030-51626-0_38

1 Introduction

Documentation and data archiving were born as tools of descriptive linguistics and then became a discipline called linguistic documentation, which aimed at preserving writings and primary data of linguistic, anthropological, archaeological, and humanistic research. For Himmelmann [1], documentation is a durable and multifunctional record of a language and its corpora allows the analysis of syntactic, semantic, discursive, pragmatic, anthropological, and sociolinguistic aspects of languages. Thus, the establishment of interdisciplinary approaches such as socio-, ethno-, psycho-, and neurolinguistics, as well as ethnographic, pragmatic, discursive, and semiotic aspects of culture allow to fully understand linguistic systems; hence, documentary and descriptive linguistics is multidisciplinary. In this way, documentation as a field of research and linguistic activity is already a goal in itself and not just an auxiliary procedure in research. Currently, documentation and archiving have transcended time and space. Bird and Simons [2] argue that data collected in research can be used for multiple purposes and can be loaded on platforms and web applications in different electronic formats [3]. In addition, Woodbury [4] determines as functions of the "Linguistics of Documentation" the elaboration, preservation, and recording of the languages of the world, and takes as possible corpora its patterns of use. In this way, it forms discursive memories that will preserve legacies of possible worlds, spaces for interaction and interference of investigative processes. On the other hand, Del Gizzo [5] reflects on the potential of the archives and the implications of mixing documents from different axes and meanings in the contemporary world. This possibility implies establishing new enunciations that require new and renewed scenarios for reading, as well as disarming the fixed senses and the rules of authority for the selection of texts or speeches; thus, altering and modifying the conditions of ordering, juxtaposing, and sequencing different disciplinary knowledge. In this way, technology helps to safeguard the archiving task through an infinite number of possibilities available today.

In this context, in 2018, the University of Azuay signed an Agreement for Scientific and Technological Cooperation between CONICET Argentina, the Peruvian Amazon Research Institute, the University of Chile, the National University of Formosa, the Center for Anthropological Studies of the Catholic University of Paraguay, the National University of San Juan, Argentina; and the Pontifical Catholic University of Peru to contribute to the documentation, preservation, and dissemination of linguistic and cultural heritage.

The main objective of this research was to provide access to the resources and materials produced by the universities that belong to the network to be exchanged for scientific and educational purposes through the creation of a website called 'Documentation and Scientific Archiving.' This site will allow access to the corpora of various disciplines. Data must be accessible to both researchers and site users, regardless of the technical specifications of the linguistic documentation, but considering the needs of the academic and scientific community. For this reason, all material collected and documented has a standard digital support and will be available to academics and scientists from anywhere in the world. It is very important that the copyright requirements are in effect in all data exchanges that the network activates.

2 Methodology

The process of developing the digital repository followed three stages: i) Identification of the need for a digital repository, ii) Definition of the requirements for the digital repository, and iii) Development of the digital repository using Software Engineering techniques as can be seen in Fig. 1.

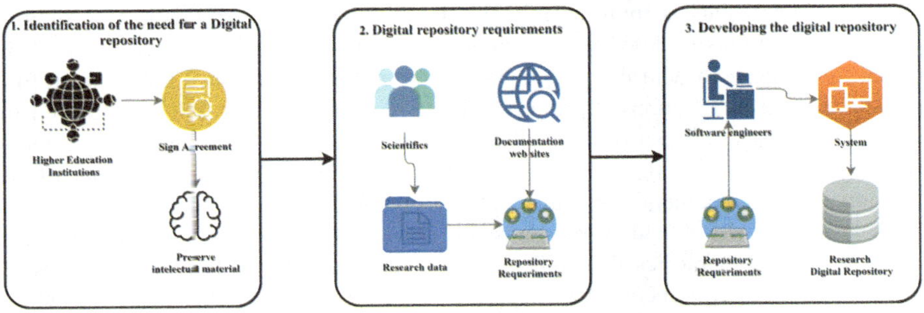

Fig. 1. Scenarios for Digital repository development.

2.1 Identification of the Need for a Digital Repository

The Agreement of Scientific and Technological Cooperation with international universities is an important factor that has highlighted the need for a digital repository that allows the preservation and sharing of research results. The authors carried out an initial review of websites oriented to the linguistic field and that allow researchers to store, maintain and share their data to place themselves in the existing digital repositories environment. However, after analyzing these websites, it was identified that several of them, despite performing the aforementioned tasks, did not provide an interface oriented to the exchange of material that could be used by Latin American researchers. Additionally, and as a major factor, it was pointed out the fact that these websites are oriented solely to the area of linguistics.

2.2 Digital Repository Requirements

In several meetings, the research team of the University of Azuay and their counterparts in different local universities discussed the importance of their findings as well as the need of storing and sharing the results in a web application of free access for society. The resulting data of each researcher has been analyzed and classified into disciplines, types of documents (images, audio, video, notes, musical chords, software, articles, etc.), and document formats to document each research process. In addition, the team has proposed the inclusion of a "mutual agreement" between the researcher and the software representatives to file their data for storage and subsequent publication. Finally, the development team needs some of the aforementioned initial requirements (disciplines, document types, formats, etc.), to define quality requirements such as, usability, security, flexibility, and portability [6] necessary for the development process.

It was necessary to perform a comprehensive review of 11 websites with similar purposes to support the definition of the initial requirements. The study focused primarily on two websites considered as the most significant to the development team due to their standardization in the structure of their data: Digital Repository DILA-CONICET Archive [7] and The Archive of Indigenous Languages of Latin America (AILLA) [8], which shows the categorization of a collection. Table 1 shows all the attributes identified on the websites to describe and store digital documents.

Table 1. Attributes extracted from the web sites to define the digital database.

Fields	[7]	[8]	[9]	[10]	[11]	[12]	[13]	[14]	[15]	[16]	[17]
Title	✓	✓	✓	✓	✓	✓	✓	✓		✓	✓
Author	✓	✓	✓	✓	✓	✓	✓	✓	✓	✓	✓
Identification	✓	✓	✓	✓	✓						
Place	✓	✓		✓							
Key words					✓	✓				✓	
Relation	✓										
Cited by	✓								✓	✓	
Source type	✓	✓		✓	✓	✓		✓	✓	✓	✓
Domain	✓	✓				✓					
File type	✓										
Language	✓	✓	✓								
Editor	✓									✓	
Rights	✓										
Source	✓										
Support	✓	✓		✓	✓	✓	✓	✓	✓	✓	✓
Collection	✓	✓			✓					✓	
Classification			✓								
Autonomous			✓								
Location			✓								
Alternative			✓								
Level				✓							
Created on				✓							
Gender		✓		✓							
Specification		✓			✓	✓					
Online date	✓	✓	✓	✓	✓	✓	✓	✓	✓	✓	✓

The identification of the most relevant and common characteristics offered by the websites has allowed the selection of the most suitable attributes to match the requirements identified by the local researchers. It has also helped to determine the necessary attributes for the design of the database (DB), as well as to avoid mishaps of the application in the future [18].

2.3 Developing the Digital Repository

The development of the website called "Documentation and Scientific Archives" is proposed to offer a service to researchers from various disciplines that allows the storage of different kinds of documents. On the other hand, it also includes rules to data preservation and security policies [19] that facilitate long-term maintenance.

Before the development of the repository, three possible users were considered. The authors, who will be allowed to save and share content of their own interest, users who research and synthetize relevant information and save their preferences of data stored in the web site, and the visitors, improving the usability characteristic of the repository [20].

3 Results

From the first meeting with the researchers of the University of Azuay and other universities of the city, it is evident that the disciplines that will be part of the Documentation and Scientific Archiving platform will be literature, archeology, anthropology, linguistics, administration sciences, arts, psychology, and cartography. The application will accept files as PDF documents, images, photography, audio, video, or maps. Table 2 shows this classification.

Table 2. Disciplines and file types for documents.

Discipline	File type
Latin American literature	PDF documents
Archeology and anthropology	Images and photography
Linguistics, ethnolinguistics	Audio, Video, and PDF documents
Administration sciences	Audio and PDF documents
Arts	Score in image
Psychology	PDF documents
Cartography	Maps and images

4 Discussion

From the central axes of the linguistics of the documentation field of action presented by Woodbury [4], Golluscio and Vidal [21], this research complies with some of its principles. It is distinct regarding situations, participants, channels: oral, written, or digital, discursive genres, varieties, and codes. The corpora that the website will compile is extensive and, with the help of technology, will allow to archive a large amount of information as well as to store it properly. Additionally, a team of academics contributes to its development and from their particular research cases and views, they can document instances of language use and train people to make high quality records.

On the other hand, the documents selected to be part of the website are transparent and properly coded so that academics in different areas of research can use them for varied objectives. Additionally, since linguistic researchers do not consider their documenting task completed only by the collection of corpora in different formats such as, audio or video, pdf and others, data will be available for different purposes. In addition, the materials gathered in this website are ethical, protect the ownership of the data and can be spread transparently. Above all, they are portable in the sense developed by Bird and Simons [2], so collected data can be easily transferred to new technologies that are constantly emerging. In this way, this study respects the interests of the social, cultural and academic subjects involved, especially those who produce or own the data or assign it a hereditary, ancestral, and memory value.

5 Conclusions and Future Work

The local interests of this research are the core of this issue because the textual materials that are collected come from different areas and from research projects developed by universities in the region who envision sharing their files to the academic community of Latin America and the world.

In this process, the corpus collected will guide the analytical apparatus to be used by the disciplinary models, so that the strategies of archiving can establish a sequence of the knowledge activated in these processes.

As future works have been proposed, the creation of the digital repository will be managed by the SCRUM methodology, reducing the development task complexity. This methodology employs the users' requirements and software technologies to create software ruled by incremental tasks [22]. The programming language PHP was used in the development task. This open source code has been used to create many web sites in the latest years because it uses a high-level interface. Moreover, it can be incrusted into HTML code to improve the client's side experiences [23]. In addition, to speed up its creation, a Laravel framework was used taking advantage of the concepts and criterions to organize the development of the platform [24].

Acknowledgments. This work is part of the "Lingüística de la documentación: red de archivos lingüísticos, arqueológicos y humanísticos" project, supported by the University of Azuay, with the contribution of the partners of the international "Agreement for Scientific and Technological Cooperation."

References

1. Himmelmann, N.P.: Documentary and descriptive linguistics (1998)

2. Bird, S., Simons, G.: Seven dimensions of portability for language documentation and description. Language (Baltim) **79**, 557–582 (2003)
3. Evans, N.: Review of essentials of language documentation. Lang. Doc. Conserv. **2**, 340–350 (2008)
4. Woodbury, A.C.: Defining documentary linguistics. Lang. Doc. Descr. **1**, 35–51 (2003)
5. Del Gizzo, L. El canon frente al Archivo. Avatares metodológicos de una relación complementaria. Chuy. Rev. Estud. Lit. Latinoam. **5**, 45–69 (2018)
6. Barbacci, M.: Software quality attributes: modifiability and usability. Software Engineering Institute, Carnegie Mellon University Pittsburgh PA, 15213 (2004)
7. CAICYT - CONICET: Repositorio Digital Archivo DILA (2019)
8. University of Texas at Austin: Archive of the Indigenous Languages of Latin America (AILLA) (2019)
9. Lewis, M.P., Simons, G.F., Fennig, C.D. (eds.) Ethnologue Language World, pp. 233–262 (2013, 2015)
10. University of London: Endangered Languages Archive at SOAS University of London (2019)
11. Universidad de Chile: Repositorio Digital de la Universidad de Chile (2019)
12. Instituto de Investigaciones de la Amazonía Peruana: Repositorio Digital IIAP (2019)
13. Universidad Nacional de la Formosa: Editorial de la Universidad Nacional de la Formosa (2019)
14. Centro de estudios Antropológicos: Revista Estudios Paraguayos (CEADUC-UC) (2019)
15. Pontificia Universidad Católica del Perú: Directorio de investigadores PUCP (2019)
16. Universidad del Azuay: Repositorio Institucional de la UDA (2019)
17. Department of Empirical Linguistics: Essentials of Language Documentation (2019)
18. Nwosu, K.C., Thuraisingham, B., Berra, P.B.: Multimedia Database Systems: design and Implementation Strategies. Springer, Heidelberg (2012)
19. Tang, J., Cui, Y., Li, Q., Ren, K., Liu, J., Buyya, R.: Ensuring security and privacy preservation for cloud data services. ACM Comput. Surv. **49**, 13 (2016)
20. Lew, P., Olsina, L., Zhang, L.: Quality, quality in use, actual usability and user experience as key drivers for web application evaluation. In: Benatallah, B., Casati, F., Kappel, G., and Rossi, G. (eds.) Web Engineering. pp. 218–232. Springer. Berlin (2010)
21. Golluscio, L., Vidal, A.: Reflections on language documentation in the Chaco (2018)
22. Cervone, H.F.: Understanding agile project management methods using Scrum. OCLC Syst. Serv. Int. Digit. Libr. Perspect. **27**, 18–22 (2011)
23. Jingling, Z., Rulin, G.: A new framework of security vulnerabilities detection in PHP web application. In: 2015 9th International Conference on Innovative Mobile and Internet Services in Ubiquitous Computing, pp. 271–276 (2015)
24. Das, R., Saikia, L.P.: Comparison of procedural PHP with Codeigniter and Laravel Framework. Int. J. Curr. Trends Eng. Res. **2**, 42–48 (2016)

Research on the Design Strategy of Green Packaging from the Perspective of Chinese Young Consumers

Jiaqian Xu(✉), Ziwei Chen, and Bing Xiao

Shanghai Jiao Tong University, Shanghai 200240, China
daphneqian@qq.com

Abstract. Facing the fact that the focus of green packaging design in China is not clear and current green packaging design can't attract young consumers, this paper attempts to propose some design strategies that meet young consumer preferences by Decision-Making Trial and Evaluation Laboratory (DEMATEL) method. By analyzing the data of 25 valid questionnaires, the scores of the prominence and influence of each factor are been calculated. Finally, a causal diagram is developed to illustrate the relationships between 9 main factors. The diagram shows that (1) the two most critical factors for green packaging design are packaging structure and the expanding use of packaging, (2) packaging's extended is greatly influenced by other factors, which should be considered in all aspects of green packaging design, (3) pattern, convenience and green message's presentation are the edge factors to be considered.

Keywords: DEMATEL · Green packaging design · Consumer behavior

1 Introduction

The concept of green packaging originated from "our common future" published by the United Nations Commission on environment and development in 1987 [1]. It can also be called environment-friendly packaging. It is a new packaging concept in line with sustainable development and the principle of 3R1D [2].

The design of green packages is an important part of promoting green consumption, however, China's current green packaging design is still in the exploration stage, with many problems. For example, the backwardness of industrial technology limits the design method to a certain extent. The concept of environmental protection floats on the surface, the homogenization of design is serious, environmental protection information cannot be well communicated, and the consumers' awareness of environmental protection is not strong, so they seldom choose green packaging products [3].

2 Purpose

Since the focus of green packaging design is not clear and the current green packaging design cannot attract consumers to buy green packaging products, this paper take the youth group of Shanghai Jiaotong University as the research object, based on the

E. Markopoulos et al. (Eds.): AHFE 2020, AISC 1218, pp. 303–310, 2020.
https://doi.org/10.1007/978-3-030-51626-0_39

perspective of consumers, proposing design strategies of green environmental protection packaging that meets consumer preferences.

3 Method

In this paper, Decision-Making Trial and Evaluation Laboratory (DEMATEL) method is used as the main research method trying to make quantitative analysis on the relationship between various factors that affect consumers' purchase behavior in green packaging design and explore a more reasonable direction of green packaging design. DEMATEL is proposed by American scholars A. Gabus and E. Fontela on a Geneva Conference in 1971. It is a systematic analysis method using graph theory and matrix tools, mainly used to solve complex practical problems with multiple influencing factors, which helps to clarify the mutual influence relationship and causality between the relevant factors. Specific calculation formula can be found in related websites [4].

4 Process

The research is divided into four stages by combining the methods of several researchers' articles [5–8]: First, list the influencing factors and determine the final 9 main factors. Second, determine the relationship between these 9 factors. Young people with experience in purchasing green packaging products will be invited to make a quantitative evaluation of the selected nine factors through a questionnaire, specifically designed for double-sided comparison DEMATEL technique. Third, based on the data from the questionnaire, calculate the direct associations between influential factors in green packaging design, and then the scores of the Prominence and influence of each factor can be calculated in DEMATEL tool and finally a causal diagram can be developed to illustrate the relationships among the 9 factors.

4.1 Determine the Influencing Factors

The factors that affect consumers' purchase behavior of green packaging products are relatively complex. To grasp these factors more comprehensively, the author uses VOSviewer[1] to analyze the literature retrieved by the keywords of green packaging design.

Combined with other materials and literature on green packaging design, and after consulting the opinions of the youth group of Chinese consumers, nine factors affecting consumers' purchase of green packaging products were determined: Color (F1), Pattern (F2), Interest (F3), Structure (F4), Convenience (F5), Expansion Use (F6), Recyclability (F7), Lightweight Design (F8), Presentation of Green Environmental Protection Signs (F9) and labeled as F1-9.

According to the attributes of these factors, they can be divided into four categories: Color, Pattern and Interest constitute the visual factors of green packaging; Structure,

[1] VOSviewer is a software tool for constructing and visualizing bibliometric networks.

Convenience and Expanding Use constitute the interactive factors of green packaging design; Recyclability and Lightweight Design constitute the material factors of green packaging design; the Presentation of Green Environmental Protection Signs is information Transmission factors. For the convenience of interviewees. At the same time, the corresponding picture cases shown in Fig. 1 are prepared for reference and distributed to the subjects before the questionnaire survey.

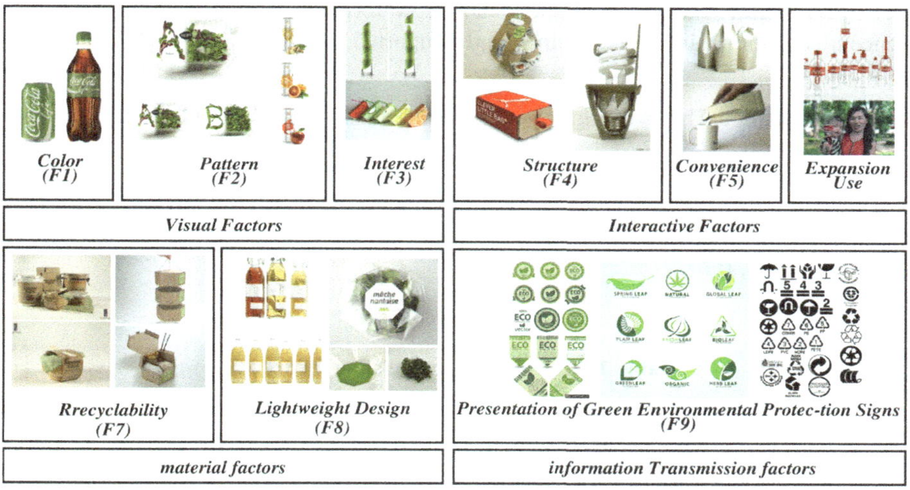

Fig. 1. Corresponding picture cases for nine influencing factors.

4.2 Questionnaire Design and Distribution

To facilitate the follow-up data processing, this paper uses Tencent Questionnaire[2] as the survey method designed a nine-pages questionnaire. Each page compares one factor with the other eight factors in twos and explains the factors in detail as much as possible in the question stem. The whole questionnaire has 72 questions, and the choice of each question is divided into four grades from "no impact" to "great impact" and is given a weight of 0–3.

4.3 Determine Respondents and Issue Questionnaires

The method of this survey combined online and offline and the interviewees need to have certain logical judgment ability and focus to judge a large number of relationships between elements. To ensure the quality of questionnaire recovery, the respondents are the people aged 20–30 with bachelor degree or above and have certain experience of green packaging product purchase.

[2] Tencent questionnaire is a free and professional questionnaire system launched by Tencent in China. It provides a variety of ways to create questionnaires, simple and efficient editing methods, powerful logic setting functions, professional data statistics and sample screening.

After determining the respondents, about 100 questionnaires were distributed through Wechat, QQ and other channels. Some of the respondents answered the questionnaire in the way of online and offline combination. Finally, 25 valid questionnaires were collected.

4.4 Acquisition and Processing of Research Data

Get Initial Data. The questionnaire data in CSV format can be directly exported from Tencent questionnaire background and imported to excel for data processing preparation.

Data Processing. The average value of the 25 questionnaires for each question was calculated by the AVERAGE function and arranged into the initial direct relation matrix M shown in Table 1. The values in the matrix marked as $M_{(i,\ j)}$ represent the average Influencing Value between the corresponding elements, which means the direct Influencing Value of factor I on element J. the values on the diagonal of the matrix are usually taken as 0.

Table 1. The initial direct relation matrix M.

Factor	F1	F2	F3	F4	F5	F6	F7	F8	F9
F1	0	2.16	1.88	1.12	0.76	1.48	1.48	0.76	1.92
F2	1.84	0	2.28	1.24	0.72	1.56	1.2	0.76	1.96
F3	1.48	1.64	0	1.84	1.36	1.96	1.28	1.2	1.36
F4	1	1.48	2.2	0	2.24	2.48	1.88	2.2	1.48
F5	0.64	0.84	1.12	1.68	0	1.96	1.72	1.8	0.88
F6	1.04	1.2	1.8	2	1.6	0	1.56	1.48	0.92
F7	1.04	1.16	1.28	2.08	1.6	1.84	0	1.8	1.24
F8	1.12	0.88	1.4	2.16	2.44	2.04	2.16	0	1.08
F9	1.96	2.08	1.44	0.92	0.8	1	1.2	0.96	0

Normalize the direct relation matrix. Sum each row of the matrix M to get $M^{max} = 14.96$ and get the standardized direct relation matrix N by Formula (1). The result is shown in Table 2. For the accuracy of subsequent calculations, three decimal places are temporarily reserved.

$$N_{(i,j)} = M_{(i,j)}/M^{max}. \tag{1}$$

The standardized direct relation matrix will tend to zero matrix after self-multiplication for many times, and self-multiplication represents the increased indirect influence among elements. Therefore, the formula of direct-indirect comprehensive relation matrix T can be deduced (2). The formula can be simplified to the Formula (3), I is a unit matrix with a diagonal value of 1 and other values of 0 and $(I - N)^{-1}$

represents the inversion of matrix N, which is the most complex step. It is usually carried out by calculation tools. There are special functions in Excel and MATLAB to deal with this calculation problem. The result is shown in Table 3.

Table 2. The standardized direct relation matrix N.

Factor	F1	F2	F3	F4	F5	F6	F7	F8	F9
F1	0	0.144	0.126	0.075	0.051	0.099	0.099	0.051	0.128
F2	0.123	0	0.152	0.083	0.048	0.142	0.080	0.051	0.131
F3	0.099	0.110	0	0.123	0.091	0.131	0.086	0.080	0.091
F4	0.067	0.099	0.147	0	0.150	0.166	0.126	0.147	0.099
F5	0.043	0.056	0.075	0.112	0	0.131	0.115	0.120	0.059
F6	0.070	0.080	0.120	0.134	0.107	0	0.104	0.099	0.061
F7	0.070	0.078	0.086	0.139	0.107	0.123	0	0.120	0.083
F8	0.075	0.059	0.094	0.144	0.163	0.136	0.144	0	0.072
F9	0.131	0.139	0.096	0.061	0.053	0.067	0.080	0.064	0

$$T = (N + N_2 + N_3 + \cdots + N_k) \quad (k = \infty). \tag{2}$$

$$T = N(I - N)^{-1}. \tag{3}$$

Table 3. The direct-indirect comprehensive relation matrix T.

Factor	F1	F2	F3	F4	F5	F6	F7	F8	F9
F1	0.310	0.474	0.513	0.459	0.394	0.512	0.458	0.381	0.442
F2	0.421	0.348	0.535	0.467	0.394	0.518	0.444	0.382	0.445
F3	0.413	0.460	0.424	0.527	0.456	0.568	0.474	0.431	0.427
F4	0.450	0.521	0.637	0.509	0.588	0.693	0.592	0.565	0.500
F5	0.328	0.373	0.447	0.481	0.343	0.526	0.462	0.435	0.360
F6	0.376	0.422	0.517	0.526	0.463	0.442	0.480	0.440	0.390
F7	0.386	0.431	0.502	0.544	0.477	0.566	0.399	0.470	0.418
F8	0.416	0.445	0.544	0.587	0.558	0.618	0.563	0.398	0.438
F9	0.398	0.437	0.451	0.411	0.363	0.446	0.409	0.359	0.299

The value $T_{(i,\ j)}$ in the comprehensive matrix represents the sum of the direct and indirect influences of factor I on factor J. The sum of each row in the comprehensive matrix T is the total influence of factor I on all other factors. It represents the Influencing Value of factor I, and the aggregate is labeled as D. The sum of each column of the comprehensive matrix T is the comprehensive influence value of factor J by all other factors, which is labeled as R_j, which is called the degree of influence of factor J, and the aggregate is labeled as R.

The sum of the Influencing Value D_i of factor I and the Affected Value R_i is called the Prominence Value of factor I, which indicates the importance of factor I in the whole system; the difference between D_i and R_i is Relation Value, which refers to the active influence ability of factor I. If the value is positive, the factor is the active cause factor, and if the value is negative, it means the factor is passive result factor. These data are listed in Table 4 (two decimal places).

Table 4. Table of Influencing Value, Affected Value, Prominence Value and Relation Value.

Factor	Influencing Value D	Affected Value R	Prominence Value D + R	Relation Value D − R
F1	3.94	3.50	7.44	0.45
F2	3.95	3.95	7.86	0.04
F3	4.18	4.18	**8.75**	**−0.39**
F4	5.06	5.06	**9.56**	0.55
F5	3.76	3.76	7.79	**−0.28**
F6	4.06	4.06	**8.94**	**−0.83**
F7	4.19	4.19	**8.47**	**−0.09**
F8	4.57	4.57	**8.43**	0.70
F9	3.57	3.57	7.29	**−0.15**

4.5 Draw a Causal Map of the Factors Affecting Green Packaging Design

Set the Cartesian coordinate system with the Prominence Value (D + R) as the horizontal axis and the Relation Value (D − R) as the vertical axis to determine the position of each element. To show more core causality, the paper uses the quartile method to divide the data in the comprehensive relation matrix into four parts and selects the top 3/4 of the data for direction drawing. The top 1/4 of the data is expressed as Solid line arrows, the rest of the data appear as dotted arrows, as Fig. 2 shown.

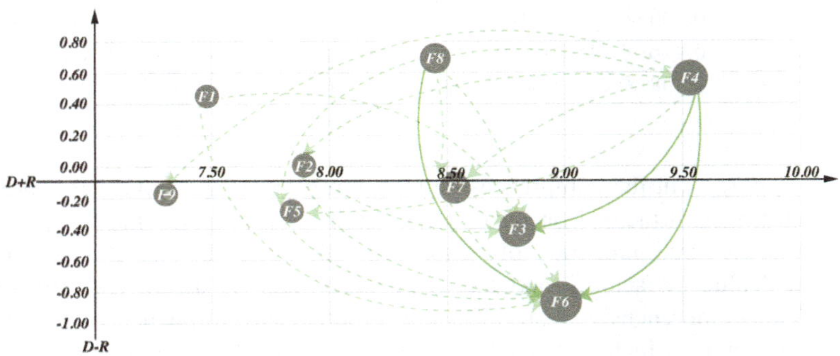

Fig. 2. Causal map of the factors affecting green packaging design.

5 Conclusions and Suggestions

5.1 Conclusion Based on Prominence Value

The factors above the average level are Interest (F3), Structure (F4), Expansion Use (F6), Recyclability (F7) and Lightweight Design (F8). These factors are the key factors influencing consumers' purchase behavior.

The two most critical factors for green packaging design are packaging structure and the expanding use of packaging. Structure (F4) is undoubtedly the most important factor affecting young consumers' purchase of green packaging products. It is not only the most central factor (9.56) but also the most positive value. It has a strong impact on Factor Interest (F3) and Expansion Use (F6) which need to be concerned. Therefore, it is suggested that the government and enterprises pay more attention to the design of structure when promoting and designing green packaging.

5.2 Conclusion Based on Affected Value

The positive factors which have a higher absolute value of the Relation Value are Color (F1), Structure (F2) and Lightweight Design (F8). These factors are the main factors influencing others. The changes in these factors have a great influence on consumers' behavior of purchasing green packaging products.

Expansion Use (F6) showed excellent performance in the Prominence Value (8.94), but a large negative value (−0.83) in the Relation Value. It means that it's greatly influenced by other factors, especially by two factors: Lightweight Design and Structure, which are the main problems to be solved in green packaging design, and other factors will be developed around them. To solve this problem, we need to make a balance between the structure and lightweight design and the extended use.

5.3 Discussion on Non-core Factors

Pattern (F2), Convenience (F5) and Presentation of Green Message (F9) are the edge factors to be considered. So, we can reduce the attention to these three factors to obtain short-term business benefits. However, it may change with time, so designers also need to keep pace with the times and revise their design direction in different stages.

References

1. Brundtland, G.H.: Our common future—call for action. Environ. Conserv. **14**(4), 291–294 (1987)
2. Lan, W., Mei, Y.: Green packaging design on the principle of 3R. Packag. Eng. **02**, 169–172 (2008). (in Chinese)
3. Wenliang, L.: Puzzling and surmounting: step out of misunderstandings surrounding green packaging. Art Des. **03**, 39–43 (2010). (in Chinese)
4. DEMATEL method. http://www.huaxuejia.cn/ism/dematelintro.php
5. Gang, L., Huang, D., Chunrong, L.: Research on preference of automobile modeling features based on decision laboratory method. China Packag. Ind. **12X**, 34–36 (2014). (in Chinese)

6. Chong, H., Lijuan, S., Jian, G.: Research on the teaching quality evaluation factors by using DEMATEL. Comput. Era **07**, 74–77 (2017). (in Chinese)
7. Xiaoguo, D.: The application of decision making trial and evaluation laboratory method in the field of design. Art Sci. Technol. **31**(09), 286–287 (2018). (in Chinese)
8. Javier, A.M., Luis, E.Q., Pedro, I.P., Astrid, M.O., Hernan, A.S.: A quantitative methodology to enhance a strategy map. Int. J. Prod. Econ. **219**, 43–53 (2020)

The Influence of Emotional Experience Relating to Communication from a Typographic Perspective

Amic G. Ho(✉)

Department of Creative Arts, School of Arts and Social Sciences, The Open University of Hong Kong, Hong Kong, People's Republic of China
amicgh@gmail.com

Abstract. Typography is not a tool that is limited to delivering messages; it is a foundation for compositional communication. Letterforms are valuable graphic elements in representing language and organising type in order to serve multiple roles in a composition. Designers organise the fundamentals of typesetting in order to use typography to explore new opportunities for interaction and storytelling. They consider the formal qualities of every typeface they select, as these typefaces create connections between the graphic form and the messages they aim to reinforce. This study reviewed the theoretical understanding of the interactions between the experience of designing and experiencing emotion. Selected masterpieces of typographic design are investigated and categorised by several features. These masterpieces and their various approaches require further analysis, in order for designers to apply them in their typographic work.

Keywords: Paradox of typography · Users experience · Usability · Typographic design

1 Introduction

Paradoxically, design elements that bring about a visual conflict can attract people's attention, an idea which many designers have applied to their work. Letterforms are valuable graphic elements in representing language, as well as organising type in order to serve multiple roles in a composition. They are regarded as a reliable approach to elevating the quality of design, and they are applicable to various projects and mediums. The consideration of a typography arrangement involves both language and composition. When considering varying approaches towards organising type, imagination, accurate copy, and the letterforms of the typeface are transformed to add further meaning to the typography design. Designers organise the fundamentals of typesetting in order to use typography to explore new opportunities for interaction and storytelling. They consider the formal qualities of every typeface they select, in order to create connections between the graphic form and the messages they aim to reinforce.

E. Markopoulos et al. (Eds.): AHFE 2020, AISC 1218, pp. 311–316, 2020.
https://doi.org/10.1007/978-3-030-51626-0_40

2 Dimensions of Experience

Before further investigate how the typography be arranged in both language and composition, there is a need to understand how the prior knowledge and experience of designing was considered in the study of design. Some study of consciousness provided a framework for a designer to understand user experience [1]. It offered a contextual perspective of the approaches users interacts with the environment around them. The research team argued that consciousness also took a role to influence participatory design critically [2, 3]. When consciousness was manipulated as an interpretive as well as the generative process in participatory design, the experience of the end-users was considered as an essential factor rather than the functions of outcomes.

3 Emotional Experience in the Paradox of Type

In order to present information effectively, a bullet point system is employed for organising information according to a certain logical sequence. After analysing previous studies, a model of how data is transformed into information and knowledge has been developed (Fig. 1). Three core elements are included within the concept of emotional experience in the paradox of type: typographic design, experience, and emotion. In the process of reading typographic design, readers use their senses to obtain information on both language and letterforms, which convey sensation and meaning [4]. Readers are often able to predict this information based on prior experience, as they have created an expectation of the typographic design presented [5, 6]. When readers experience disconfirmation, emotions are stimulated [7]. Regardless of whether readers experience satisfaction or dissatisfaction, an impression is created.

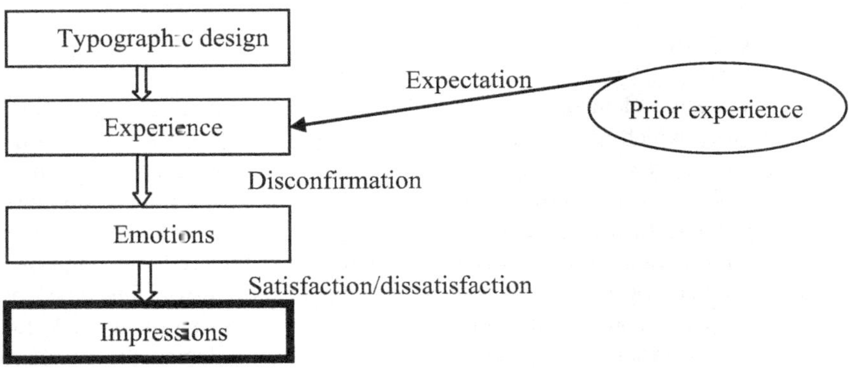

Fig. 1. Model of emotional experience in the paradox of type.

Some masterpieces of the typographic design were investigated. Several features were categorised for further analysis by designers looking to apply these approaches in their typographic work. Designers used contradictions for presenting the concept, thus solving the design problems. Then, designers decided how to organise type and other visual elements. The design can, therefore, relate back to these focal elements.

4 A Pilot Study for Exploring the Emotional Experience in the Paradox of Type

This study aimed to explore how the experience of emotions created by typographic design specifically, and its relationships with the design experience of typography. These theories developed the hypothesis of why emotion may be influenced by typographic design. The insights are then employed to formulate fundamental knowledge for investing the design experience as well as design iteration. The concepts also reflected a continuous force connecting designers and readers through the dimensions of experience includes perception, cognition as well as reaction. Within the dimensions of experience, emotions are involved. This study provided more productive knowledge of the design and the development of design projects.

4.1 Research Methods

To illustrate the paradox of type, the design concept of 'Blockter' was proposed to be an example (Fig. 2). Not only was the type selected by the designer involved in the design, but the approach applied for the presenting of the concept. Types were not restricted to only one aspect or rendered in one dimension or one texture. Designers are required to investigate methods to introduce variety to the visual elements in typographic work, and the designer should build bridges between the various styles applied. In the experimental work, the negative space was minimised as well as repositioned in order to emphasise the outline shape of the letterforms. This approach maximises the weight of each letterform, so at the same time, the position of the little negative space was redesigned to prevent it from being overly bold—the different letterforms of the typeface would provide the effect of creating a wedge through the middle of the design.

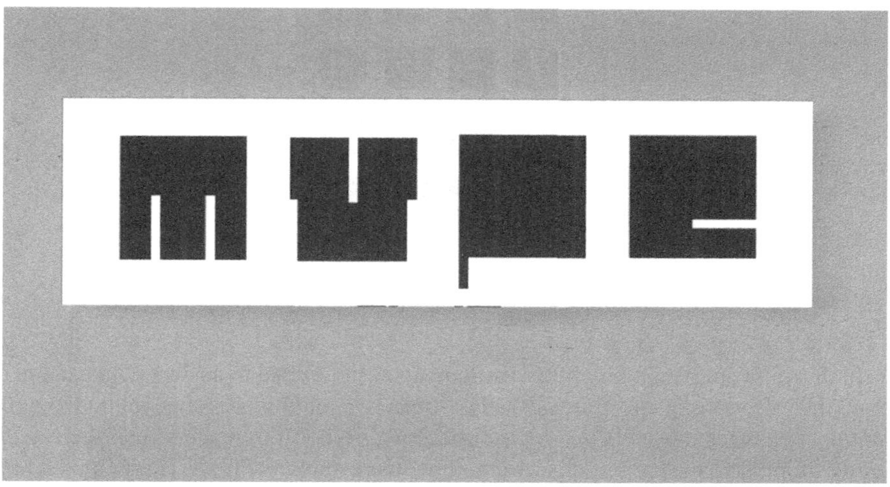

Fig. 2. The decision-makings among the design process of Blockter were directed by the knowledge and critical thinkings of its designer.

Within the design process of Blockter, decision-making processes associated with the actions performed in designing. During the decision-making process, perception and action were integrated simultaneously. In other words, designers investigated feedback, tested out prototypes, drew sketches, made calculations, evaluated shapes and forms with dimensions of experience which involved acting as well as perceiving together. Sketching (as an example action in the design process) is determined by the previous perception (the negative space of the typeface must be placed in a particular location in particular shape), via the body (the display typeface).

Most of the typographic works were intimately familiar with different typefaces and giving designers themselves the space for trying unusual things with them. The term 'paradox' was used a few times to describe this typeface design, and that was a great way to describe this kind typographic design feature in general as shown in Fig. 3. Every typeface design, even the most ideal one, is just a starting point. It's up to the designer to optimise it for their design purposes. Applying display type would be the scaffolding concept for the remaining of the design work (Fig. 4).

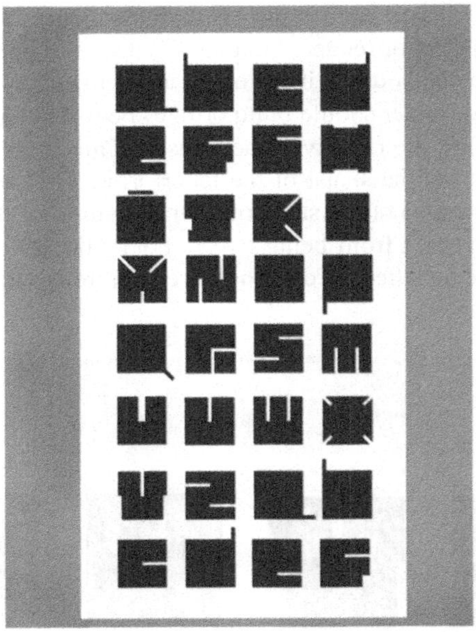

Fig. 3. Experimental typeface sample of typographic paradox.

In this pilot study, a total of fifty participants were invited to answer a questionnaire about their feedbacks of 'Blockter'. The objective of this stage is to find out the emotional changes those elicited by the paradox of type from readers' perspectives. In the questionnaire, twelve options of emotions were shown (Fig. 4). Participants were asked to identify the emotions those elicited by the paradox of type. They also asked to share if the paradox of type would provide the most effective information to them.

Please select the emotions you feel about this type:

Positive emotions	Negative emotions	Neutral emotions	Mixed emotions
Surprised	Sick	Thoughtful	Interested
Smug	suspicious	Indifferent	Puzzled

Fig. 4. Twelve options of emotions were shown for participants to identify the emotions those elicited by the paradox of type.

4.2 Research Results

The relationships between their attended information and elicited emotional changes were investigated. The relationships between emotional concern and communication design those implied in typographic paradox were briefly understood. The insight proposed in this study suggested there are emotional concerns created by the paradox of type (Table 1 and Fig. 5).

Table 1. General characteristics of participants in the questionnaire (n = 50).

Characteristics	Categories	Total
1. Gender	Male	25
	Female	25
2. Age	<20 years	2
	20–25 years	13
	26–30 years	25
	30–35 years	10

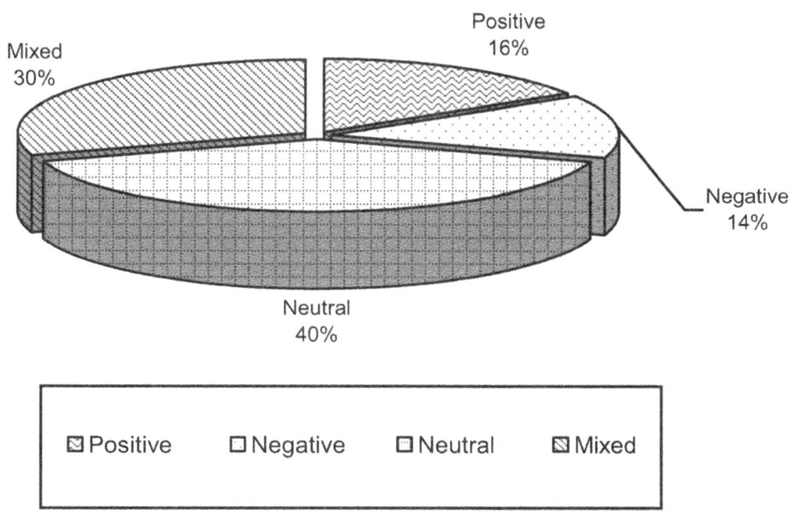

Fig. 5. Pie-chart that shown the most identified type of emotions.

5 Conclusion

The study of consciousness critically influences participatory design. Since consciousness has been manipulated into an interpretive element of design as well as the generative process in participatory design, the experience of end-users has been considered as an important factor, as well as the function of the outcome. At a similar period, researches about emotions and even treatments for initiatives moved away from traditional diagnostic research directions; but turned to transdiagnostic topics. Most psychotherapeutic studies were organised and conducted for investigation about the experience. Some scholars regarded emotion involved in generating concepts, organising, and providing their services. The discussions of the participants in this study revealed the previous finding from some prior studies of emotion. Based on the fundamental knowledge provided by psychotherapeutic research, it was not difficult to understand certain experience emotions, irrespective of the reason. Changes related to cognition, perception, personal experience, agency, and embodiment were reviewed across the psychiatry literature, which provided an opportunity to develop further an interdisciplinary framework for better understanding the bordering on the experience of designing and the experience of emotions. This study reviewed changes that could theoretically be discovered in the interaction of experiences of designing and the experiences of emotions. Some masterpieces of the typographic design were investigated and were categorised into several features for further analysis of some approaches for designers to apply in their typographic work.

References

1. Frauenberger, C., Good, J., Keay-Bright, W.: Phenomenology: a framework for participatory design (2010)
2. Pienkos, E., et al.: Hallucinations beyond voices: a conceptual review of the phenomenology of altered perception in psychosis. Schizophr. Bull. **45**(1), 67–77 (2019)
3. Hamm, J.A., Leonhardt, B.L., Ridenour, J., Lysaker, J.T., Lysaker, P.H.: Psychosis psychological, social and integrative approaches phenomenological and recovery models of the subjective experience of psychosis. Discrepancies Implic. Treat. **10**(4), 340–350 (2018)
4. Parnas, J.: A disappearing heritage: the clinical core of schizophrenia. Schizophr. Bull. **37**(6), 1121–1130 (2011)
5. Leonhardt, B.L., et al.: Recovery and serious mental illness: a review of current clinical and research paradigms and future directions. Expert Rev. Neurother. **17**(11), 1117–1130 (2017)
6. Sterzer, P., Mishara, A.L., Voss, M., Heinz, A.: Thought insertion as a self-disturbance: an integration of predictive coding and phenomenological approaches. Front. Hum. Neurosci. **10**, 502 (2016)
7. Yanagisawa, H., Miyazaki, C., Nakano, S.: Kansei modeling for multimodal user experience. Visual Expectation Effect on Product Sound Perception (2016)

Study on Thermal Comfort of Vehicle Interior Environment Based on Thermal Manikin

Wang Rui, Li Wei(✉), Zhao Chaoyi, and Gao Jianfeng

Ergonomics Laboratory, China National Institute of Standardization, No. 4 Zhi Chun Road, Haidian District, Beijing, China
{wangrui,liwei,zhaochy,gaojf}@cnis.ac.cn

Abstract. The average daily stay time of city people in the vehicle is more than one hour. For some office workers in megalopolis, the stay time in the vehicle will be longer. Therefore, the thermal comfort of vehicle air conditioning system has an important impact on the life of residents. Using the warm manikin to test the comfort of the interior environment of the vehicle under different working conditions, test the steady-state conditions and compare with the comfortable equivalent space temperature of different parts of the human body, and give the air conditioning design optimization program. The results show that the cooling comfort in summer is relatively good, and the discomfort areas are mainly concentrated in feet, shins and other parts. In some cases, the thighs are overheated due to strong solar radiation, and the equivalent space temperature exceeds the comfort area.

Keywords: Automobile · Vehicle interior environment · Thermal comfort · Thermal manikin

1 Introduction

The research on human thermal comfort first appeared in the field of architecture. The research on human thermal comfort in architecture has been relatively mature, and there are corresponding thermal comfort evaluation methods and criteria. As a non-uniform thermal environment, the thermal comfort evaluation method in the building is not suitable for the passenger compartment. The factors that affect the thermal comfort of the human body in the car are divided into objective factors and human factors. The objective factors include air temperature, air velocity, relative humidity and evaluation radiation temperature. The human factors include clothing thermal resistance and human activity. Therefore, the research on the thermal comfort of the human body in the car is analyzed from two aspects: the experimental research and the evaluation method of the thermal comfort of the human body.

The change of the thermal environment inside the car directly affects the thermal comfort of human body, which is different from the indoor thermal environment of the building. Therefore, it is necessary to study the characteristics of the thermal environment inside the car. Ishihara [1] et al. studied the distribution of air velocity through visual experiments and compared it with the results of numerical simulation. Without considering the solar radiation, the influence of foot blowing on the thermal environment in the car is studied. The results show that blowing foot has a great influence on the distribution

E. Markopoulos et al. (Eds.): AHFE 2020, AISC 1218, pp. 317–324, 2020.
https://doi.org/10.1007/978-3-030-51626-0_41

of air flow in the car. Ishihara and Shibata [2] et al. used a combination of particle tracing and pulse laser technology to obtain the velocity distribution in the passenger compartment, the flow field distribution in the vehicle with the dummy model is also measured. Moos and Currie [3] studied the thermal comfort of a roadster, analyzed the thermal comfort of people in the car when the roof is opened in summer and winter. Ozeki [4] et al. studied the spectral difference between the solar radiation at the same intensity and the fluorescent lamp used in the test, and studied the influence of the vehicle glass on the thermal comfort of the human body in passenger compartment. Zhang [5] et al. used two test conditions to test two vehicles respectively, collected the changes of the thermal environment parameters in the vehicle, used fluent to simulate the thermal environment in the vehicle, and compared the simulation results with the test results. Kaynakli [6] et al. carried out an experimental study on the thermal environment in the car during the refrigeration and heating stages, and the mathematical model is used to calculate the heat exchange between human body and thermal environment, and to predict the thermal feeling of the passengers in the car. Kilic [7] et al. measured the thermal environment in the car during the cooling stage of the air conditioner in summer, obtained the data of the relative humidity, air velocity, air and surface temperature in the car, measured the human skin temperature, and conducted a subjective questionnaire survey on the thermal feeling of the human body. And according to the data of two cars, the change characteristics of the thermal environment inside the car were analyzed, and the change of the thermal comfort of human body was studied. Alahmer [8] et al. by controlling the relative humidity of the air in the environment cabin, obtained the local thermal feeling, local comfort, overall thermal feeling and the three-dimensional Berkeley model is used to simulate the thermal environment inside the car, and the influence of indoor relative humidity on human thermal comfort is analyzed; and analyzed the impact of the indoor relative humidity on the thermal comfort of the human body; calculated the human body's thermal comfort with the PMV-PPD method The results show that in the heating and cooling stage of air conditioning, the control of relative humidity level is helpful for A/C system to achieve faster thermal comfort. Chakroun [9] et al. studied the temperature characteristics of the air inside the car parked in the sun, and tested the temperature characteristics of the car under different shielding arrangements.

The warm manikin can simulate the heat exchange process between human body and environment in the set environment, and evaluate the thermal comfort of the tested environment scientifically, so as to avoid the influence of individual differences in human body experiments, and it has high accuracy and good repeatability. It is recognized as an essential means of ergonomics research. The warm manikin is an instrument to simulate the heat and humidity exchange between human body and environment. Its body size is similar to that of ordinary adults, and it is composed of head, chest, back, abdomen, buttocks, upper limbs, hands, lower limbs and feet. In this paper, warm manikin is tested by using the comfort of the interior environment of the vehicle under different working conditions, and the test results are compared with the equivalent space temperature of different parts of the human body under steady-state working conditions. The air conditioning design optimization scheme is given. The results show that the cooling comfort in summer is relatively good, and the discomfort areas are mainly concentrated in feet, calves and other parts. In some cases, the thighs are overheated due to strong solar radiation, and the equivalent space temperature exceeds the comfort area.

2 Test Method

In order to simulate the real environmental conditions when the vehicle is running, the test is carried out in the vehicle environment wind tunnel laboratory, and the laboratory diagram is shown in Fig. 1. The temperature, humidity and wind velocity of the external environment are simulated by wind tunnel, and the solar simulation light source is used to create the thermal radiation environment in the test area.

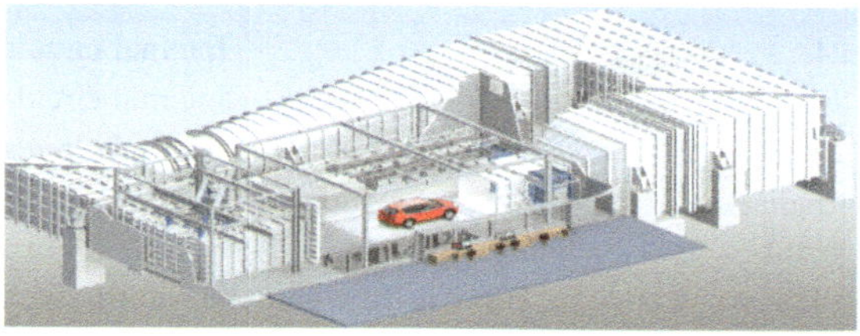

Fig. 1. Schematic diagram of wind tunnel laboratory

During the test, the vertical incidence of the simulated sun was 1050 W/m^2, the ambient temperature is 35 °C, and the relative humidity is 40%. The test starts when the vehicle is standing in the environment for more than 10 h. the working condition of the air conditioner is set as: 26 °C, automatic, internal circulation, the wind direction blows to the face, and the wind velocity is 2. In the test process, start the vehicle, turn on the air conditioner, and drive the vehicle at a constant speed at a speed of 60 km/h. The equivalent space temperature changes of the current environment driving position and co driving position are tested with the warm body dummy, and the test process is shown in Fig. 2.

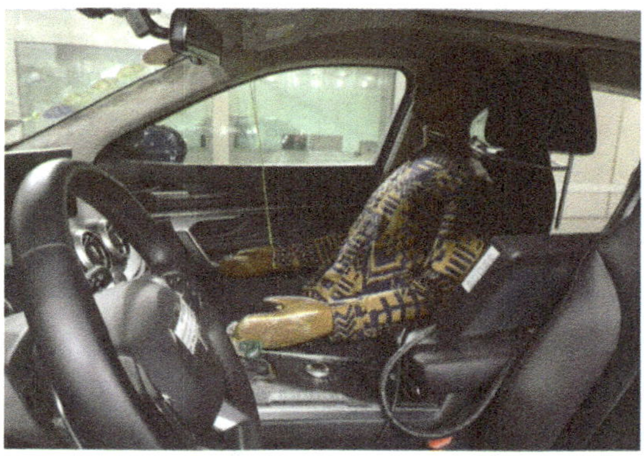

Fig. 2. Measurement of the equivalent space temperature in the vehicle by using the warm manikin

After the air conditioner is turned on, the equivalent space temperature gradually decreases with time, and t he temperature non-uniformity of the local small space in the car gradually expands, becoming a typical non-uniform environment. Therefore, the law of the equivalent space temperature change of the uneven environment and different parts of the body is tested by using the warm body manikin.

3 Test Result

This test only tests the thermal comfort of the interior environment under refrigeration condition. The three test conditions are shown in Table 1.

Table 1. Test conditions

Conditions	Setting temperature °C	Air velocity grade	Internal/external circulation
Condition 1	24	2	Internal circulation
Condition 2	25	4	Internal circulation
Condition 3	25	4	Internal circulation
Conditions	**Irradiation W/m^2**	**Wind direction**	
Condition 1	1050	Blow to face and chest at the same time	
Condition 2	1050	Blow face	
Condition 3	1050	Blow face	

The equivalent space temperature test results of three test conditions are shown in Fig. 3.

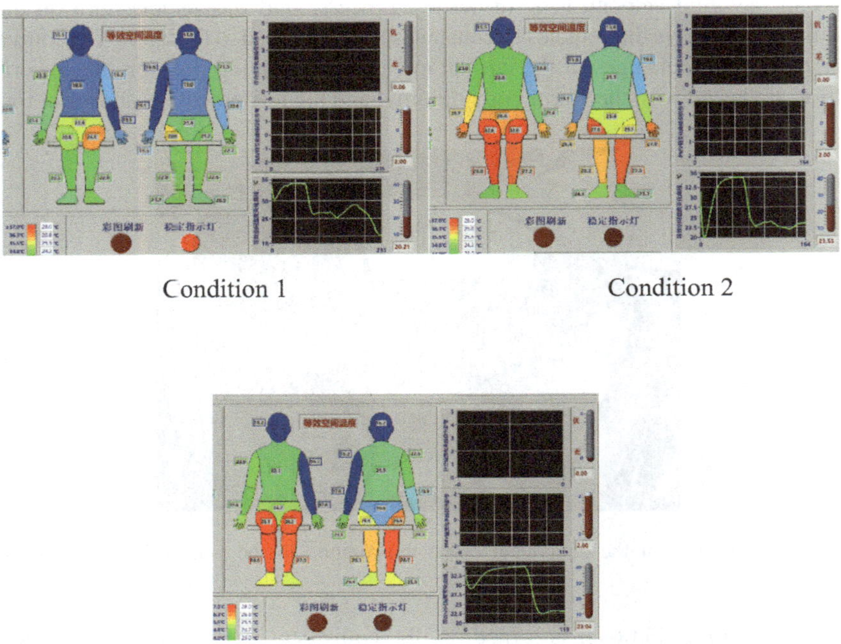

Condition 1

Condition 2

Condition 3

Fig. 3. Test results of equivalent space temperature under different test conditions

The test results show that the cooling rate of condition I is slow, while that of condition II and condition III is very fast, because the wind stall is low, there are only two gears, and the equivalent space temperature has dropped to the comfort zone about 10 min after the air conditioning is turned on. Due to the fact that the direction of air flow is the same for the head and chest, the final equivalent space temperature of the whole body is low and the body feels cold. However, the equivalent space temperature of the whole body after the stability of condition 2 and condition 3 is at the beginning of 23 °C, which belongs to a more comfortable range. It can be seen from the temperature test results that the equivalent space of the local part of the body blows towards the face under condition 2 and condition 3, which directly results in lower details of the equivalent space temperature of the head, far lower than other parts of the body. The equivalent space temperature of the head under condition 2 is only about 15 °C, while the equivalent space temperature of the head under condition 3 is slightly higher, but also about 18 °C, far lower than the comfortable temperature of the human body. The equivalent space temperature of face and chest is lower when face and chest are winded at the same time in the condition 1, and the equivalent space temperature of face is not about 15 °C, while chest is also lower, only 18 °C, far lower than 22 °C in the 2 and 3 conditions. The leg equivalent space temperature condition I is also lower, because the air supply direction of condition 1 is lower, which is beneficial for cold air to be sent to the legs, while the wind direction of condition 1 and condition 3 only blows to the head and face, resulting in high equivalent space temperature of the legs, and the legs are in a hot state.

4 Comfort Evaluation

The steady-state test results of equivalent space temperature of different parts of the body obtained from the test of warm body manikin are counted and listed in Table 2.

Table 2. Test results of equivalent space temperature of the whole body Unit °C

Condition	Right foot	Left foot	The right calf	The left calf	Right thigh	Left thigh	Right hand	Left hand	The lower right arm
1	20.9	23.2	22.5	22.4	22.0	24.8	22.7	19.5	20.9
2	23.3	24.3	28.2	26.7	29.2	30.1	27.0	26.4	25.1
3	25.5	25.4	28.6	26.8	27.4	26.8	24.3	23.1	21.2

Condition	The lower left arm	Right upper arm	Left upper arm	Upper back	Chest	Face	Head	The whole body
1	17.4	22.4	17.9	19.0	18.8	15.1	13.0	20.2
2	20.3	21.4	17.8	21.1	22.8	15.5	13.8	23.5
3	17.4	23.2	15.9	21.3	22.1	18.2	16.7	23.0

In terms of the whole body equivalent space temperature, he equivalent space temperature of condition 1 is the lowest, which is far away from the comfortable equivalent space temperature range in summer. The human body must feel cold under this condition. In terms of the local equivalent space temperature, due to the direct blowing of the air conditioning outlet flow, the equivalent space temperature of the head is the lowest, and the left arm and left hand are lower, while the right arm and right hand are relatively better. Lower body because did not receive air conditioning direct blow. Its equivalent space temperature is relatively high. The difference of equivalent space temperature in whole body between working condition 2 and working condition 3 is small, but the difference of equivalent space temperature in local area is relatively large due to the different air outlet modes. The equivalent space temperature of the head (scalp and face) in condition 2 is about 3 °C lower than that in condition 3, while the equivalent space temperature of the right hand and right lower arm in condition 2 is about 3 °C higher than that in condition 3. In order to be able to objectively evaluate the current environmental comfort, according to the comfort evaluation method in ISO 14505-2 series of international standards for ergonomics of thermal environment – assessment of internal thermal environment of vehicles, the interior environment created by three working conditions was evaluated in detail. The evaluation results are shown in Fig. 4.

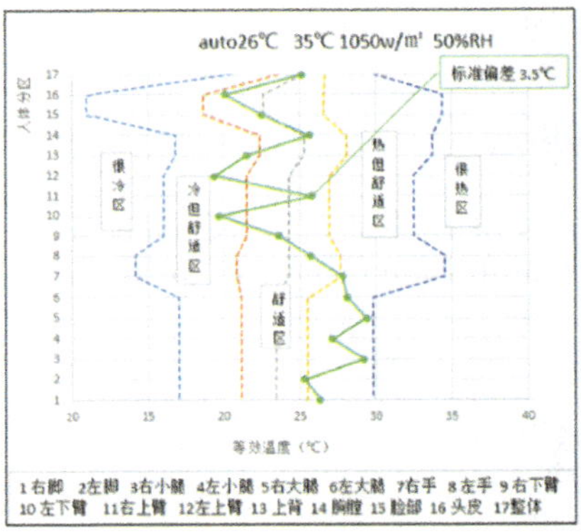

Fig. 4. The evaluation results of condition 1

Compared with the comfort evaluation chart, it can be seen that in the test results of condition 1, the lower body has been beyond the comfort zone, are a hot but comfortable state, upper body in cold but comfortable area, the whole body to see thermal comfort, but each part of the body are not in a very cold or very hot area, the overall comfort is relatively comfortable (Fig. 5).

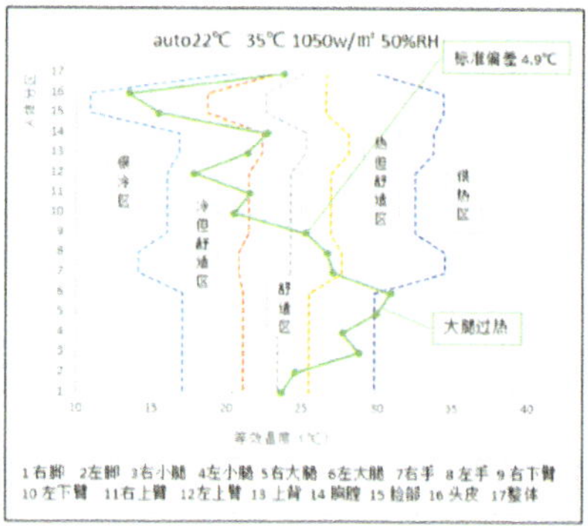

Fig. 5. The evaluation results of condition 2

According to the comfort evaluation chart, it can be seen that in the test results of working condition 2, the trend of upper body cold and lower body hot is more obvious. The head and face have almost exceeded the cold but comfortable zone, while the left and right thighs of the lower body have entered the hot zone, which is obviously uncomfortable. Other parts are relatively comfortable, and the overall thermal feeling of the whole body is in the comfort zone (Fig. 6).

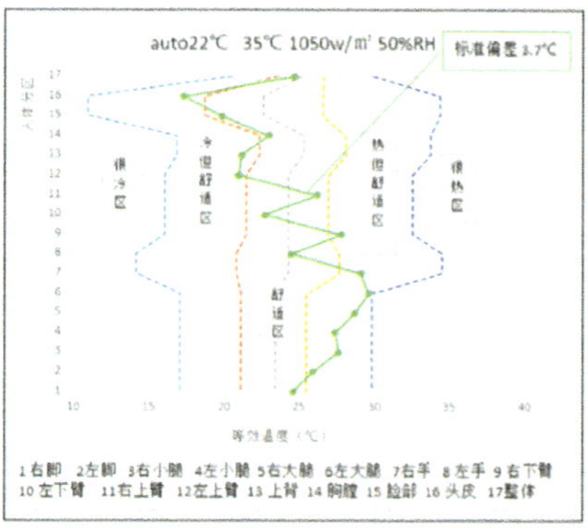

Fig. 6. The evaluation results of condition 3

Compared with the comfort evaluation chart, it can be seen that in the test results of condition 3, the upper body is warmer than the other two conditions, the comfort is better, the lower body hot trend is not so obvious, there is no any parts of the body beyond the comfort zone (including cold but nice and hot but comfortable), the body of the overall thermal sensation in a comfort zone.

5 Conclusion

Using the warm manikin to test the comfort of the interior environment of the vehicle under different working conditions, test the steady-state conditions and compare with the equivalent space temperature of different parts of the human body, and give the air conditioning design optimization scheme The results showed that the comfort of refrigeration in summer was relatively good, and the discomfort areas were mainly concentrated in the feet, shins and other parts. In some working conditions, due to strong solar irradiation, the thigh was overheated, and the equivalent space temperature exceeded the comfort zone.

Acknowledgments. This research is supported by "Special funds for the basic R&D undertakings by welfare research institutions" (522018Y-5984) and "2017 National Quality Infrastructure (2017NQI) project" (2017YFF0206605).

References

1. Ishihara, Y., Hara, J., Sakamoto, H., et al.: Determination of flow velocity distribution in a vehicle interior using a visualization and computation techniques. In: International Congress & Exposition (1991)
2. Ishihara, Y., Shibata, M., Hoshino, H., Hara, J., Kamemoto, K.: Analysis of interior airflow in a full-scale passenger-compartment model using a aser-light-sheet method. In: International Congress & Exposition (1992)
3. Moos, O., Currle, J.: Evaluation of thermal comfort in convertibles. In: SAE 2002 World Congress & Exhibition (2002)
4. Ozeki, Y., Takabayashi, T., Tanabe, S.: Effects of spectral properties of glass on thermal comfort of car occupants. Hydrocolloids Part 2, Fundamentals and Applications in Food, Biology, and Medicine, pp. 349–355 (2003)
5. Zhang, H., Dai, L. Xu, G., et al.: Studies of air-flow and temperature fields inside a passenger compartment for improving thermal comfort and saving energy. Part I: test/numerical model and validation. Appl. Therm. Eng. **29**(10), 2022–2027 (2009)
6. Kaynakli, O., Pulat, E., Kilic, M.: Thermal comfort during heating and cooling periods in an automobile. Heat Mass Transf. **41**(5), 449–458 (2005)
7. Kilic, M., Kaynakli, O.: An experimental investigation on interior thermal conditions and human body temperatures during cooling period in automobile. Heat Mass Transf. **47**(4), 407–418 (2011)
8. Alahmer, A., Abdelhamid, M., Omar, M.: Design for thermal sensation and comfort states in vehicles cabins. Appl. Therm. Eng. **36**(1), 126–140 (2012)
9. Chakroun, W., Al-Fahed, S.: Thermal comfort analysis inside a car. Int. J. Energy Res. **21**(4), 327–340 (2015)

Research on Clothing Warmth Preservation Based on Thermal Manikin

Wang Rui, Li Wei(✉), Gao Jianfeng, and Zhao Chaoyi

Ergonomics Laboratory, China National Institute of Standardization,
No. 4 Zhi Chun Road, Haidian District, Beijing, China
{wangrui,liwei,gaojf,zhaochy}@cnis.ac.cn

Abstract. The thermal manikin is a kind of equipment which can simulate the data of human body environment heat exchange and be used for the comfort test of clothing and so on. In this paper, the thermal resistance of the thermal manikin is measured according to the heat flow, surface temperature and environment temperature. The results show that the thermal resistance of the waist and abdomen is very close to that of the mountain suit. The thermal resistance of the local position is even better than that of the mountain suit, but the leg part of the thermal resistance value is only half of the mountain suit, which seriously affects the thermal resistance of the whole suit. Combined with the clothing surface temperature collected by infrared thermal imager, the test results are verified. It provides the basis for the design of clothing warmth preservation.

Keywords: Thermal manikin · Clothing thermal resistance · Thermal clothing

1 Introduction

Since the research of thermal manikin, more than 100 kinds of thermal manikin have been made at home and abroad [1, 2]. According to the purpose of thermal manikin, it can be divided into dry thermal manikin, breathing thermal manikin, sweating thermal manikin, numerical thermal manikin, immersion thermal manikin, etc. Its development process mainly experienced three stages: the first stage was based on the concept of "clo", i.e. clothing thermal resistance put forward by American scholar Gagge and his research team in 1941. The first generation of thermal manikin was developed by the US military Climate Research Institute, which is a single-stage thermal manikin and can only be used for static clothing thermal resistance test [3]. The early thermal manikin generally served for the military, and later extended to other fields. However, at present, thermal manikin are still important tools for testing military supplies. The second stage of thermal manikin is segmented thermal manikin, which was born in the 1960s. Compared with single segment thermal manikin, the advantages of each segment of thermal manikin are that each segment can be controlled independently and heated independently to simulate some simple posture of the human body. The third generation of thermal manikin is sweat thermal manikin [4], which was made by Glodenman and Professor Mecheels [5] in 1970s. Compared with the previous two generations, the third generation can not only simulate the skin temperature of human

E. Markopoulos et al. (Eds.): AHFE 2020, AISC 1218, pp. 325–332, 2020.
https://doi.org/10.1007/978-3-030-51626-0_42

body, measure the static and dynamic thermal resistance of clothing, but also simulate the sweating of human body under the control of computer and software. Based on the fact that human sweat is divided into gas and liquid, sweating thermal manikin is divided into sweat gland and non sweat gland. The appearance of sweating thermal manikin makes it possible to quantitatively evaluate the thermal performance (wet resistance and heat insulation performance) of clothing. Nowadays, it is widely used in the evaluation of special clothing (aerospace, fire protection, etc.) and the evaluation of human-clothing-environment heat exchange.

Thermal manikin is used to simulate the heat and humidity exchange process between human body, clothing and environment. The performance state of manikin can be described by surface temperature, heat production, thermal resistance, humidity resistance and other indicators. Before the experiment, it is of great significance to use accurate and reasonable test scheme and parameters to grasp the test rule and accurately evaluate the thermal and wet comfort performance of clothing.

2 Clothing Thermal Resistance

According to the heat balance theory of human body and the heat and humidity transfer principle of clothing, when the sweating thermal manikin, clothing and environment are in heat balance, the heat output of the manikin is equal to the heat dissipation through the clothing, i.e.,

$$Q = Qt + Qe \tag{1}$$

In this formula, Q is the heat production flow of the manikin (W); Qt is the non evaporation heat dissipation flow of the human body through the clothing surface (W); Qe is the evaporation heat dissipation flow of the human body through the clothing surface (W).

The resistance of heat flow caused by temperature gradient in garment layer is called thermal resistance, and the calculation formula is

$$R\alpha = \frac{\mathrm{A} \times (\mathrm{T}skin - \mathrm{Tabm})}{Qt} \tag{2}$$

In this formula, $R\alpha$ is the thermal resistance of clothing (°C·m2/W); Tskin is the skin temperature of manikin (°C); Tabm is the ambient temperature of manikin (°C); A is the surface area of manikin (m2); Qt is the non evaporation heat dissipation flow of manikin (W).

The total thermal resistance Ra of clothing consists of two parts, one is the effective thermal resistance value RCf of clothing, the other is the thermal resistance Rcto of the air layer on the human body surface, which is the thermal resistance measured by the naked manikin under the same environmental conditions, that is, $R\alpha = RCf + Rcto$.

Considering the influence of the increase of body surface area on the heat transfer characteristics of clothing, the basic thermal resistance of clothing is defined as

$$RCf = Rt - (Radm/fd) \tag{3}$$

In the formula, clothing area coefficient fd is the ratio of clothing surface area to human body net surface area.

The total thermal resistance of clothing refers to the thermal resistance from the skin surface to the environment, including the influence of the increase of the body surface area and the resistance of the air layer on the body surface; the effective thermal resistance of clothing refers to the thermal resistance from the skin to the clothing surface, including the influence of the increase of the body surface area; the basic thermal resistance of clothing refers to the thermal resistance from the skin to the clothing surface, excluding the influence of the increase of the body surface area after the clothing. Therefore, for the thermal insulation performance of clothing, it is necessary to define the type of thermal resistance and test indicators, in order to accurately characterize and compare with each other.

There are three modes to measure clothing thermal resistance with thermal manikin: constant skin temperature, constant heat flux and thermal comfort equation adjustment. At the same time, there are three modes to calculate the thermal resistance of clothing: parallel mode, serial mode and global mode. Table 1 lists the details of the three modes [11].

Table 1. Calculation Model of clothing thermal resistance measured by thermal manikin

Pattern	Calculation formula	Applicable mode
Parallel mode	$\frac{1}{I_t} = \sum f_i \times \left(\frac{A_{s,i}}{T_{sk,1} - T_0}\right) = \sum f_i \times \left(\frac{1}{I_{T,1}}\right)$	Constant skin temperature
Serial mode	$I_t = \sum_i \left(f_i \times \frac{T_{sk,i} - T_0}{O_{s,i}}\right) = \sum_i f_i \times I_{T,i}$	Constant heat flux
Global mode	$I_t = \frac{\sum (f_i \times T_{sk,i}) - T_0}{\sum (f_i \times O_{sk,i})}$	All modes

Pattern	Feature
Parallel mode	Compared with the serial model, the measured value is more stable
Serial mode	It is widely used, but the measured value is on the high side (especially for clothing with obvious local thermal resistance difference)
Global mode	Wide range of application

Note: It refers to the total thermal resistance of clothing (clo value);
f_i refers to the ratio of the surface area of parts to the surface area of manikin;
$O_{s,i}$, refers to the sensible heat flow (W/m);
$T_{sk,i}$ refers to the average skin temperature (°C);
T_0 refers to the ambient temperature (°C);
$I_{t,i}$ refers to the total thermal resistance of clothing (clo value)

In the calculation of clothing thermal resistance, in addition to calculating the total thermal resistance of clothing, it is also very important to calculate the local thermal resistance of clothing, because the local thermal resistance can truly express the thermal insulation performance of clothing to the covered part of the body. For the same fabric clothing, the local thermal resistance can also distinguish the subtle differences caused by the clothing structure, which is more conducive to improving the clothing structure design. The test standards and comfort evaluation of the thermal and wet comfort of the manikin mainly include the clothing thermal resistance, wet resistance test and comfort evaluation. At present, the testing technology of thermal manikin is quite mature. Table 2 is the international testing standard of thermal manikin.

Table 2. International test standards for thermal manikin

Thermal manikin test standards	(1) ISO7920, evaluation of thermal performance of clothing; (2) ASTM f1291, standard test method for measuring thermal resistance of clothing using thermal manikin; (3) EN-ISO 15831-2003, basic thermal resistance of clothing using thermal manikin
Application standard of thermal manikin test	(1) ENV342, standard for cold clothing; (2) EN 511, protective gloves (warm artificial hand); (3) ISO DIS14505, evaluation of the thermal environment inside the vehicle
Criteria requiring thermal manikin test results	(1) ISO 7730, test standard of PMV and PPD index; (2) ISO DIS 7933, thermal environment, analysis of human thermal resistance by prediction of thermal stress; (3) ISO DIS 11079, interpretation of human cold feeling by clothing thermal resistance and local cooling effect in cold conditions

3 Study on Thermal Resistance and Warmth Retention of Warm Clothing in Severe Cold Environment

The city suit consists of 3 coats and 2 pants. The test conditions are as follows: environment temperature −15 °C, relative humidity 50%, average wind velocity 0.07 m/s, and environmental conditions as shown in Fig. 1.

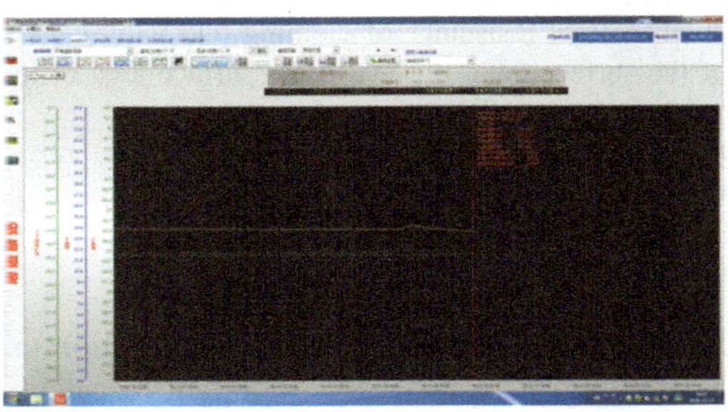

Fig. 1. Environmental parameters of city suit test

The thermal resistance of the clothing after the test is stable is shown in Fig. 2, the total thermal resistance of the whole body is 3.077clo, which is about 0.7clo lower than the mountain suit; except for the head, feet and hands, the thermal resistance of the clothes is 4.674clo, which is only about 0.1clo lower than the mountain suit. According to the calculation of thermal resistance of clothing in cold environment, the applicable environment in jogging state is shown in Table 3. The clothing can ensure that people can walk normally (1.5Mat) for 3 h without cold in the windless environment of −11.7 °C (wind velocity < 0.05 m/s); if the environment wind velocity is 4 m/s, people can walk normally (1.5Mat) for 3 h without cold in the environment of 1 °C.

Table 3. Applicable environmental temperature of city suit for 3 h in walking state

Environmental wind velocity (m/s)	Environmental temperature (°C)
0.05	−11.7
4	1

The thermal resistance values of each part of the body are listed in Table 4.

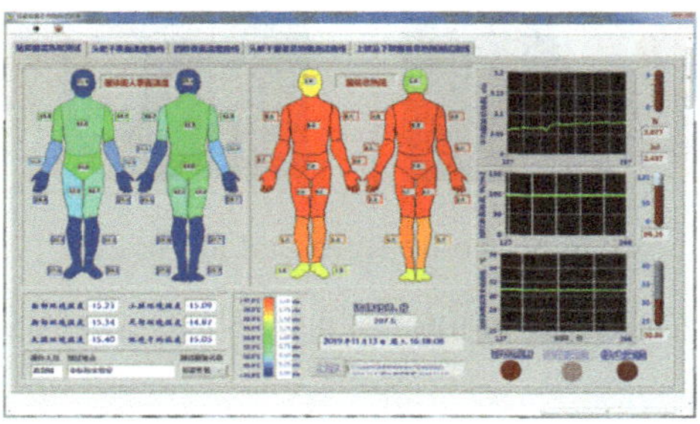

Fig. 2. Thermal resistance test results of city suit

Table 4. Test results of thermal resistance value of each part of city suit

Position	Clo value	Position	Clo value
Face	1.6	The back side of head	1.3
Left frontal upper arm	6.4	Left rear upper arm	4.5
Front chest	5.0	Back	5.9
Right frontal upper arm	5.8	Right rear upper arm	4.9
Left frontal forearm	3.6	Left rear forearm	3.3
Abdomen	7.3	Waist	7.0
Right frontal forearm	3.7	Right rearl forearm	3.8
Left front thigh	4.5	Left reart thigh	4.7
Right front thigh	4.0	Right rear thigh	5.2
Left front calf	1.8	Left rear calf	1.8
Right front calf	1.8	Right rear calf	1.7
Left foot	1.5	Left hand	2.4
Right foot	1.5	Right hand	2.4

The results show that the thermal resistance of the waist and abdomen is close to that of the mountain suit, and the thermal resistance of the front chest, back, arm and thigh is also slightly different from that of the mountain suit. The thermal resistance of the local position is even better than that of the mountain suit, but the thermal resistance of the leg part is only half of that of the mountain suit, which seriously affects the thermal resistance of the whole suit.

After stabilization, use infrared thermal imager to collect the surface temperature of different parts of clothing, and the results are shown in Fig. 3.

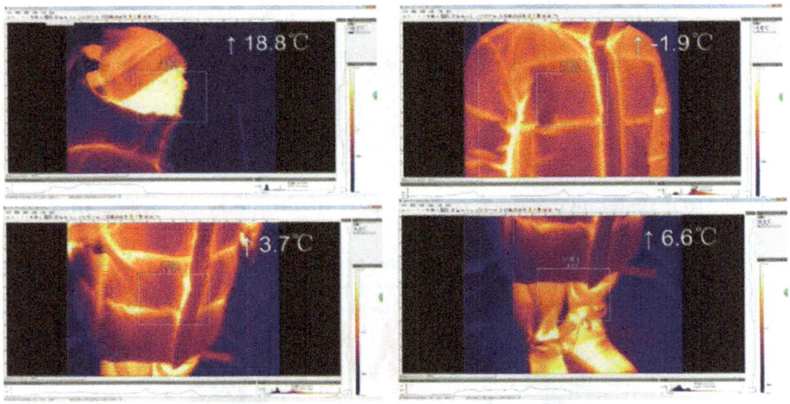

The pictures are: head; upper body; waist and thigh; calf

Fig. 3. Temperature distribution of city suit test clothing surface

In order to test the difference of thermal resistance of clothing in different environments, grasp the thermal resistance of clothing in potential application environment, the thermal resistance of urban suit clothing was tested at −20 °C, the test results are shown in Fig. 4.

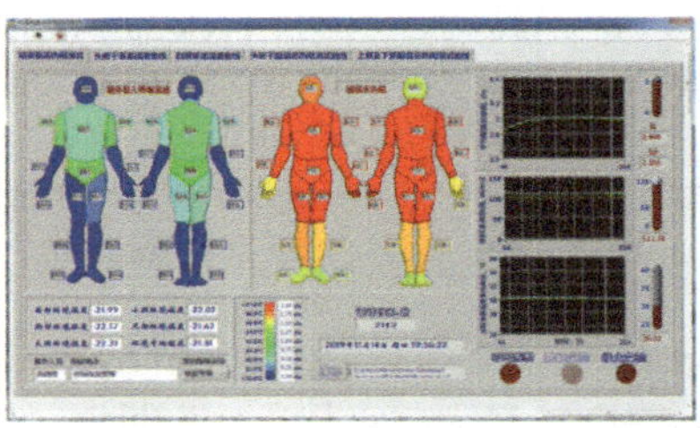

Fig. 4. Test results of thermal resistance of city suit at −20 °C

The total thermal resistance of city suit at −20 °C is 2.999clo, which is slightly smaller than that at −15 °C. The test results show that the environmental temperature has little effect on the thermal resistance of clothing. In order to further confirm the change of thermal resistance of clothing with the environment temperature, the thermal resistance of the whole set of urban suits is tested under the environment of 5 °C. The test results are shown in Fig. 13. The total thermal resistance of the whole body of the urban suits under 5 °C is 3.055clo, which is very small compared with the thermal resistance of the suits under −20 °C and −15 °C, and the variation is less than 3%, which is equivalent to the measurement error of the thermal resistance of the clothing tested by the thermal manikin. The test results further shows that the environmental temperature has little effect on the thermal resistance of clothing.

4 Calculation of Thermal Resistance of Required Clothing

Under the condition of no wind, the environment is in the state of windless (wind velocity 0.1 m/s), and the calculation results of the required thermal resistance are shown in Fig. 18 below. When the necessary thermal resistance of the clothing base is about 4.32clo, and the thermal resistance of the clothing base is 3.75clo, it can ensure that there will be no discomfort within 3 h of the current environment exposure.

According to the working condition of clothing, the method of evaluating human stress in cold environment by using thermal resistance of clothing needs, the environmental temperature −20 °C, radiation temperature −20 °C, wind velocity is 4 m/s, relative humidity is 50%, human metabolic rate is 1.5 M (about 90 W/m^2), slow walking speed of 0.17 m/s are calculated. Parameters of air permeability and clothing surface emissivity are not specified, but the parameters have little impact on the results, so according to the default The recognition value is given, and the calculation results under this working condition are shown in Fig. 19 below: the results show that when the basic thermal resistance of clothing reaches to the thermal neutral state is about 6.2clo, and the basic thermal resistance of clothing is 5.3clo, there will be no discomfort within 3 h of the current environment exposure.

5 Conclusion

In this paper, the thermal resistance of the thermal manikin is measured according to the heat flow, surface temperature and environment temperature. The thermal resistance of a warm clothing is measured. The results show that the thermal resistance of the waist and abdomen is very close to that of the mountain suit. The thermal resistance of the front chest, back, arm and thigh is also slightly different from that of the mountain suit. The thermal resistance of the local position is even better than that of the mountain suit Suit, but the leg part of the thermal resistance value is only half of the mountain suit, which seriously affects the thermal resistance of the whole suit. Combined with the clothing surface temperature collected by infrared thermal imager, the test results are verified. Finally, the thermal resistance value of the garment and its applicable environmental conditions are given. It provides the basis for the design of clothing warmth preservation.

Acknowledgments. This research is supported by "Special funds for the basic R&D undertakings by welfare research institutions" (522018Y-5984)and "2017 National Quality Infrastructure (2017NQI) project" (2017YFF0206605).

References

1. Goldman, R.F.: Tolerance time for work in the heat when wearing CBR protective clothing. Military Med. **128**, 776 (1963)
2. Mecheels, J., Umbach, K.H.: The textile and clothing-physiological requirements to be met by therapeutic thermal segments. Schriftenr Zentralbl Arbeitsmed Arbeitsschutz Prophyl Ergonomie **6**, 205–211 (1981)
3. Richards, M.G.M. Mattle, N.G.: A sweating agile thermal manikin (SAM) developed to test complete clothing systems under normal and extreme conditions. RTO-MP **76**(4), 1–7 (2002)
4. Wissler, E.H.: Steady-state temperature distribution in man. J. Appl. Physiol. **16**(4), 734–740 (1961)
5. Wang, H., Ma, C., Lv, H.: Research status and development direction of sweating thermal manikin. Techn. Textiles **35**(10), 1–7 (2017)
6. Le, Z.-X., Qian, X.-M., Xing, J.-J.: Application of thermal manikin in clothing heat-moisture comfort. Prog. Textile Sci. Technol. **12**, 50–52 (2016)

Ergonomic Test of Human Skin Contacting with Hot Surface

Wang Rui, Li Wei(✉), Zhao Chaoyi, and Gao Jianfeng

Ergonomics Laboratory, China National Institute of Standardization, No. 4 Zhi Chun Road, Haidian District, Beijing, China
{wangrui, liwei, zhaochy, gaojf}@cnis.ac.cn

Abstract. When human body operates some heating equipment, the surface temperature will affect the working efficiency and safety of workers. This study included the critical temperature of skin in contact with a hot solid surface and the method of assessing the risk of burns, that is, the application of the data provided by ergonomics in the process of risk assessment. Through the test results, how to choose a reasonable temperature limit value is put forward, and how to establish the same temperature limit value for different products with the same risk is proposed. The time of intentional and unintentional contact is different. Considering the distribution of human response time, when a healthy adult inadvertently contacts the hot surface, 0.5 s is the safe contact time, while the intentional contact time is longer. It is very necessary to choose the contact time, which can best represent the actual situation of contacting hot products.

Keywords: Ergonomics · Contact surface temperature · Hot surface

1 Introduction

At present, with the improvement of people's living standards, people put forward higher requirements for the hot surface contact of the working environment. There are more researches and relevant specifications on the hot surface that may be contacted in the work abroad [1, 2], but there are less researches on this aspect in China, and the relevant standards and regulations are relatively lacking [3]. In recent years, people pay more and more attention to the protection of human body in the process of labor. Some scholars are gradually carrying out the research work on the influence of hot surface temperature and time length in the process of human-computer interaction. In order to better protect workers, and provide the basis for ergonomics of human-computer interaction system that will contact the hot surface during the operation process, we intend to test the human contact hot surface reaction, and study its evaluation method.

When human body operates some heating equipment, the surface temperature will affect the working efficiency and safety of workers. In order to prevent people from burning when contacting the hot product surface, the relevant contact surface temperature limit value must be specified. How to choose a reasonable contact surface temperature limit is very important. Under different tasks, the contact time with the heating surface is also different. In addition, some high-temperature objects in the working area do not need to be directly touched by hand to operate, but unintentional

E. Markopoulos et al. (Eds.): AHFE 2020, AISC 1218, pp. 333–338, 2020.
https://doi.org/10.1007/978-3-030-51626-0_43

touch can also cause burns. Therefore, the surface temperature of high-temperature objects that may be touched in the working area also needs to be limited to ensure the work and safety of operators.

Mechanical surfaces that are hot and easily accessible during operation are a source of burn risk [4]. Contact with the hot surface may occur consciously, for example, by manipulating the machine handle; it may also be done unconsciously by the person next to the machine [5]. In this paper, the general guidelines for the design of mechanical safety are specified, which take into account the measures to prevent injury, including thermal injury. In order to assess the risk of burns caused by hot surfaces, it is necessary to know the factors and effects of burns caused by skin contact with hot surfaces. The most important factors are: surface temperature; surface material components; time of skin contact with surfaces. There are other secondary factors, but they have little impact on the test results. In this paper, the data for assessing the risk of burns when the skin is in contact with a hot surface are specified. In other standards or rules, it is necessary to specify the thermal surface temperature limits, and these data can also be used. The data specified in this paper are based on scientific research and reflect the characteristics of human skin in contact with hot surfaces.

2 Test Equipment

The temperature control equipment with feedback function is used for the test, as shown in Fig. 1. The five touch bars of the equipment are hollow design, and the central control part is electric heating bar. The electric heating bar is bonded with the outer surface material by the heat-conducting adhesive. A temperature measuring point is arranged on the outer surface of the material to monitor the surface temperature and control the heating power of the heating rod as a control signal. Before each test, the surface temperature needs to be calibrated to ensure the accuracy of the test data.

Fig. 1. Hot surface contact test equipment

3 Test Method

Hot surface endurance test is mainly carried out for five materials, namely: wood, nylon, steel, aluminum, stone.

During the test, two methods of grasping and touching are used. When grasping, hold the sample with hands, maintain a certain grip strength, start timing when grasping, release when feeling hot and unbearable, stop timing when releasing, and record the holding time. If the holding time is longer than 1 min, stop the test, and write the time for 1 min. The test table is shown in Table 1 below (taking steel as an example), and the interval after each test is long enough to ensure that the thermal feeling result of the next test is not affected by the last test.

Table 1. Statistical table of thermal feeling test data (steel)

Temperature °C	40	43	46	49	52	55	58	61
Time S								

4 Test Results

The occurrence of burns depends on the temperature of the skin and the time when the skin temperature rises. The relationship between skin temperature, influence time and burn has been studied and established. When the hot surface of the product is burned, it is not feasible to measure the skin temperature with a simple method. This value is not the skin temperature, but the hot surface temperature of the product. The surface temperature of the product can be simply measured with appropriate measuring tools.

The surface temperature of the burn when the skin contacts the product depends on the material of the product and the contact time. Figure 2 shows this relationship. It can be seen that several materials have similar thermal conductivity and burn critical temperature values.

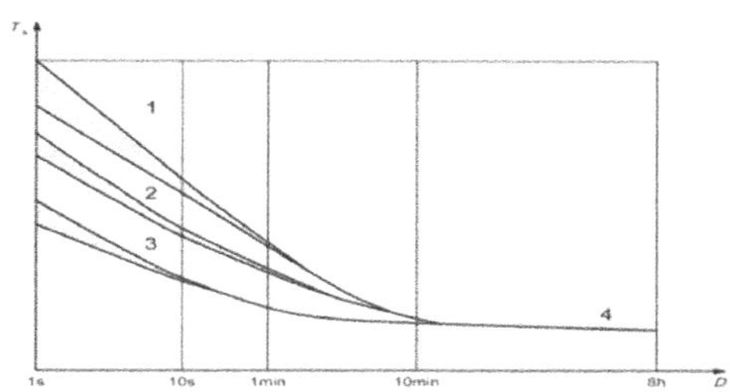

D contact time *Ts* surface temperature

1 Plastic 2 ceramics 3 metal 4 Burn threshold

Fig. 2. Relationship between burn critical temperature and contact time when skin contact surface is burned

A point on the critical temperature curve of burn indicates: a special period of contact, the surface temperature is the temperature between the uninjured and the slight burn when the skin contacts the hot surface. Generally, lower than the critical surface temperature will not cause burns, higher than the critical surface temperature will cause burns. Figure 2 illustrates the relationship but does not provide a clear burn threshold. In the short contact time, the critical temperature of burn is not plotted as a straight line as shown in Fig. 2. A problem needs to be considered. In a short contact time, the related problem of critical temperature between unburned and burned is not complete. The critical temperature of burn depends on many factors, including skin thickness at the contact point, skin surface humidity, skin surface pollutants (such as oil), and the contact force of a variety of heat conductive materials, which results in the measurement error of critical temperature of burn. However, these factors are insignificant compared with the effects of different combinations of materials with different heat conduction properties. Compared with the short contact period, the long contact period has less uncertainties. Therefore, the critical temperature of long-term contact burn is determined, and the effect of different material combinations does not exist. The data given assume that the surface temperature remains constant during contact through product quality or heating sources. These cases have been studied under the worst conditions that meet the requirements.

Figure 3 shows the burn critical value of pottery, glass pottery, glass, porcelain and stone. The burn critical value of marble and concrete is lower than that of the material in Fig. 3, and the burn critical value of glass is higher than it.

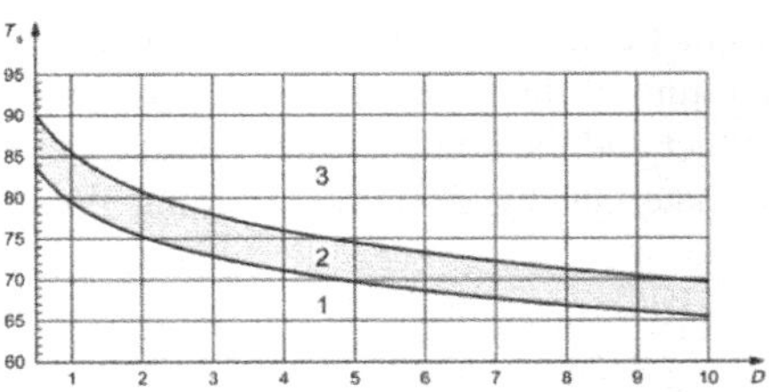

D contact time, s *Ts* surface temperature, °C

1 Unburned 2 Burn threshold 3 Burn

Fig. 3. Burn threshold for skin contact with hot smooth surfaces of pottery, glass, and stone.

Figure 4 shows the burn critical value of plastic (nylon, acrylic glass, polytetrafluoroethylene, hard plastic).

Note the thermal conductivity of plastics with different chemical components is different, and Fig 4 shows the burn critical value of most solid plastics. However, the material with heat conduction performance is obviously different from the material here, and its burn critical value cannot be used.

Figure 5 shows the burn threshold of wood. The burn critical value of softwood with low moisture content is higher than that of the material in the figure, and that of hardwood with high moisture content is lower than it.

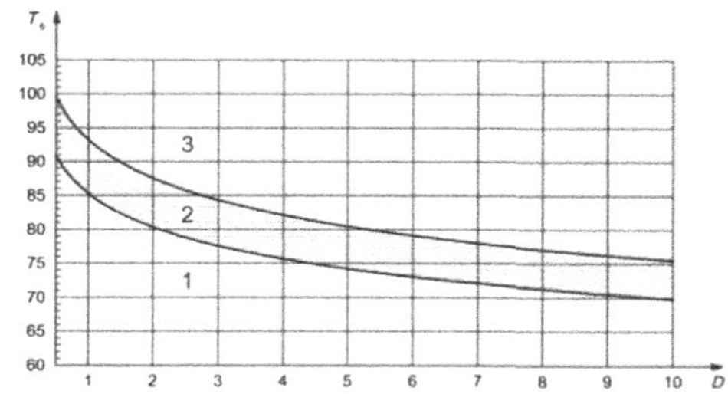

D contact time, s *Ts* surface temperature, ℃

1 Unburned 2 Burn threshold 3 Burn

Fig. 4. Burn threshold for skin contact with a smooth hot surface of plastic.

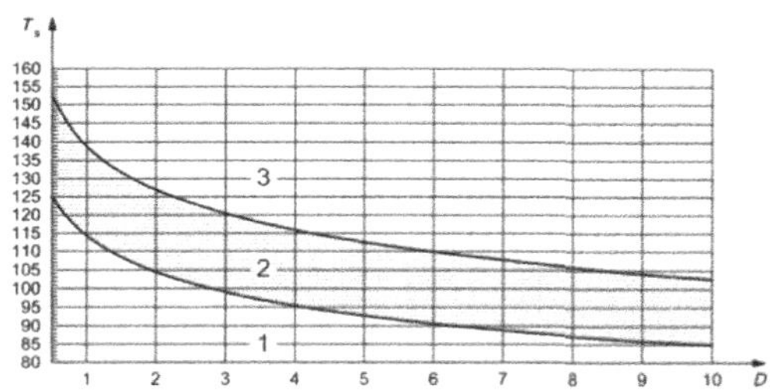

D contact time, s *Ts* surface temperature, ℃

1 Unburned 2 Burn threshold 3 Burn

Fig. 5. Burn threshold for skin contact with a smooth hot surface of wood

In order to evaluate the burn risk of non procedural personnel contacting hot surfaces, the following steps shall be followed:

——Identification of accessible hot surfaces
——Job analysis
——Measurement of surface temperature
——Selection of usable burn critical value
——Comparison of surface temperature and burn critical value
——Determination of burn risk
——Reassessment

There are some special cases that differ from their steps for a reason. For example, if the number of surfaces to be considered can be reduced by this process, the job analysis can be performed prior to the identification of the accessible hot surfaces, or the burn threshold can be selected prior to the measurement of the surface temperature. Pay attention to one or several hot surfaces of the product, and collect necessary information about the hot surface of the whole product. Related properties are:

——Accessibility of surfaces
——Rough estimate of surface temperature (hot, moderate, cold)

——Composition of material surface
——Surface texture
——All operating conditions of the product (including the worst conditions), such as the maximum surface temperature.

5 Conclusion

This study includes the critical temperature value when the skin contacts with the hot solid surface and the method to evaluate the burn risk, that is, the application of the data provided by ergonomics in the risk assessment process. Data can be further applied to specific hot surface temperature limits. This temperature limit may be specified in the product's standards or rules in order to prevent people from burning when contacting the hot product surface. Through the test results, how to choose a reasonable temperature limit value and how to establish the same temperature limit value for different products with the same risk are proposed. Touching a hot surface may be intentional, such as operating a motor, gas engine, or tool, or when a person inadvertently approaches a hot object. The time of intentional and unintentional contact is different. Considering the distribution of human response time, when a healthy adult inadvertently contacts the hot surface, 0.5 s is the safe contact time length, while the intentional contact time is longer. It is very necessary to choose the contact time, which can best represent the actual situation of contacting hot products.

Acknowledgments. This research is supported by "Special funds for the basic R&D undertakings by welfare research institutions" (522018Y-5984)and "2017 National Quality Infrastructure (2017NQI) project' (2017YFF0206605).

References

1. Siekmann, H.: Recommended maximum temperatures for touchable surfaces. Appl. Ergon. **21** (4), 69–73 (1990)
2. Siekmann, H.: Determination of maximum temperatures that can be tolerated on contact with hot surfaces. Appl. Ergon. **20**(4), 313–317 (1989)
3. GB/T 18153-2000. Safety of machinery–Temperatures of touchable surfaces–Ergonomics data to establish temperature limit values for hot surfaces
4. EN 71-1:1998, Safety of toys — Part 1: Mechanical and physical properties
5. ISO 7010:2003. Graphical symbols — Safety colors and safety signs — Safety signs used in workplaces and public areas

Author Index

E. Markopoulos et al. (Eds.): AHFE 2020, AISC 1218, pp. 339–340, 2020.
https://doi.org/10.1007/978-3-030-51626-0

The manufacturer's authorised representative in the EU is Springer Nature Customer Service Centre GmbH, Europaplatz 3, 69115 Heidelberg, Germany. If you have any concerns regarding our products, please contact ProductSafety@springernature.com

Printed and bound by CPI Group (UK) Ltd, Croydon, CR0 4YY
30/09/2026
02225509-0009